**CANADIAN SOCIETY OF
EXPLORATION GEOPHYSICISTS
50TH ANNIVERSARY**

DEPTH IMAGING
OF
FOOTHILLS SEISMIC DATA

EDITED BY
LAURENCE R. LINES, SAMUEL H. GRAY, AND DON C. LAWTON

Cover photograph courtesy of Gary Maclean

ISBN: 0-9692354-1-0

Contents

Preface .... *vii*

Chapter 1 – Introduction

    Editors' Introduction *1*

    Foothills Depth Imaging – An Exploration Perspective
        G. Maclean *2*

    Seismic Modeling of an imbricate thrust structure from the foothills of the
    Canadian Rocky Mountains
        S. H. Lingrey *6*

Chapter 2 – Prestack Migration Concepts

    Editors' Introduction *21*

    Kirchhoff migration using eikonal equation traveltimes
        S. H. Gray and W. P. May *23*

    Seismic imaging: Use the right tool for the job
        S. H. Gray *31*

    Simple prestack migration of crosswell seismic data
        C. L. Liner and L. R. Lines *35*

    Speed and accuracy of seismic migration methods
        S. H. Gray *47*

    Smiles and frowns in migration/velocity analysis
        J. Zhu, L. R. Lines and S. H. Gray *84*

    Comparison of Kirchhoff and reverse-time migration methods with applications to
    prestack depth imaging of complex structures
        J. Zhu and L. R. Lines *94*

Chapter 3 – Migration in Structurally Complex Areas

    Editors' Introduction *105*

Interpretative seismic imaging in structurally complex areas
    S. H. Gray *106*

Migration from topography: improving the near-surface image
    S. H. Gray and K. J. Marfurt *108*

Migration from topography: experience with an Alberta foothills data set
    L. R. Lines, W. Wu, H. Lu, A. Burton, and J. Zhu *115*

Wavefield extrapolation by nonstationary phase shift
    G. F. Margrave and R. J. Ferguson *122*

Chapter 4 – Case Histories

Editors' Introduction *135*

Migration of a multi-offset VSP: A case study in NE British Columbia
    M. A. Slawinski and J. M. Parkin *136*

Prestack depth migration of an Alberta foothills data set – The Husky experience
    W. Wu, L. R. Lines, A. Burton, J. Zhu, W. Jamison, and R. P. Bording *145*

An imaging comparison of three depth migration algorithms on Foothills data sets
    L. Yan and L. R. Lines *152*

Crooked line, rough topography: advancing toward the correct seismic image
    G. Maclean, S. H. Gray, and K. Marfurt *156*

Numerical seismic modeling and imaging of exposed structures
    M. G. Kirtland Grech, D. C. Lawton, and D. A. Spratt *159*

Computational experiments with the Spratt Foothills Model
    Larry R. Lines *173*

Chapter 5 – Anisotropic Migration

Editors' Introduction *183*

Slip-slidin' away – some practical implications of seismic velocity anisotropy on depth imaging
    D. C. Lawton *185*

Anisotropic prestack depth migration
    J. M. Leslie and D. C. Lawton *193*

A refraction seismic field study to determine the anisotropic parameters of shales
    J. M. Leslie and D. C. Lawton *198*

Anisotropic true-amplitude migration
    T. Fei, J. Dellinger, G. E. Murphy, J. Hensley, and S. H. Gray *204*

Prestack depth migration in TI media: examples with numerical and physical modeling data
    N. Dai, S. Cheadle, and J. H. Isaac *206*

Comparison of structural imaging in anisotropic media using P-wave and S-wave data
    M. G. Kirtland Grech, J. H. Isaac, and D. C. Lawton *208*

Image mispositioning due to dipping TI media: a physical seismic modeling study
    J. H. Isaac and D. C. Lawton *215*

Imaging structures below dipping TI media
    R. W. Vestrum, D. C. Lawton, and R. Schmid *224*

Nonstationary phase shift (NSPS) for TI media
    R. J. Ferguson and G. F. Margrave *232*

Examples of prestack depth migration in TI media
    R. J. Ferguson and G. F. Margrave *259*

Chapter 6 – Conclusions and Future Work *275*

ACKNOWLEDGEMENTS OF REPRINT PERMISSION

There are a number of papers in this book that have been reprinted from CSEG and SEG journals. These papers are listed at the beginning of each chapter. We thank the CSEG and SEG journals and the authors of these papers for permission to reprint them here.

We also thank the SEG and Steve Lingrey for permission to print his case history (p. 6) from "Seismic Modeling of Geologic Structures" (edited by Stuart Fagin).

We are grateful for the permission of Geophysical Press to reprint the paper by Liner and Lines (p. 35) from the Journal of Seismic Exploration.

Finally, we thank the Consortium for Research in Elastic Wave Exploration Seismology (CREWES) and the Foothills Research Project (FRP) at the University of Calgary for permission to print from past research reports.

Preface

This book, *Depth Imaging of Foothills Seismic Data* is published by the Canadian Society of Exploration Geophysicists (CSEG) as part of the CSEG Fiftieth Anniversary Project in the year 2000. The foothills depth imaging project was initiated by Larry Lines and Sam Gray in 1998 as a CSEG Superfund Project at the University of Calgary. The goal of the project was to provide a course and a book for Canadian universities outlining the techniques and applications of depth imaging in foothills exploration. The book provides an outline for a short course on the subject.

The structural geology of the Western Canadian Foothills is dominated by a series of thrust faults, complex folds and steeply dipping formations. Seismic imaging is an essential tool in the petroleum exploration of such complicated geology, and due to structural complexity and steep dips, migration is a process which is necessary for the correct positioning of seismic reflection events. Migration is defined as a processing step for placing anisotropic reflectors in their true subsurface location. More specifically, seismic prestack depth migration is the preferred method for the following reasons:

1. Prestack migration does not use the common midpoint stacking, which will smear reflections from steeply dipping layers.
2. Depth migration can accommodate the lateral seismic velocity variations found in foothills geology, whereas time migration methods do not.
3. Foothills rock formations, especially shales, may exhibit anisotropic behavior. That is, seismic velocities will depend on the direction of wave travel. Failure to accommodate this behavior may result in mispositioning of reflectors.

In this set of course notes, we look at the depth imaging of foothills structures from Western Canada through the description of theory and applications.

Chapter 1 describes seismic responses for foothills models to emphasize the complexity of seismograms. Chapter 2 introduces the concepts of prestack depth migration in terms of two techniques – Kirchhoff migration and reverse-time depth migration. Chapter 3 explains how prestack migration must be revised to account for complex topography with large elevation changes. Chapter 4 includes a number of case history examples from Alberta and British Columbia. Chapter 5 describes how the failure to consider seismic anisotropy can cause mispositioning of seismic events, and shows how algorithms can be revised to handle anisotropic behavior. Chapter 6 summarizes the state of the art in present studies and discusses possible future directions.

In completing this book, the editors thank the CSEG, especially Jack Pullin and Susan Eaton, for their support of the project. We thank the Foothills Research Project (FRP) personnel, McAra Printing, and Linda St. Pierre of FRP for their invaluable assistance.

References made in the introductory material to articles appearing in this volume are set in italics.

Chapter 1 - Introduction

This first chapter contains an introductory discussion by *Maclean (1999)* on the need for prestack depth migration in foothills petroleum exploration. The discussion explains exploration requirements in practical terms, and it links advanced technological developments to these practical needs.

This chapter also presents a modeling study, originally from *Lingrey (1991),* in which synthetic seismograms are computed for imbricate thrust structures typical of the Canadian Rocky Mountains. These models and their seismic responses give the reader an appreciation of the complexity of the region and the seismic imaging challenges.

References:

Lingrey, S.H., 1991, Seismic Modeling of an imbricate thrust structure from the foothills of the Canadian Rocky Mountains: in Fagin, Stuart W., ed., Seismic modeling of geologic structures, Society of Exploration Geophysicists, 111-125.

Maclean, G., 1999, Foothills Depth Imaging – An Exploration Perspective: unpublished manuscript.

Foothills Depth Imaging – an Explorationist's Perspective

Gary Maclean

There is no more important tool for a structural geophysicist than prestack depth imaging. In the subsequent portions of this chapter, reasons for this somewhat bold statement will be outlined. Perspectives from exploration line management, and the extended team of which the geophysicist is a part, will hopefully demonstrate the importance of prestack depth imaging. In conclusion, one of many strategies for producing a correctly imaged seismic section will be discussed.

Management perspective

For the exploration line managers or team leaders, the focus of exploration is to understand the risk of projects for which they are accountable. Entrusted with significant capital dollars, they must take responsibility for the success or failure of the project insofar as senior management is concerned. Risk for structural plays involves timing or product migration, reservoir, seal, and most importantly from a geophysical perspective, trap definition. Trap definition encompasses more than an estimation of the real extent of the structure. It also involves an understanding of risks associated with closure, seal failure mechanisms, the geometry of the structure and estimation of the transition zone above the water line.

Typically, success or failure for a "foothills" play is not determined by whether gas is present. Encountering gas in a structure would fall into the technical success category. The true measure of success is the correct estimation of original gas in place (OGIP) that is subsequently verified by production. The cost of acquiring seismic and drilling in structural plays is significant but substantially less than pipeline and gas processing infrastructure (or processing fees from a mid-stream operator). Numerous cases can be cited where enormous capital has been deployed for infrastructure only to find that the basic assumptions regarding resource size were in error.

Trap definition

At some point in a play's evolution, seismic data is acquired, interpreted and mapped in time. In general, for stratigraphic plays, this is typically the end point for a geophysicist's contribution to the effort. Structural plays require additional effort beyond this. With the potential for wide variation in surface geology, and surface seismic velocities, the demonstration of four-way closure as well as an estimate of column height must be done in the depth domain. Recent work by active foothills explorers has also demonstrated a requirement to apply more rigor to the calculation of volumetrics for the structure. This rigor involves the calculation of layered rock volumes *versus* a simple net pay estimate, the incorporation of the losses related to the transition zone above the water line, and the weighting of seal failure mechanisms into the final calculation. All this points to an absolute requirement of a correct image in depth for the seismic that defines the structure.

Drilling

Road and location costs aside, the drilling of a foothills well in Western Canada can range from $1MM up to $15MM (in addition to drilling "train wrecks"). Quite often, drilling on an exploration wildcat well is done with less than a full understanding of the geology prior to spud. Recent advances in drilling technology have increased the variance of the final cost of a well depending on the methods used. Perhaps the single most important cost saving technology is that of air-drilling. Penetration rate increases in the order of 3-fold are not uncommon. Costs associated with drilling are generally a direct function of the days spent over the hole and air-drilling can make a substantial difference over directional drilling with drilling fluids. In air drilling, the bit tends to want to go in a direction that is perpendicular to the bedding plane of the rock. With a properly imaged seismic section in depth, drilling engineers can work from the target back to surface to find a location where air-drilling techniques can be utilized. Sometimes several different surface locations need to be evaluated due to surface constraints on placing a well site.

As different seismic markers are encountered during the actual drilling operation, the seismic can be refined by re-migrating the data with the new information. The behavior of the bit will also determine which seismic reflectors can be "believed" and those which are not related to the subsurface geology. Once the borehole is logged via wireline methods, the information regarding dip analysis (most commonly done with borehole imaging), velocities, and other petrophysical properties can be directly tied to the seismic. This allows for a common mindset of what has actually been encountered by the drill-bit for all the team members such as log analysts, drilling engineers, production engineers, and geoscientists.

Technology requirements

There are several components required for successful, and somewhat painless, depth imaging. These involve accurate migration algorithms, computational horsepower, high quality preprocessing of the input data, appropriate analysis tools and experienced interpretation skills. It is safe to say that not all seismic depth migration algorithms in use today are equal in their accuracy. In particular, the correct handling of topographic variation is not common to all applications. With surface elevation changes in the order of 1000 meters in certain areas, the effect of topography on the final image can be substantial. The correct treatment of amplitudes within the algorithm is also key to successful imaging.

A decade ago, there were a handful of computers available in North America that could handle the computational requirements of prestack depth imaging in 2-D, let alone 3-D prestack depth imaging. With advances, the computational power of high-end workstations put at least 3-D prestack depth imaging capability on the desk of the geoscientist. Due to high data volumes and massive computational requirements, 3-D prestack depth imaging remains in the high-end server and mass data storage domain.

Preprocessing of the data involves placing the correct survey (geometry) information in the header, signal processing for noise removal and signal enhancement, and most importantly for foothills data, correct static estimation. Although it is common for both a seismic processor and a geophysical interpreter to work jointly to produce a prestack depth image, it can be argued that it is more efficient for the interpreter to possess the skills required to run the migrations and perform the analysis themselves.

Of equal importance to an accurate migration algorithm is the requirement for analysis tools. These tools take many forms but the superior ones have the ability to interact with the velocity model and the migrated output across a number of common depth point gathers, a graphical display of both the velocity model and ray tracing paths, and imbedded migration parameter selection.

Imaging strategy

Assuming all the tools and skills are in place, a strategy is required to efficiently image the data. One of the first aspects of the strategy is to understand the limitations of the seismic data that is being imaged. The quality of the output seismic section is directly related to the quality of the data being input into the process. Noise, sampling of the data and the quality of the preprocessing can put severe restrictions on the ability to produce a seismic section that gives a true picture of the subsurface geology. In particular, understanding the limitations on resolving dip and velocity due to sampling are key.

The uppermost part of the geological section represents the biggest challenge to successful imaging. Unfortunately, this is the section that is the most poorly sampled by seismic acquisition. Prior to commencing pre-stack imaging for foothills data, all statics should be applied and the data then brought back to topographical surface. In general, it is beyond the data quality of most seismic to expect to extract a velocity-depth model for the very near surface utilizing migration techniques. Although statics are simply vertical time shifts, it has been found that the removal of the effect of the high frequency component of the near surface geology improves the ability to successfully image the data without significantly affecting the quality of the final output.

It is absolutely key to obtaining a proper image that as much of the surface geology be incorporated into the velocity model as possible. This requires strong interaction between the geophysicist and geologist on the team. If possible, the seismic line should be walked by both members of the team, in order to gain surface information and structural style knowledge. A geological cross-section can then be constructed to incorporate this information into the model. With these constraints in place, iterative imaging can then be performed to extract the near surface velocities. The ability to correctly determine these velocities is hampered by the lack of offset in the input data.

If sonic velocity logs are available in the regional vicinity of the seismic line being imaged, one can extract a background "shale velocity gradient". An estimate of the variation of velocity as a function of depth for the clastic section can be input into the model as a first pass. Typically, there should not be any velocities slower than those present in this gradient model. Subsequent iterations of migrations should be analyzed with that thought in mind. These iterations progress down through the depth section until the deepest reflector of interest is imaged.

The interpretation aspect of the pre-stack depth imaging process can best be seen when a variety of hypothetical velocities are placed into the model and a gross understanding of the potential geology can be evaluated.

In conclusion, the practice of pre-stack depth imaging in structural areas can be seen as a key part of the interpretation process and vital to the understanding of the technical risk of the prospect. The product of the process is one that can be used as a common link by the members of a multidisciplinary team. Drilling exploratory foothills wells incurs large capital expenditures in high-risk settings with hope of a very substantial reward. Armed with the tools of prestack depth imaging, the structural geophysicist can make a very large contribution to the success of their team.

Seismic Modeling of an Imbricate Thrust Structure from the Foothills of the Canadian Rocky Mountains

*Steven Lingrey**

ABSTRACT

Seismic structural modeling provides a useful means toward understanding the complex geometry associated with detached-style fold and thrust deformation. Routine seismic methods do not tightly constrain details of structural interpretation, but methods incorporating geometric and seismic modeling allow these details to be inferred. Geometric models measure spatial elements of the folds and faults, and check them for internal consistency, usually against an assumed condition of material balance. Seismic models measure the effects of ray-path bending through a complex, structurally defined velocity field. The seismic model predicts the presence (or absence) of reflections and their pattern on unmigrated sections. These patterns in the modeled data can be checked against the patterns observed in the real data. Iteration between the two models allows the interpretation to converge toward a mutually acceptable solution.

Structural analysis of a seismic profile across the Quirk Creek gas field from the Foothills belt of the Canadian Rocky Mountains is used to illustrate this iterative method of seismic interpretation. The internal geometry of thrust sheets, essentially opaque on the basis of routine examination of a migrated seismic profile, is developed with the aid of geometric and seismic modeling techniques. As with many model-based approaches, proposed solutions in a given geometric or seismic model are nonunique. Used in combination, however, they each check solutions independently and thus can more narrowly constrain the range of acceptable interpretations. The incorporation of synthetic seismic modeling greatly improves the accuracy of interpretation for the structurally complicated Quirk Creek gas field.

INTRODUCTION

Structural interpretations of reflection seismic data in the Foothills belt of the Canadian Rocky Mountains of southern Alberta have typically relied heavily on concepts of structural style (Bally et al., 1966; Dahlstrom, 1970; Harding and Lowell, 1979; Sheriff and Geldart, 1983). This reliance on structural style is necessary in view of the fact that seismic images of fold and thrust structures are typically incomplete and of uneven quality. Parts of the stratal geometry may be

*Esso Resources Canada Ltd.

clearly shown while other parts show either a lack, or a confusing overabundance of reflection signals. Migrated results of uniform quality are difficult to obtain. Thus, generalizations of structural style have been necessary to complete interpretations in areas where imaging is poor.

In general, characteristic patterns of seismic reflections reveal the large scale (regional) geometry of detached-style fold and thrust deformation (e.g., the several seismic illustrations in Bally, 1983). In detail, however, the internal geometry of individual fold and fault structures is not as well defined. Barring difficulties in acquisition or statics, the regions of poor seismic image can be attributed to the effects of ray-path bending caused by the structure itself. Routine processing typically does not adequately account for this effect. The main purpose of this paper is to show how synthetic seismic modeling provides a means for assessing this effect and improving the interpretation.

A structural interpretation of a seismic profile across the Quirk Creek gas field from the Foothills belt of the southern Canadian Rocky Mountains is presented (Figure 1). High effort seismic processing (refraction statics corrections, velocity sweeping for optimal VNMO, and recursive migration) provides a high quality data base. Yet characteristically, the seismic reflections incompletely image the structure due to the complexity of folds and faults. Standard interpretation methods are augmented with the inclusion of geometric and seismic modeling analysis. Geometric models are incorporated to test the interpretation for material balance. Seismic models are incorporated to test the interpretation for reflection imaging potential.

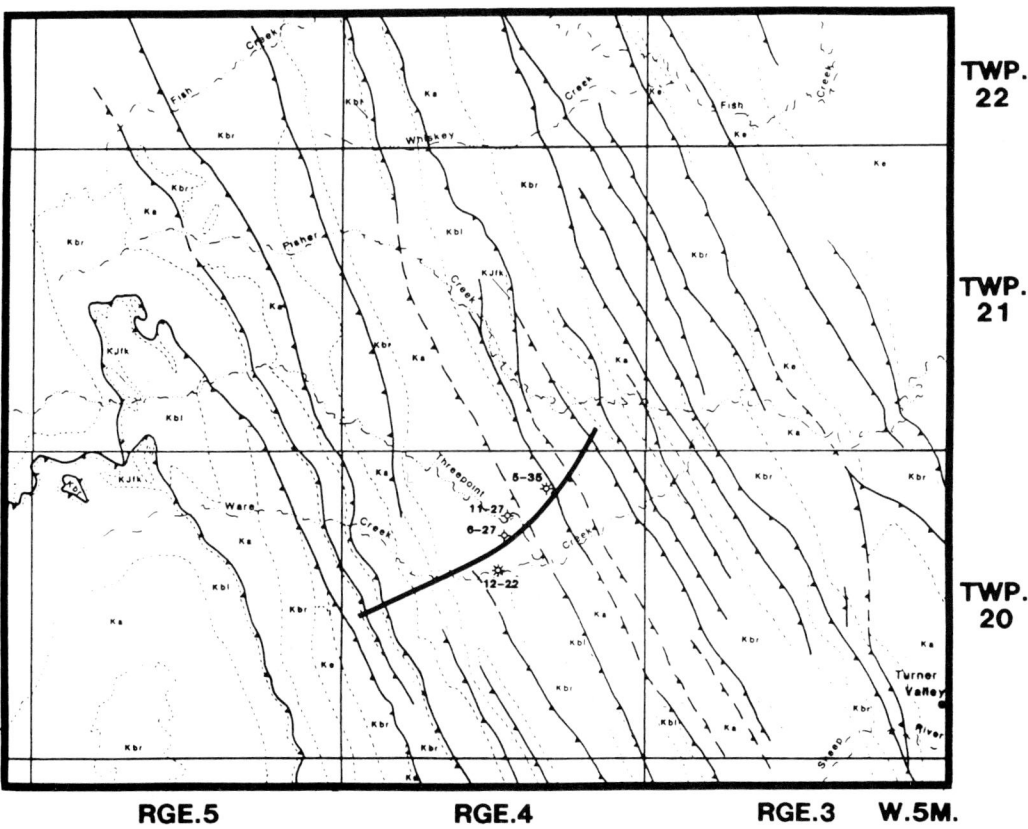

Fig. 1. Location map showing the surface geology for the Foothills Belt of the Canadian Rocky Mountains southwest of Calgary, Alberta and the location of the seismic profile.

METHOD OF SEISMIC INTERPRETATION

Since full seismic imaging of the complicated structural geometry associated with fold and thrust deformation in the Quirk Creek structure cannot be produced, routine seismic interpretation methods are insufficient to achieve an accurate interpretation. Two modeling procedures provide an opportunity to remedy this limitation (Figure 2). First, a geometric model of the seismic interpretation can be constructed and this depth image can be analyzed for material balance. Second, a seismic model (using the balance-tested geometric model) can be run to predict the ray-path distortion and its effect on the patterns of reflections. Modifications in the structural geometry of the interpretation can be tried and assessed on the basis of the fit between observed and predicted (i.e., modeled) reflection patterns. Iteration between the geometric and seismic modeling allows modifications to converge on the strongest interpretation. In the end, the structural interpretation is more completely constrained by the seismic data, even in areas where reflections are missing or improperly migrated.

Geometric modeling involves measurements of certain spatial elements, generally to check for a condition of material balance. In its simplest form, a balanced section assumes conservation of bedlength and this status can be tested by measuring, and comparing for equality, successive pretectonic stratal horizons. In a more advanced form, a kinematic analysis is used to compare pre- and post-deformational profiles. A forward model starts with an undeformed profile and computes the deformed state; an inverse model (or palinspastic restoration) starts with a deformed profile and computes the undeformed state. For a two-dimensional (2-D) analysis, plane strain is assumed. In this study, a palinspastic restoration is used to demonstrate the material balance of the structural interpretation.

The mechanism of deformation is important to geometric modeling as this defines the mathematical process by which the strain is calculated. In this

PROCEDURE FOR INTERPRETATION

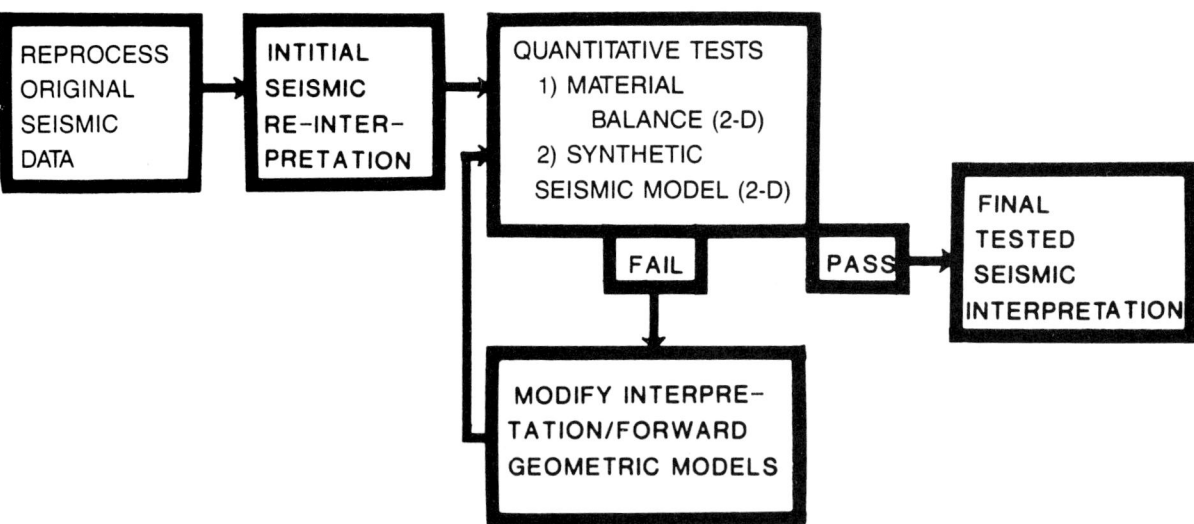

Fig. 2. Flow chart for the method of interpretation in areas of complex structure.

analysis, a flexural-slip mechanism is used to inversely model the folding deformation (Donath and Parker, 1964). As such, bedding surfaces approximate planes (lines in 2-D) of no finite strain within the strain ellipsoid. The principal constraints on flexural-slip deformation on the geometry of folds in fold and thrust belt settings were outlined in Dahlstrom (1969). Recently, Suppe (1983; 1985) has trigonometrically defined two mechanisms of detached-style, flexural-slip folding and thrust faulting: fault-bend and fault-propagation folding. In the process of inferring structural interpretations, the patterns of folds and faults were calculated by means of graphic experiments using Suppe's formulations (forward models).

Geometric models were constructed and restored with the aid of commercially available computer software (Kligfield et al., 1986). This software enables precise representation and kinematic restoration of flexural-slip geometry.

Synthetic seismic modeling involves the analysis of ray-path bending through a complex velocity field (Taner et al., 1970; May and Hron, 1978). Because the pattern of reflections in a time display is sought, and not their amplitude character, the model profile defines relatively thick layers representing averaged interval velocities. The object of the analysis is to calculate the traveltime of a zero-offset ray-path to and from a given reflective horizon. This can be plotted as a synthetic time section showing the pattern of reflection arrivals for an unmigrated, stacked section.

The synthetic result can be compared with the reflections present in the real section. Two modifications to improve the fit between synthetic and real data are possible: (1) the geometry of the layer boundaries in the model can be changed, or (2) the average interval velocities can be changed. In this study, there are several wells containing velocity information on the Mesozoic and Paleozoic rock units; the average sonic velocities used are displayed in Table 1. The values of these velocities are relatively uniform and not too dependent on depth. This uniformity allows the modification of the seismic model to critically test, principally, for the geometry of the layer boundaries.

Seismic models were constructed using the tested geometric models. These depth profiles were input into an interactive modeling system for computation of ray-tracing and plotting of resultant reflection arrivals on synthetic time sections.

STRUCTURAL INTERPRETATION

Figure 3 is a migrated seismic profile situated across the southern part of the Quirk Creek gas field located near the external margin of the Foothills belt of the Canadian Rocky Mountains of Alberta. The approximate location of the line is shown in Figure 1. The structure at Quirk Creek is a thrust faulted, anticlinal trap involving imbricate repetitions of upper Paleozoic carbonate rocks. This Paleozoic level structure is not represented at the surface. No upper Paleozoic rocks crop out at the surface and no conspicuous pattern of fold closure appears in overlying Cretaceous strata at the surface. A stratigraphic chart shows the formations, lithology, and average interval velocities for the sedimentary rocks in this part of the Foothills (Figure 4).

Table 1. Interval velocities used for depth conversion and seismic modeling

Stratal units	Interval velocity
Upper Cretaceous/Lower Cretaceous	13,600 kft/s (4.15 km/s)
Lowermost Cretaceous/Jurassic	15,000 kft/s (4.57 km/s)
Mississippian	21,500 kft/s (6.55 km/s)
Devonian/Cambrian	21,000 kft/s (6.40 km/s)

The goal of this seismic interpretation aims at a more detailed description of the Paleozoic structural geometry of the Quirk Creek trap. The analysis utilizes: (1) the recently recognized principals of structural deformation in detached fold and thrust terranes (Dahlstrom, 1970; Boyer and Elliot, 1982; Suppe, 1983, 1985; Woodward et al., 1985), particularly the importance of the presence of several bedding-parallel detachments (duplex style imbrication), (2) an evaluation of material balance by means of computer-aided geometric modeling (Hossack, 1979; Ramsay and Huber, 1987; Kligfield et al., 1987; Marshack and Woodward, 1988), and (3) an evaluation of the final depth interpretation by means of seismic structural modeling.

Seismic interpretation of the line across the southern part of the Quirk Creek structure was aided by four wells and surface geology taken from Geological Survey of Canada maps (Figure 1; Hume, 1931, 1941; Hume and Beach, 1942; Hage, 1946). The results of the structural interpretation are illustrated in Figure 5.

Structural interpretation of the seismic data is influenced by the accepted importance of bedding-parallel detachment faults. For this part of the Foothills and at the structural level of the Quirk Creek structure, four stratigraphic horizons of detachment are important (Figure 5): (1) a basal detachment near the base of the Cambrian, (2) a detachment within the lower part of the Mississippian Banff Formation, (3) a detachment within the Jura-Cretaceous Fernie-Kootenay Formations, and (4) a detachment near the top of the Lower Cretaceous Blairmore Group. Additional detachments within the Upper Cretaceous section are important, but not critical to the interpretation presented.

The importance of multiple surfaces of bedding detachment is shown in Figure

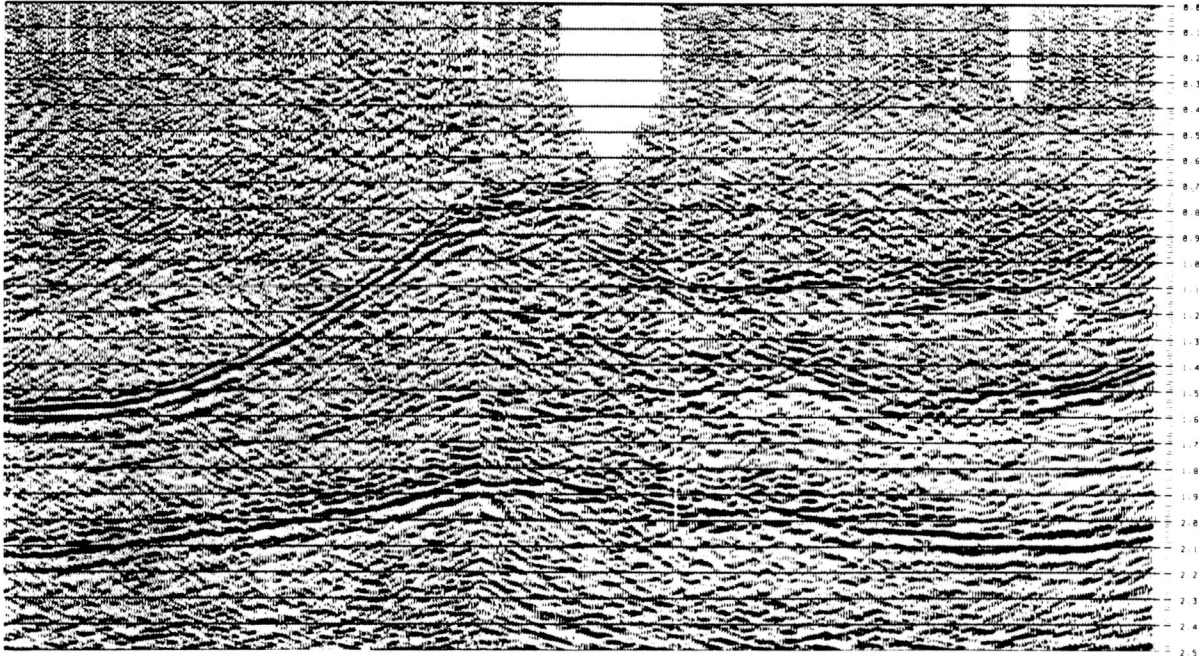

Fig. 3. Migrated seismic profile across the southern end of the Quirk Creek gas field. Structural features at the Paleozoic level are outlined by: (1) a band of high amplitude reflections (Jura-Cretaceous, near top of Mississippian) located at 1.6 s on the left edge, continuous and rising to 0.8 s in the center, and discontinuous and returning to 1.5–1.6 s on the right edge, and (2) a band of high amplitude, lower frequency reflections (mid-Cambrian) continuous across the length of the section located at 2.2 s on the left edge, gently rising to 1.9 s in the center, and located at 2.1 s on the right edge. The upper Paleozoic is imbricately faulted by thrusts, but the lower Paleozoic is autochthonous and unfaulted. Note also the band of high amplitude reflections between 1.0 and 1.1 s on the right half of the seismic line. These reflections arise from the same stratal reflectors described in (1), but at this location do not overlie the Mississippian. The base of these reflections marks the Fisher Mountain thrust fault which overlies repeated Lower Cretaceous Blairmore Group strata.

AGE		FORMATION OR GROUP		LITHOLOGY	THICKNESS	AV. VELOCITIES IN ft/s
CRETACEOUS	UPPER	EDMONTON	UPPER BRAZEAU		1500 - 2000	11,000 / 12,000
		BEARPAW			0 - 150	
		BELLY RIVER	LOWER BRAZEAU		1200 - 3500	12,000 / 13,000
		WAPIABI	ALBERTA GROUP		1250 - 1500	12,500 / 13,000
		CARDIUM			300 - 400	13,000 - 14,000
		BLACKSTONE			800 - 1000	13,000 / 14,000
	LOWER	BLAIRMORE			1100 - 1400	14,000 / 16,000
		KOOTENAY			0 - 1200	14,000 - 16,000
JURASSIC		FERNIE			50 - 500	13,500 - 14,000
MISSISSIPPIAN		MOUNT HEAD	RUNDLE		200 - 300	21,000
		TURNER VALLEY			250 - 350	
		SHUNDA			250 - 300	
		PEKISKO			250 - 300	
		BANFF			500 - 600	
DEVONIAN	UPPER	PALLISER			800 - 1000	21,000
		FAIRHOLME			1000 - 1300	
CAMBRIAN	U?	ARCTOMYS ?			0 - 100 ?	20,000
	MIDDLE	PIKA			150 - 400	
		ELDON			500 - 900	
		STEPHEN			200 - 250	
		CATHEDRAL			200 - 800	
PE		HUDSONIAN				

Fig. 4. Table of formations for the southern Alberta Foothills of the Canadian Rocky Mountains showing sites of bedding detachment and average interval velocities. Paleozoic interval velocities are insensitive to depth of burial; Mesozoic interval velocities only mildly increase with depth of burial. (modified from Gordy et al., 1975).

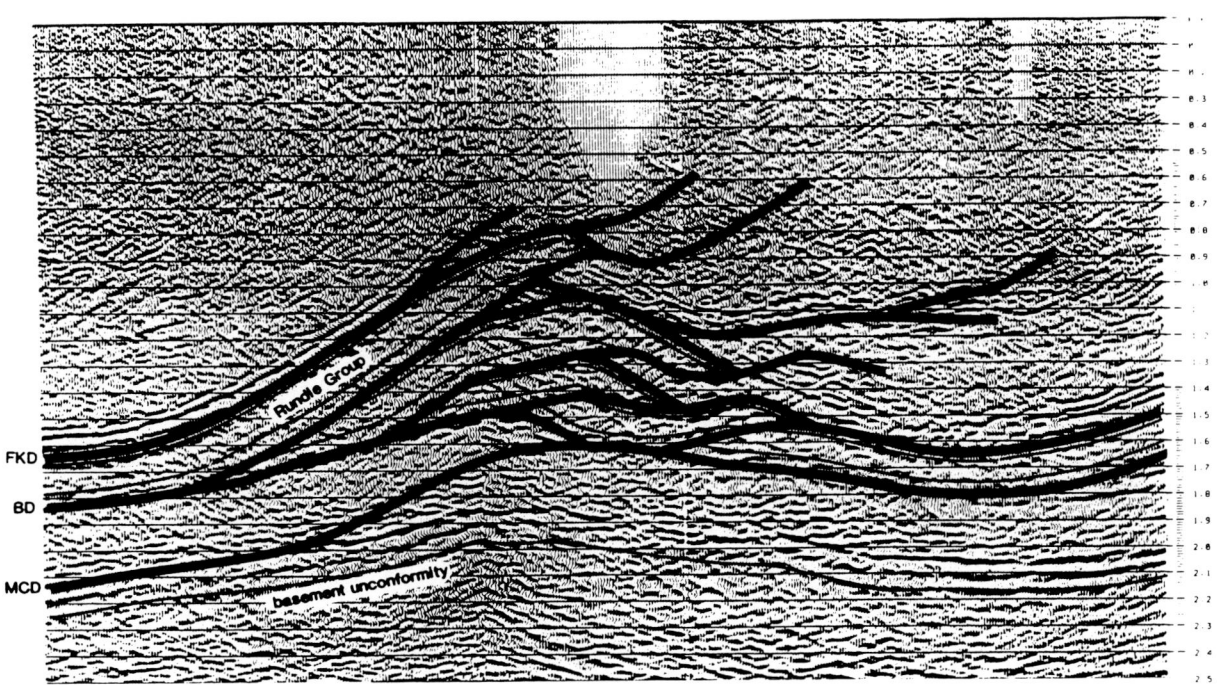

Fig. 5. Final interpretation of the seismic profile showing structural development of thrust sheets above Cambrian and Banff detachments. Structural deformation above the Fernie-Kootenay and Blairmore detachments is not interpreted. Time converted well control is from wells listed in Appendix 1.

6. Many interpretations have emphasized a basal detachment, but have neglected or minimized the higher detachments. The structural interpretation presented here for the Quirk Creek structure emphasizes a duplex-style imbrication of the Mississippian strata between the Banff and Fernie-Kootenay detachments.

The gross features of the trap image well on routinely processed seismic data as a band of three to four, high amplitude reflections that mark a position near the base of the Blairmore Group and the top of the Kootenay Formation. While not precisely at the top of the Paleozoic (Mississippian Mount Head Formation), these reflections are close enough (25–50 ms) to reliably delimit the top of Paleozoic at most places. These reflections define an asymmetric, northeast-vergent fold in profile. The seismic dip line (Figure 3) shows the top of Paleozoic as deep on the southwest (1.6–1.7 s), high on the crest (0.8–0.9 s), then deep again on the northeast (1.55–1.65 s). The structure shows a smoothly defined crest and southwest-dipping backlimb, and a raggedly defined, northeast-dipping forelimb. Reflections do not explicitly define the shape of the forelimb, yet do constrain the region of its occurrence. While much of the external aspect of the structural culmination is clearly discernible, several deep wells into the structure demonstrate a complicated internal structure consisting of several repetitions of the Mississippian section. This internal structure is *not* clearly imaged in the seismic profile and thus cannot be interpreted unambiguously.

A deeper, single reflection of high amplitude marks a position near the base of the Banff (detachment). The continuity of this reflection is poor, but the time dip clearly shows the reflection to be more flat-lying than the top of Mississippian. This reflection pattern is indicative of a floor thrust (lower detachment) above which the overlying strata have been tilted by imbrication. This reflection shows a long wavelength anticlinal warp on the time section, but with a relatively small structural relief (220–260 ms). A small wavelength step in this reflection occurs just east of the crest of the gentle anticline. As is discussed later, the gentle anticline is an artifact (velocity anomaly) on the time section. The sharper,

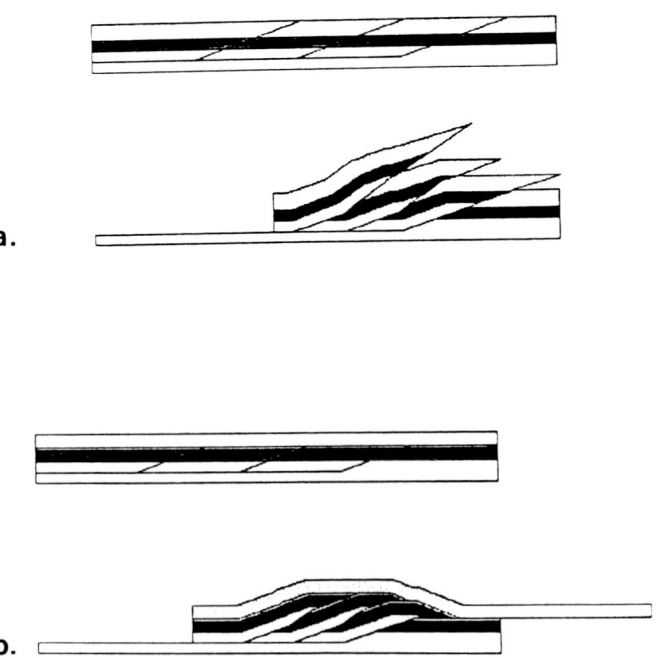

Fig. 6. Structural style of imbrication involving: (a) single basal bedding detachment, and (b) two levels of bedding detachment (duplex style imbrication). Note the absence of tectonic thickening on the left and right edges of the diagram.

east-descending step is interpreted to be the fold at the leading-edge of a thrust coming from the deeper, basal detachment.

An even deeper band of two to three, high amplitude (lower frequency) reflections mark a position near the upper part of the Cambrian section. In contrast to the near top Paleozoic reflections and the near base of Banff reflection, these Cambrian reflections are more or less continuous, indicating little or no disruption by thrust faults (i.e., these strata are largely below the basal detachment and the detachment is interpreted to lie at the top of this package of reflections). These reflections also display a gentle anticlinal warp on the time section with a structural relief of 200–250 ms.

The pattern of these prominent reflection bands demonstrates unambiguously that the Quirk Creek structure is a product of duplex style imbrication. The Fernie-Kootenay detachment (the base of the upper reflections band) is the folded roof thrust of the duplex. The duplex is relatively simple being confined for the most part to imbrications between the Banff detachment (floor thrust) and the Fernie-Kootenay detachment (roof thrust). In detail, the deeper duplex imbrication between the Cambrian and Banff detachments slightly modifies (folds) the shallower duplex.

While the recognition and tracing of the external shape of the duplex deformations can be made with some confidence, the internal geometry of the individual thrust sheets comprising the duplex are less clearly defined. Two types of reflection patterns are critical for the definition of these thrust sheets (Figure 7). The first type of pattern is that associated with the reflection terminations marking the leading edge of a thrust sheet. These terminations indicate the position of the hanging-wall ramp. The second type of pattern is that associated with the reflection terminations marking the trailing edge of a thrust sheet. These terminations indicate the position of the foot-wall ramp. Many reflection patterns on the dip lines are suggestive of these structural elements. To confirm or deny these inferred thrust ramps and the overall structural interpretation, two tests have been made in the process of achieving the final interpretation. These tests are: (1) geometric modeling that involved checks on 2-D material balance (palinspastic restoration), and (2) seismic structural modeling that involved zero-offset ray tracing to confirm the magnitude of pull-up effects and to predict imaging attributes on unmigrated seismic data.

GEOMETRIC MODEL

Initial structural interpretations of the seismic data are tested by means of geometric models. This process involves: (1) depth conversion of the seismic interpretation, (2) geometric analysis of the interpretation (checks for material

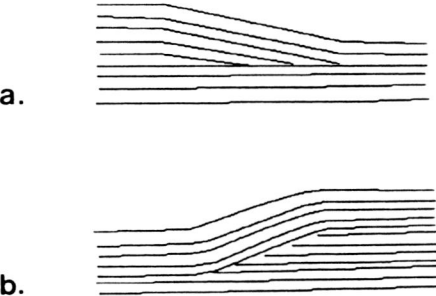

Fig. 7. Line drawing showing the pattern of reflections that are indicative of: (a) leading-edge of thrust sheet, and (b) trailing-edge of thrust sheet.

balance), and (3) modifications of the interpretation to remedy areas of imbalance in the interpretation. The computer program allowed the impact of changes in the geometry of the interpreted thrust sheets to be assessed, thereby suggesting changes that could improve areas of imbalance. For the Quirk Creek seismic line, three cycles of interpretation modification were involved.

Depth conversion of the interpreted seismic line in this part of the Foothills Belt of the Canadian Rocky Mountains is aided by: (1) the platformal character of the Paleozoic strata which shows a very gradual primary thickening (approximately 0.3 degrees) of stratigraphic layers to the west, (2) the fairly uniform interval velocities characteristic of the Mesozoic strata, and (3) deep well control, one with check-shot information.

To the northeast and southwest of the Quirk Creek structure, there are zones in which the Paleozoic has not been duplicated across thrust ramps. In these zones, the top of the Paleozoic in depth can be calculated using the average interval velocity for Mesozoic strata treated as one layer. This is fortunate, because the necessity to recognize and compensate for the complexly deformed Mesozoic layers is obviated. By connecting the calculated depth in front (northeast) of the Quirk Creek structure to the calculated depth in back (southwest), a regional gradient in depth for the top of the Paleozoic is established. Using known thicknesses of Paleozoic strata (checked for internal consistency with the time thickness in the seismic interpretation), the regional gradient of the autochthonous basement can be determined.

Estimations of the appropriate depths across and within the Quirk Creek structure can be either calculated downward using the interval velocities, or calculated upward using the net amount of duplicated Paleozoic section above the regional gradient. These calculations are well constrained; since for each individual thrust sheet interpreted, the horizontal dimension of its leading- and trailing-edge can be measured directly and the vertical dimension between the Banff and Fernie-Kootenay detachments can be inferred from the nearly uniform stratigraphic thickness. Forward geometric models using the idealized fault-bend fold behavior were used in some cases to anticipate the spatial details of imbricate thrust patterns (Suppe, 1983).

The depth conversion of the initial interpretation made for the Quirk Creek seismic line is shown in Figure 8 with its palinspastic restoration. Significant gaps occur between the restored thrust sheets indicating a material imbalance in this interpretation. Accepting the appropriateness of the assumptions (plane strain, flexural-slip folding), the implication of this analysis is that the interpretations need to be modified to remove the spatial gaps between the thrust sheets. As the external geometry of the Quirk Creek structure is rather tightly constrained by the seismic data, modifications are directed toward the internal geometry of the thrust sheets.

In the following discussion the seismic model for this geometric model points out a significant discrepancy between the calculated and observed time anomaly in unfaulted Cambrian reflections underlying the structure. The implication of this analysis is that too much Paleozoic and not enough Mesozoic is interpreted in the core of the structure.

The depth conversion of the final interpretation made for the Quirk Creek seismic line is shown in Figure 9 with its palinspastic restoration. This result follows two reinterpretations in which modifications were made and tested to see if the state of material balance improved. The restoration shows no significant gaps. While not necessarily a unique solution, this interpretation is demonstrably consistent with the assumed mechanism of deformation. When joined with the observation that the seismic model for this interpretation well matches the observed seismic reflection patterns in the unmigrated section, the accuracy of the final interpretation is considered to be optimized.

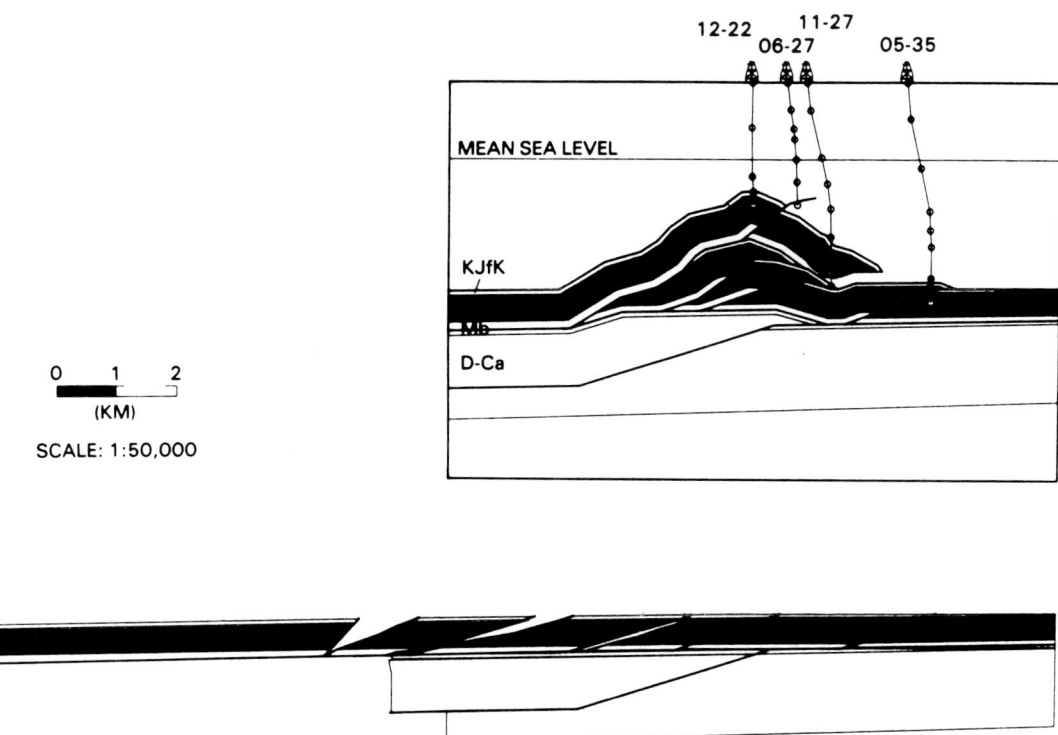

Fig. 8. Depth converted structural profile of initial interpretation for seismic line across southern Quirk Creek gas field showing palinspastic restoration.

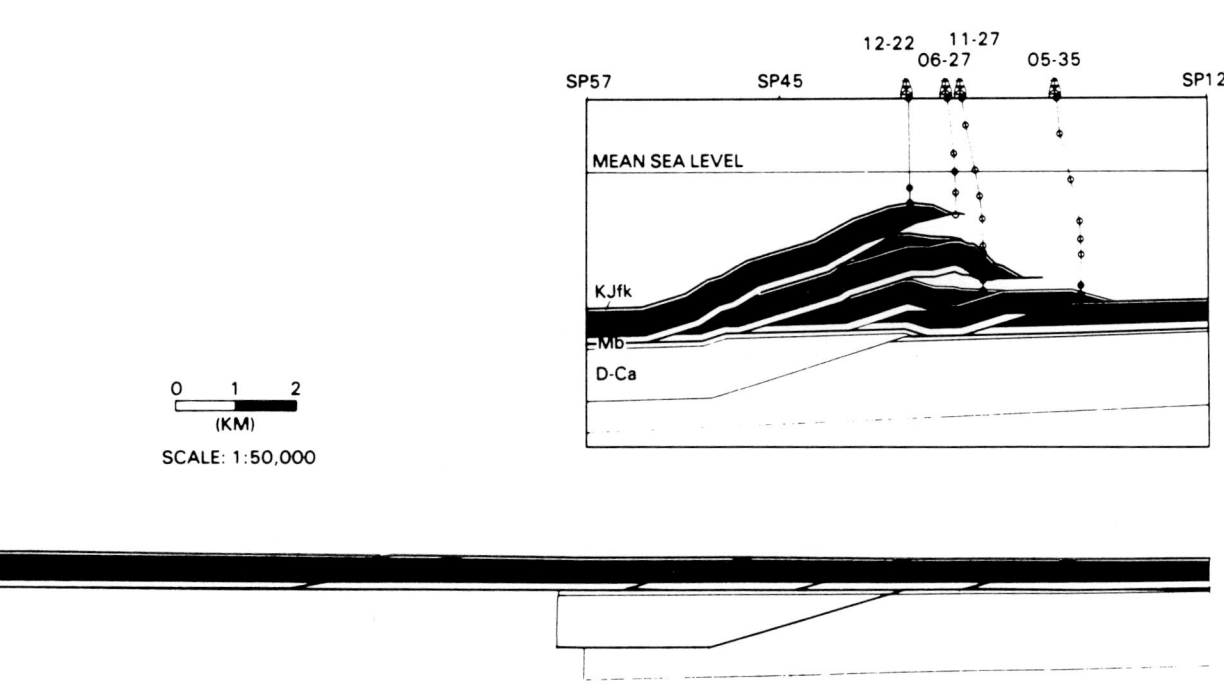

Fig. 9. Depth converted structural profile of final interpretation for seismic line across southern Quirk Creek gas field showing palinspastic restoration.

SEISMIC MODEL

Initial structural interpretations of the seismic data, in addition to the geometric models, are tested by means of synthetic seismic models. This process involves: (1) depth conversion of the seismic interpretation (the same depth converted model used for geometric analysis can be input for synthetic seismic analysis), (2) definition of model layers and their appropriate average interval velocities, (3) computation of the zero-offset ray-paths using the general acquisition parameters (seismic datum, shot-point spacing, etc.) of the real seismic line, (4) generation of a plot of reflection arrivals at the proper time scale for comparison with the real seismic line, and (5) modification of the interpretation to remedy areas of mismatch in the unmigrated reflection image. Small modifications to one or two of the layer geometries in the model allow the impact of interpretation changes to be assessed. These modifications could be tried interactively using the modeling program. For the Quirk Creek seismic line, two of the three geometric models generated were input as seismic models.

The seismic model of the initial interpretation made for the Quirk Creek seismic line is shown in Figure 10 showing the average interval velocities assigned the 18 layers. The model calculates reflection arrivals for the top of Paleozoic, the top of the Devonian (equivalent to the Banff detachment), and the base of Cambrian (about 150 ms below the Cambrian reflections). The predicted reflection patterns are shown in Figure 11, overlaying the unmigrated seismic data. As is the case with the geometric model of this interpretation, several problems are apparent. An important attribute of the model is the strong time "pull-up" of the spatially flat-lying top of basement reflector. This attribute constitutes the most conspicuous discrepency between the predicted and observed reflections. The model predicts a maximum time pull-up of 300 ms, while the observed pull-up of Cambrian reflections is 220 ms.

Estimation of the expected pull-up effect due to the imbrication of high velocity Paleozoic carbonates (interval velocity = 6.4–6.5 km/s; 21.5 kft/s) into lower velocity Mesozoic clastics (interval velocity = 4.11–4.27 km/s; 14.0 kft/s) is shown in Figure 12. Imbrication result from displacement along thrust faults repeats part or all of the upper Paleozoic section. The repetition creates a tectonically thickened section in the region of the thrust ramp; up-dip and down-dip of the ramp, the section may show normal stratigraphic thickness. As such, traveltimes beneath a region of thickened carbonate strata are less than those beneath a region

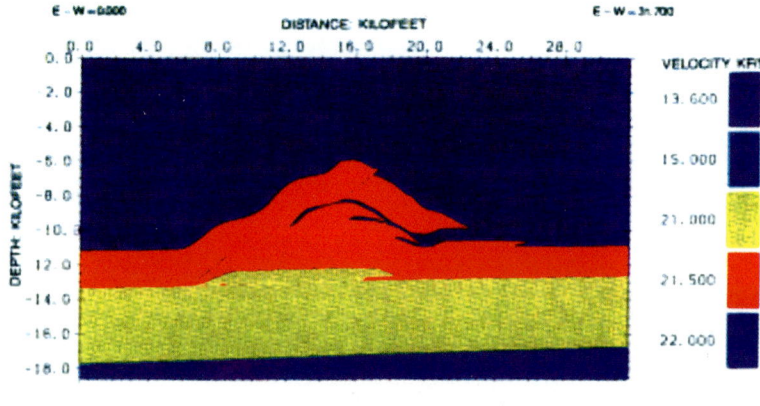

INITIAL QUIRK CREEK MODEL FOR LINE 60636

Fig. 10. Seismic model of initial interpretation (same structural interpretation shown in Figure 8) showing average interval velocities of model layers.

of normal thickness for an equivalent depth. Thus a repetition of 550 m (1800 ft) of Paleozoic carbonate create the effect of lowering the traveltimes (velocity pull-up) for reflections beneath the repetition by 90–100 ms (Figure 12). Duplications of Lower Cretaceous clastics (interval velocity of some layers as high as 4.57 km/s; 15,000 kft/s) can also cause velocity pull-up, but the magnitude of repetition must be over three times that of a Paleozoic repetition to produce an equivalent effect. Modifications to the initial interpretation that include more Mesozoic strata

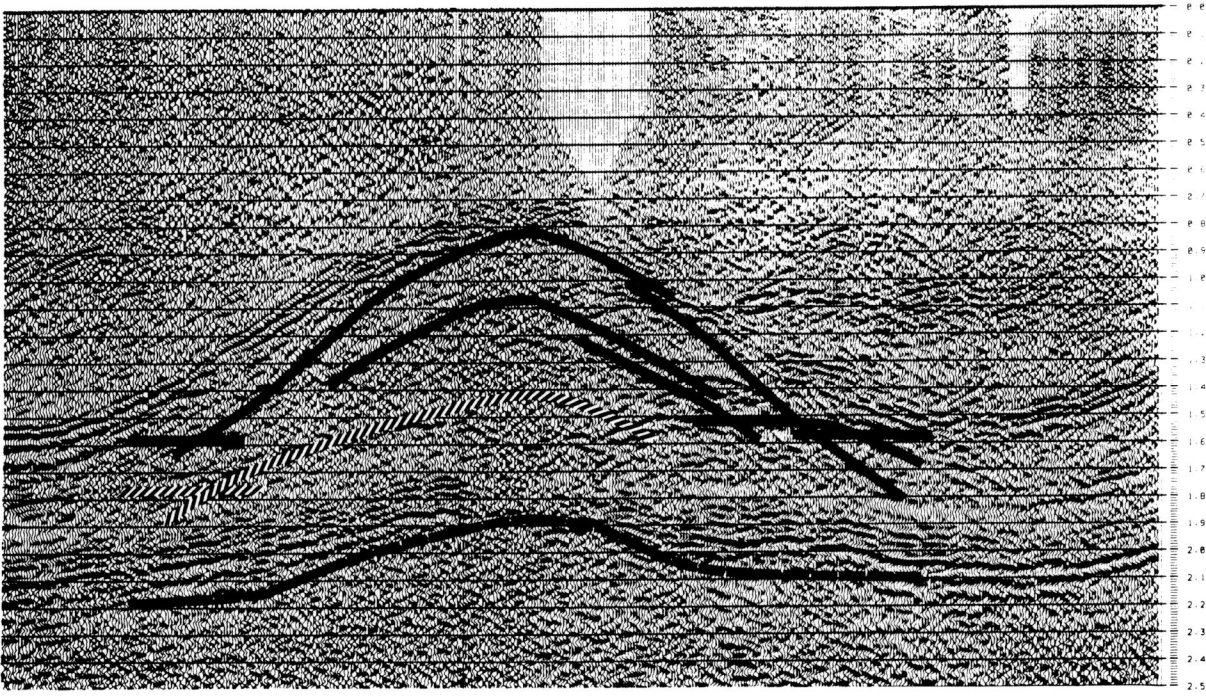

Fig. 11. Plot of reflection arrivals calculated for zero-offset ray-paths through model shown in Figure 10.

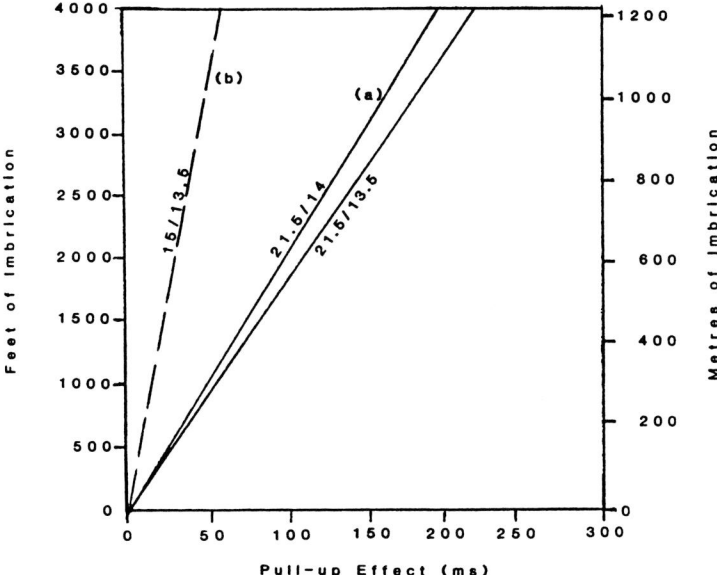

Fig. 12. Graphic plot of the calculated magnitude of time pull-up when replacing: (a) high-velocity carbonates and (b) high-velocity clastics.

in internal imbrication of the Quirk Creek structure should improve the fit of the model.

In the process of seismic reinterpretation, the graph shown in Figure 12 was used to estimate the amount of pull-up for any interpreted amount of Paleozoic imbrication. Excess Paleozoic thickness (the observed tectonic thickening in time of Paleozoic strata) was measured and the calculated pull-up was determined. The consistency between the structural interpretation and the observed velocity pull-up effect was thereby maintained. Changes in reinterpretation needed to preserve the magnitude and location of the pull-up.

The seismic model for the final interpretation is shown in Figure 13. This model incorporates reinterpretations in which modifications were checked against material balance and estimations of velocity pull-up. The predicted pattern of reflection arrivals now constitutes a good match with the real data (Figure 14). In addition to the velocity pull-up simulated in the Cambrian and Devonian reflections, the patterns characterizing the near top Mississippian reflections approximate the seismic image observed. To reiterate, the accuracy of this final interpretation is considered optimal, given the harmony between results of the independent checks of the geometric and seismic models.

To contemplate more sophisticated seismic models, and to recognize that tighter constraints on the interpretation are possible, is interesting. Gather modeling could check the quality of VNMO correction and the adequacy of assumed hyperbolic moveout patterns. Diffraction arrivals could be plotted to identify diffraction signals in real data that are not adequately suppressed in stacking and migration. Migration of synthetic sections could evaluate the effectiveness of the migration algorithm on the data set. Finally, 3-D seismic models could predict sensitivity to 2-D aquisition methods.

CONCLUSIONS

Interpretations of fold and thrust belt structures on reflection seismic data have frequently broken down when tested by well penetrations. The routine methods of aquisition, processing, and interpretation seem to be insufficient to the problem of predicting the often complex structural geometry. Geometric and seismic models provide a means for extending routine methods to cope with the complex geology.

A geometric model can bring in constraints on an interpretation in the form of strain analysis. Certain geometric attributes of the interpretation, appropriate for

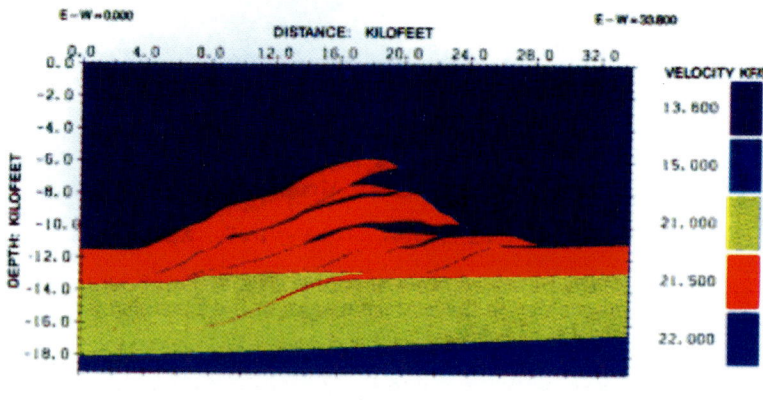

Fig. 13. Seismic model of final interpretation (same structural interpretation shown in Figure 9) showing average interval velocities of model layers.

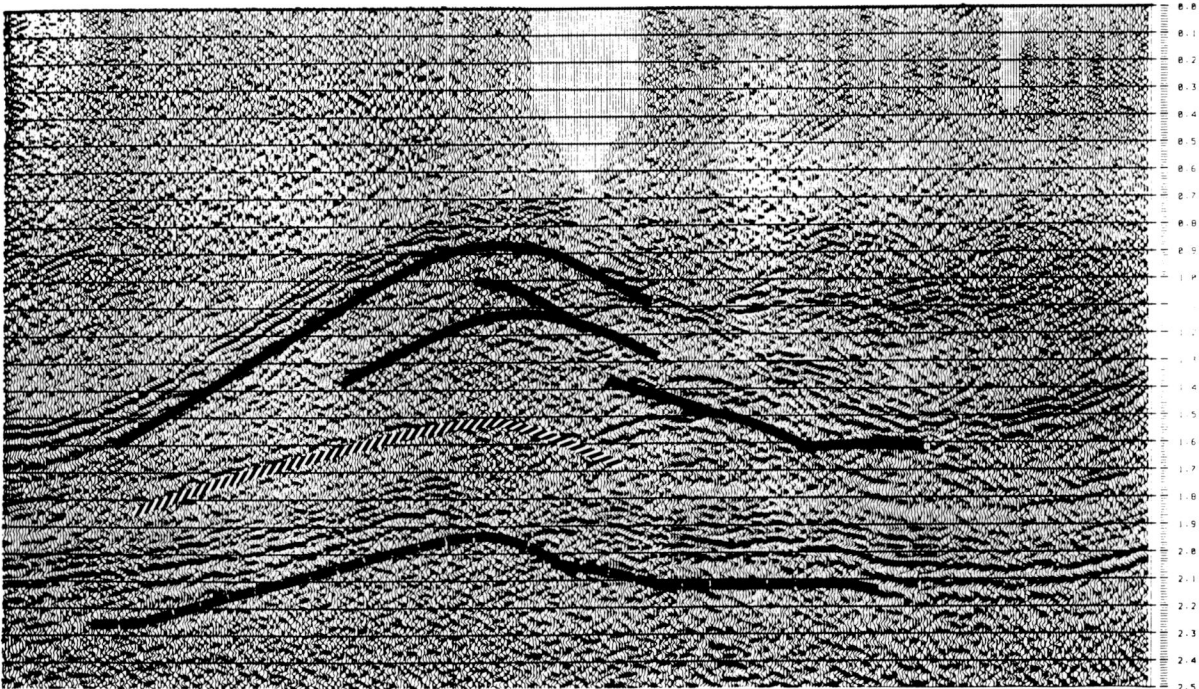

Fig. 14. Plot of reflection arrivals calculated for zero-offset ray-paths through model shown in Figure 12.

the assumed mechanism of deformation, can be checked and if need be, modified. A seismic model can bring in constraints on an interpretation in the form of ray-path analysis. The imaging potential of unmigrated, stacked sections can be predicted. Provided the interval velocity properties of the sedimentary rocks being deformed are known, the seismic model predicts the effect of interpreted structural geometry on the ray-paths. Seismic attributes that may or may not be constructive to the standard reflection image can be identified.

Each of the models can lead to solutions that are not necessarily unique. Each model, however, works independently. Used in combination the models can more narrowly constrain the range of acceptable solutions. The opportunity therefore exists to greatly improve the accuracy of interpretation in structurally complicated areas.

REFERENCES

Bally, A. W., 1983, Seismic expression of structural styles, Volume 3—Tectonics of compressional provinces/strike slip tectonics: Am. Assn. Petr. Geol. Studies in Geology Series No. 15.

Bally, A. W., Gordy, P. L., and Stewart, G. A., 1966, Structure, seismic data and orogenic evolution of the southern Canadian Rocky Mountains: Bull. Can. Petr. Geol., 14, 337–381.

Boyer, S. E., and Elliott, D., 1982, Thrust systems: Bull., Am Assn. Petr. Geol., 66, 1196–1230.

Dahlstrom, C. D. A., 1969, Balanced cross sections: Can. J. Earth Sci., 6, 743–757.

——, 1970, Structural geology in the eastern margin of the Canadian Rocky Mountains: Bull. Can. Petr. Geol., 18, 332–406.

Donath, F. A. and Parker, R. B., 1964, Folds and folding: Geol. Soc. Am. Bull., 75, 45–62.

Hage, C. O., 1946, Dyson Creek: Geol. Surv. Canada Map 827A.

Harding, T. P. and Lowell, J. D., 1979, Structural styles, their plate tectonic habitats, and hydrocarbon traps in petroleum provinces: Bull., Am. Assn. Petr. Geol., 63, 1016–1058.

Hossack, J. R., 1979, The use of balanced cross-sections in the calculation of orogenic contraction: A review: J. Geol. Soc. London, 136, 705–711.

Hume, G. S., 1931, Turner Valley sheet: Geol. Surv. Canada Map 257A.

———, 1941, Fish Creek: Geol. Surv. Canada Map 667A.

Hume, G. S. and Beach, H. H., 1942, Bragg Creek: Geol. Surv. Canada Map 654A.

Kligfield, R., Geiser, P., and Geiser, J., 1986, Construction of geologic cross sections using microcomputer systems: Geobyte, Spring issue, 60–66.

Marshack, S. and Woodward, N., 1988, Introduction to cross-section balancing: *in* Marshack, S. and Mitra, G. (eds.), Basic methods of structural geology: Prentice-Hall, Inc., 303–332.

May, B. T. and Hron, F., 1978, Synthetic seismic sections of typical petroleum traps: Geophysics, **43**, 1119–1147.

Ramsay, J. G., and Huber, M. I., 1987, The techniques of modern structural geology, Volume 2, Folds and fractures: Academic Press, 309–700.

Sheriff, R. E. and Geldart, L. P., 1983, Exploration seismology, Volume 2, Data-processing and interpretation: Cambridge University Press, 221.

Suppe, J., 1983, Geometry and kinematics of fault-bend folding: Am. Jour. Sci., **283**, 684–721.

———, J., 1985, Principals of structural geology: Prentice-Hall, 537.

Taner, M. T., Cook, E. E., and Neidell, N. S., 1970, Limitations of the reflection seismic method, Lessons from computer simulations: Geophysics, **35**, 551–573.

Woodward, N., Boyer, S. E., and Suppe, J., 1985, An outline of balanced cross-sections: Univ. Tenn. Dept. Geol. Sciences Studies in Geol. 11, 2nd ed., 170.

Chapter 2 – Prestack Migration Concepts

Chapter 2 describes the concepts and features of prestack depth migration. One of the most straightforward ways to understand prestack migration is to consider the possible subsurface reflection points for given source and receiver positions. The surface defining the possible subsurface reflecting points for a seismic sample is termed an aplanatic surface. In the case of constant velocity, the surface is an ellipse since the sum of distance from source to scatterer and scatterer to receiver is constant. Figure 2-1 from *Liner and Lines (1994)* in this chapter shows the aplanatic surface for a given trace which is the response of a diffracting scatterer. For a single trace, there is considerable ambiguity in defining the location seismic scatterer. However, as more of the aplanatic surface traces are added to the image, the image point becomes well defined since the seismic responses will constructively interfere at the imaging point, as shown by Figure 2-2. To be more generally correct, the amplitudes should be weighted by an obliquity factor before summation (Scales, 1995).

In the general case of variable velocity, the traveltimes from source or receiver to subsurface scattering points are defined by raytracing solutions. Wavefront mapping can be done by using the eikonal equation *(Gray and May, 1994)* or by wave equation modeling (Chang and McMechan, 1986). In order to do this accurately, by any method, we must have a good estimate of the seismic velocity model. A cynic would say that to find the image (the answer), we need to know the answer in terms of a model. In practice, things are not as demanding as that. We can iteratively apply the prestack migration process itself to estimate the answer. Two criteria are generally used to estimate velocity. The first is the focusing criterion that as the velocity estimate approaches the correct answer, the depth image becomes "focused". This process of adjusting velocity and monitoring the image is similar to the process of adjusting the focal length on a camera to improve the image. A second criterion and one that is discussed by *Zhu, et al. (1998)* involves the fact that in a depth migration the depth image of a scattering point should not depend on the source-receiver offset if the correct velocity is used in imaging. In other words, the common image gathers should not have "smiles" or "frowns" if the velocity is correct. Features of prestack depth migration and the accompanying velocity analysis are discussed in this chapter. A comparison of depth migration methods is given by *Gray (1997), Gray (1998)* and *Zhu and Lines (1998)*.

References:

Chang, W.F., and McMechan, G.A., 1986, Reverse-time migration of offset vertical seismic profiling data using the excitation-time imaging condition: *Geophysics*, 51, 67-84.

Gray, S.H., 1998, Speed and accuracy of seismic migration methods: unpublished report, 36 p.

Gray, S.H., 1997, Seismic imaging: Use the right tool for the job: *The Leading Edge*, 1585-1588.

Gray, S.H., and May, W.P., 1994, Kirchhoff migration using eikonal equation traveltimes: *Geophysics*, 59, 810-817.

Liner, C.L., and Lines, L.R., 1994, Simple prestack migration of crosswell seismic data: *Journal of Seismic Exploration*, 3, 101-121.

Scales, J.A., 1995, Theory of seismic imaging: Springer Pub. Co. and Samizdat Press.

Zhu, J., Lines, L.R., and Gray, S.H., 1998, Smiles and frowns in migration/velocity analysis: *Geophysics*, 63, 1200-1209.

Zhu, J., and Lines, L.R., 1998, Comparison of Kirchhoff and reverse-time migration methods with applications to prestack depth imaging of complex structures: *Geophysics*, 63, 1166-1176.

Kirchhoff migration using eikonal equation traveltimes

Samuel H. Gray* and William P. May*

ABSTRACT

The use of ray shooting followed by interpolation of traveltimes onto a regular grid is a popular and robust method for computing diffraction curves for Kirchhoff migration. An alternative to this method is to compute the traveltimes by directly solving the eikonal equation on a regular grid, without computing raypaths. Solving the eikonal equation on such a grid simplifies the problem of interpolating times onto the migration grid, but this method is not well defined at points where two different branches of the traveltime field meet. Also, computational and data storage issues that are relatively unimportant for performance in two dimensions limit the applicability of both schemes in three dimensions. A new implementation of a gridded eikonal equation solver has been designed to address these problems. A 2-D version of this algorithm is tested by using it to generate traveltimes to migrate the Marmousi synthetic data set using the exact velocity model. The results are compared with three other images: an F-X migration (a standard for comparison), a Kirchhoff migration using ray tracing, and a Kirchhoff migration using traveltimes generated by a commonly used eikonal equation solver. The F-X-migrated image shows the imaging objective more clearly than any of the Kirchhoff migrations, and we advance a heuristic reason to explain this fact. Of the Kirchhoff migrations, the one using ray tracing produces the best image, and the other two are of comparable quality.

INTRODUCTION

In spite of greater imaging accuracy offered by other migration methods, Kirchhoff migration promises to remain a method of choice for prestack migration, especially in three dimensions, for some time to come. Kirchhoff migration is unique in its ability to migrate input traces selectively onto a prespecified output volume, and this capability allows a target-oriented 3-D prestack migration to be performed hundreds of times faster by the Kirchhoff method than by competing methods. A key issue for Kirchhoff depth migration is the choice of an algorithm for generating traveltime maps. One such algorithm is ray tracing followed by interpolation of traveltimes onto the regular grid used in the migration. Another is based on solving the eikonal equation directly on the regular grid. The eikonal equation, in two space dimensions

$$\left(\frac{\partial \tau}{\partial x}\right)^2 + \left(\frac{\partial \tau}{\partial z}\right)^2 = \frac{1}{c^2}, \quad (1)$$

has primarily been used to analyze traveltime along individual raypaths in a medium with acoustic velocity $c(x, z)$. Recently, however, Reshef and Kosloff (1986) and Vidale (1988) proposed using the eikonal equation to compute traveltime fields directly on a regular grid as an alternative to methods based on tracing many raypaths. These workers solved the eikonal equation numerically, by finite differences, at each grid point. This procedure eliminates several issues associated with ray tracing, such as: shooting versus two-point ray tracing, how many rays to trace, and the type of velocity model (gridded or interface-based) to use. A further reason to study gridded solutions of the eikonal equation is for traveltime calculations in three dimensions, where even the simplest ray-based methods can be horribly complex. On the other hand, some questions about this method need to be answered. How well does the method work in complicated velocity fields where several raypaths can intersect at a depth point (caustics)? Will the anticipated failure of the method at caustics cause instability of solutions beyond caustics? Is the method efficient and, if not, can it be made efficient without significant loss of accuracy? What are the extra requirements of the method in three dimensions?

One purpose of this paper is to examine critically the use of gridded eikonal solvers in computing traveltimes for Kirchhoff depth migration. Since the effects of small pointwise errors in a traveltime field are likely to be fully compensated by the averaging process of migration, we are more concerned with the aggregate accuracy and stability

---

Manuscript received by the Editor April 27, 1993; revised manuscript received October 15, 1993.
*Amoco Production Company, P.O. Box 3385, Tulsa, OK 74102.
© 1994 Society of Exploration Geophysicists. All rights reserved.

properties of the method than with detailed features of individual traveltime tables. Our criterion for assessing these properties (at least for migration) will be the correctness of the image produced by using the traveltime tables. To this end, the Marmousi model of the Institut Français du Petrole (IFP) (Versteeg and Grau, 1991) provides an ideal test. This synthetic data set, generated using a very complicated geologic model, provides a challenge for any migration method, even when the correct velocities are used. We shall compare several migrated images produced using different published methods: one using an F-X (recursive) integral migration method, one using Kirchhoff migration with ray-generated traveltime tables, and one using Kirchhoff migration with traveltime tables generated by Vidale's (1988) gridded eikonal solver. We include the F-X migration in the comparison as a benchmark result. The great accuracy of F-X migration, except at dips steeper than those present in the Marmousi model, makes it a useful standard for comparison, even though it lacks the speed and versatility of Kirchhoff migration.

The second purpose of the paper is to present a modification of two popular eikonal solvers that is well-suited to three dimensions. By combining features of Vidale's (1988) and van Trier and Symes's (1991) methods, we have developed a method which is very fast and has reasonable memory requirements. We include in our comparison a Kirchhoff migrated image using the traveltime tables generated by this method.

THE MIGRATION COMPARISON

We show the migration results in Figures 1–4. These are stacked sections from migrating the Marmousi data set, where the exact velocity model was used for all the migrations. Horizontal and vertical scales are identical in these figures. For the Kirchhoff migrations, the velocity (actually, the slowness or inverse velocity) model was smoothed over a square box whose sides had a length approximately equal to one wavelength at the dominant frequency of the seismic data. This was necessary for traveltime calculation by either ray tracing or the eikonal equation, both of which require some smoothness of the velocity function. We show final stacks, rather than unstacked migrated sections, for comparison because the complexity of the seismic data and the velocity model caused poor imaging quality on individual sections, even for the near offsets. All migrations were performed on a Cray-2 computer. Even on this machine, where F-X migration is easier to optimize than Kirchhoff migration, the Kirchhoff migrations ran significantly (about 20 times) faster than F-X migration. In the Appendix we explain the accuracy and speed differences between F-X migration and Kirchhoff migration.

Figure 1 shows an F-X migration, where the input records were plane-wave records synthesized from the original shot records (Whitmore and Garing, 1990). This migration is the best of the four. It shows most clearly the fault faces in the upper part of the section and the imaging objective, which is the flat event centered at a depth of about 2440 m at shotpoint #140. The objective was modeled as a structural high representing a hydrocarbon-bearing sand with a large acoustic impedance contrast.

Figures 2–4 show the Kirchhoff migrations. These migrations operated on input shot records, producing output common offset records. Amplitude terms for the migrations shown here were adapted from the constant-velocity, com-

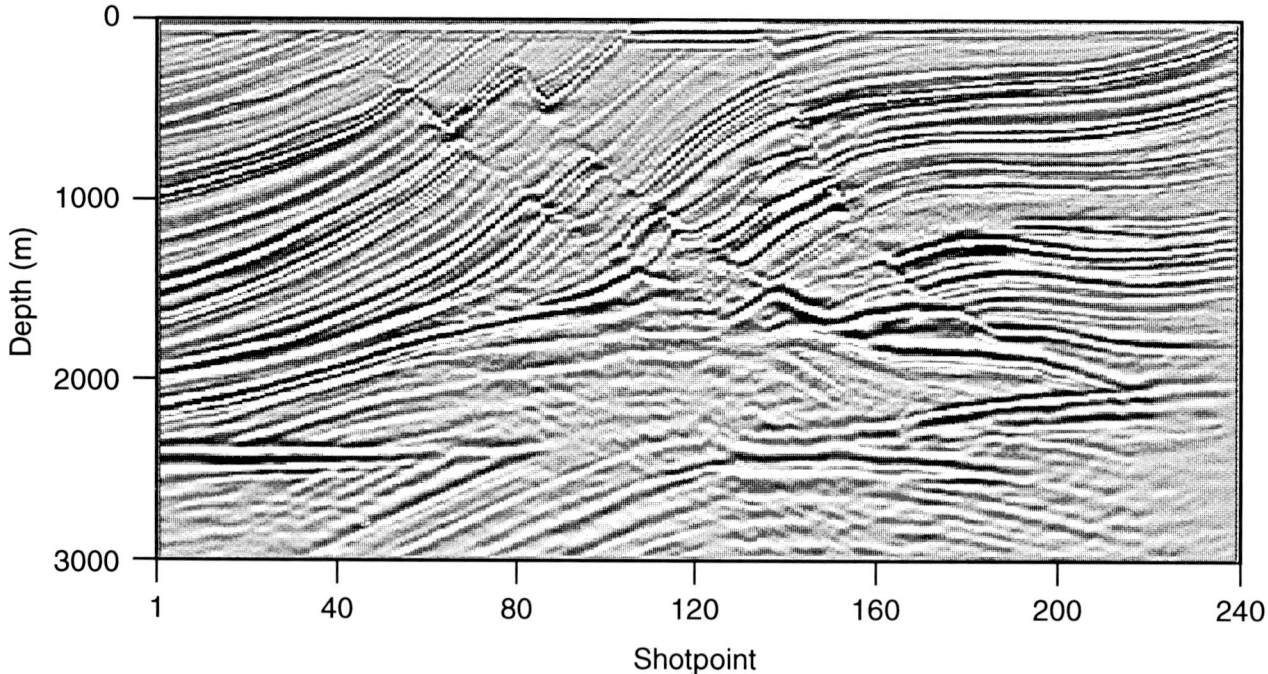

FIG. 1. Marmousi migration: stack of individual migrated images. F-X migration.

mon-offset migration formula found in Bleistein et al. (1987). (In that formula [their equation (63)], the first square root was evaluated using the law of cosines. The resulting product, with distance factors approximated by the product of upper-surface velocity and time, was easily tabled for each offset.) This represents a compromise between the complete absence of amplitude terms and the added expense of amplitude terms from dynamic ray tracing.

In Figure 2, traveltimes were obtained by ray tracing. The traveltime between any source or receiver point and any depth point was specified to be the minimum time along all raypaths between the two points. This image is nearly as good as the F-X migrated image, except near the objective, which appears weaker than on the F-X migration. Attempting to make the objective appear stronger by applying a gain to the migrated data also increases the amplitude of noisy events near the objective. In this case, the choice of the minimum time branch of the traveltime function is the cause of the inaccuracy. We verified this by producing another Kirchhoff migration, using instead the maximum amplitude branch from dynamic ray tracing, but using the same amplitude terms as the previous migration. This migration produced a slightly stronger image of the objective (showing that some energy associated with the objective propagated along raypaths that were not the "fastest"), at the expense of some clarity in the fault blocks higher in the section. However, contrary to results reported by Geoltrain and Brac (1993), the image using minimum traveltimes is acceptable, in that the objective is visible even if relatively weak.

Next, in Figure 3 we show the Kirchhoff migration using traveltimes obtained by Vidale's (1988) eikonal equation solver. This gridded eikonal-based Kirchhoff image is nearly as good as the ray-based Kirchhoff migration, but is clearly inferior to the F-X migration. The most notable difference between Figures 2 and 3 is visible at the objective which, in Figure 3, appears less continuous than in Figure 2. This accuracy difference can be explained provisionally by the relative superiority in accuracy and stability of ray calculations versus gridded eikonal calculations beneath a complex overburden. Alternatively, the eikonal equation solver has lost some accuracy because of its confusion, beneath caustic locations, in attempting to reconcile several traveltime branches into one. Geoltrain and Brac (1993) present an explanation similar to this in much greater detail. A separate problem for migrations using gridded eikonal solvers is their inability to control the propagation angle as easily as ray tracers, which merely disable, or kill, rays that exceed a prespecified dip. Unless this problem is attacked (as it has been here), say by controlling the maximum allowable difference between computed traveltimes in adjacent cells, it will affect a migration by producing smile-shaped artifacts. Having mentioned these two disadvantages of gridded eikonal solvers relative to ray tracing, we must also emphasize that Vidale's solver has helped to provide a reasonable seismic image, given the complexity of the underlying velocity model.

It is difficult to compare in detail the computer time spent computing traveltimes from ray tracing and from Vidale's gridded eikonal solver. First, the ray tracing traveltime calculation is a FORTRAN subroutine, while Vidale's eikonal equation solver (written by Vidale) is a C subroutine. Second, while the ray tracing itself vectorizes fairly well on the Cray-2, the interpolation of the ray traveltimes does not vectorize well, and the gridded eikonal solver involves recursive calculations that do not vectorize at all. However, we observe that, in computing traveltimes on the migration

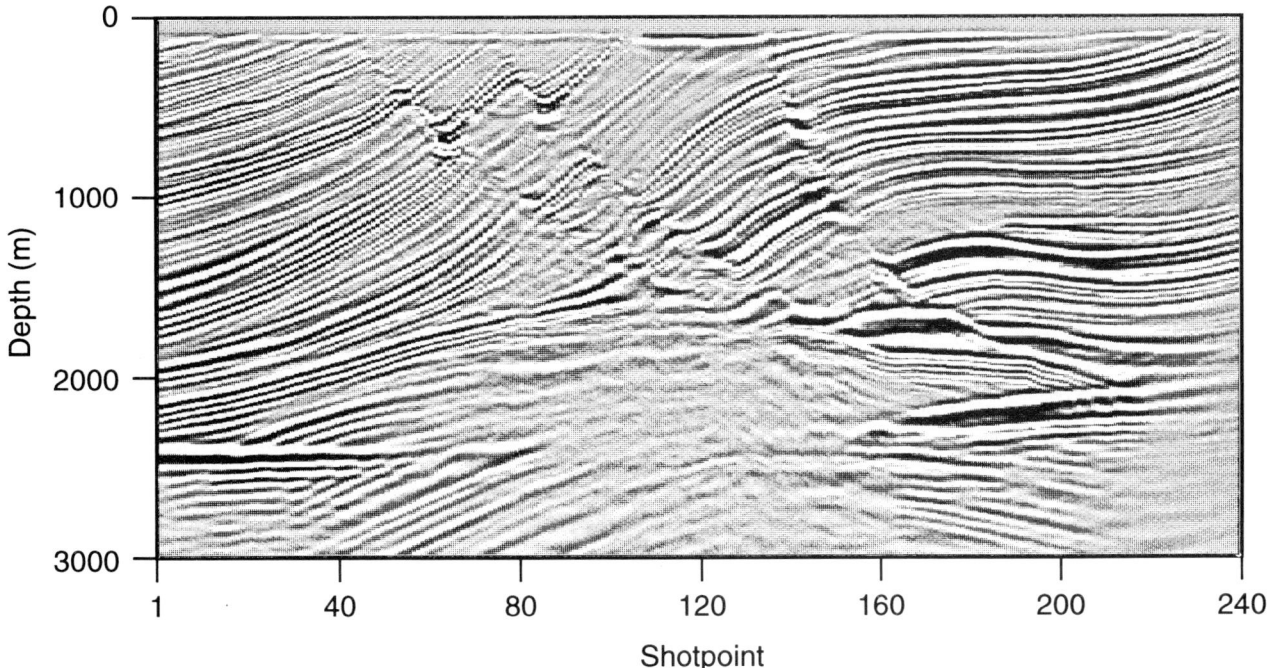

FIG. 2. Marmousi migration: stack of individual migrated images. Kirchhoff migration with traveltimes computed by ray tracing.

grid points, the ray-tracing subroutine touches every grid point with a linear interpolation (Gray, 1986), while the eikonal solver touches every point with a square root operation (Vidale, 1988). Thus, the total number of poorly vectorized or scalar (unvectorized) operations for the two algorithms is comparable. In fact, for both the ray-based and the eikonal-based migrations, the traveltime calculations took approximately one-third the total run time.

Finally, in Figure 4 we show the Kirchhoff migration using traveltimes obtained by a hybrid eikonal solver that we describe in the next section. The image quality is comparable to that of Figure 3, with some differences both shallow and deep. Although the two gridded eikonal equation-based migrations are of lower quality than those shown in Figures 1 and 2, they are still reasonably good. Given typical migration velocity errors of a few percent on the average, it would be difficult to justify the claim that any one of the migration methods discussed here will image significantly better or worse than any of the others.

A HYBRID EIKONAL EQUATION SOLVER

In the previous section we mentioned that both the ray tracer/interpolator and the eikonal equation solver need to touch each image point with some operation to fill the migration grid with traveltimes. We also noted that, even on a vector computer, the two methods were similar in run time on a 2-D migration. In three dimensions, however, the situation can be expected to change. In two dimensions, one can expect to fit into computer memory at least enough of the velocity model to compute traveltimes for the migration aperture of a trace, but in three dimensions this aperture can hold tens of millions of samples. The 3-D extension of the ray tracer used to produce the image in Figure 2 is not impacted by large memory requirements (the ray tracing proceeds depth by depth, using a single slice of the 3-D velocity model to compute a single slice of traveltimes), although the problem of interpolating ray traveltimes onto the migration (x, y)-grid on a depth slice is extremely complicated. However, both the eikonal equation solvers of Vidale (1988) and of van Trier and Symes (1991) compute traveltimes on a grid that expands outward from the source point. In two dimensions, the faces of this grid are lines of constant x and constant z (Vidale, 1988) or circles whose radii are at constant distances from the source point (van Trier and Symes, 1991). Thus, the 3-D extensions of these methods will require having in computer memory planes of constant x, constant y, and constant z (Vidale, 1988) or shells whose radii are constant distances from the source point (van Trier and Symes, 1991). The problem of organizing these grids for fast transfer from disk to memory is at least as formidable as, though very different from, the problem of interpolating ray traveltimes in three dimensions. Finally, another issue that may become more important in three dimensions than in two dimensions is the vectorizability of the algorithm. The eikonal solver of van Trier and Symes (1991) vectorizes well and permits a fast and regular interpolation of traveltimes from the expanding grid to the migration grid, but Vidale's method is inherently recursive, making it, in our opinion, unsuitable for routine application to 3-D migration where very many 3-D volumes need to be filled with traveltimes. The difficulty of these problems has motivated us to develop an alternative method which, like ray tracing, requires access only to depth slices from the 3-D velocity model and which, like the solution of van Trier and

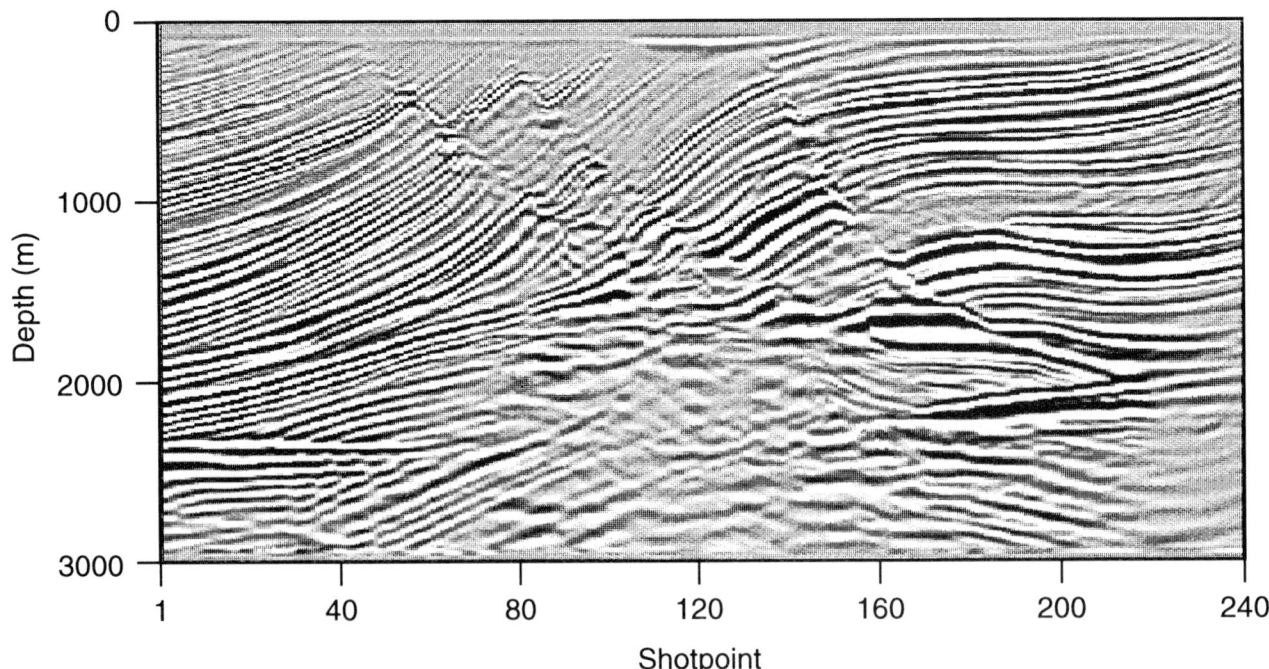

FIG. 3. Marmousi migration: stack of individual migrated images. Kirchhoff migration with traveltimes computed by Vidale's eikonal equation solver.

Symes, allows fast interpolation of traveltimes in three dimensions. Our alternative, illustrated in Figure 5, necessarily represents a compromise between accuracy and speed.

Like the ray tracer, our eikonal solver operates on depth slices from the velocity model. It computes traveltimes within a cone whose apex is at the source point, and does not attempt to track extremely high-angle reflection and refraction events. Since most of these events should not usually be migrated anyway (refractions because they are not reflections and high-angle reflections because they are usually aliased), we do not consider this to be a serious limitation. In Figure 4, near-surface propagation angles up to 60 degrees were retained. Within the cone, all propagation angles are tracked, often leading to imaged dips that exceed the angular limits of the cone. To extrapolate traveltimes from one depth to the next, we proceed not on a rectangular grid, but on an angular grid. Using traveltimes at the present depth z along a set of equally spaced points that represent a set of vectors from the source point with a corresponding set of angles, we first compute angular derivatives of those traveltimes, just as with van Trier and Symes's method. Because the points at depth z are not all the same distance from the source point, it is necessary to interpolate or extrapolate the times at neighboring points onto the same radius when approximating the angular derivative at the point (x, z) by forward and backward differences (Figure 5). For example, the forward difference approximation to $\partial \tau / \partial \theta$ at (x, z) is approximated by

$$\frac{\partial \tau}{\partial \theta} \approx \frac{\tau(r, \theta + \Delta\theta) - \tau(r, \theta)}{\Delta\theta}$$

$$\approx \frac{[\tau(z, \theta + \Delta\theta) - \tau_r(z - \Delta z, \theta + \Delta\theta)\Delta r] - \tau(z, \theta)}{\Delta\theta},$$

where r is the distance from source point to depth z along direction θ, $r + \Delta r$ is distance from the source point to z along direction $\theta + \Delta\theta$, and τ_r is the computed radial derivative at angle $\theta + \Delta\theta$ from depth $z - \Delta z$. The backward difference approximation to $\partial \tau / \partial \theta$ at (x, z) has a similar expression, involving quantities computed along direction $\theta - \Delta\theta$ rather than $\theta + \Delta\theta$. Having these differences, we choose one of them as van Trier and Symes did to ensure that our calculations remain "upwind." Then we use the eikonal equation in polar coordinates,

$$\left(\frac{\partial \tau}{\partial r}\right)^2 + \frac{1}{r^2}\left(\frac{\partial \tau}{\partial \theta}\right)^2 = \frac{1}{c^2}, \qquad (2)$$

to compute the radial derivative, and we use the radial derivative to propagate the traveltimes to the next depth $z + \Delta z$. In contrast to van Trier and Symes, we propagate traveltimes, not their derivative, from one depth to the next. We justify this by noting, like Vidale, that our governing equation is for traveltimes. Differentiating that equation does not change the physics, and it obscures the motivation slightly for choosing one or the other difference approximation to the derivative of traveltime to maintain the upwind

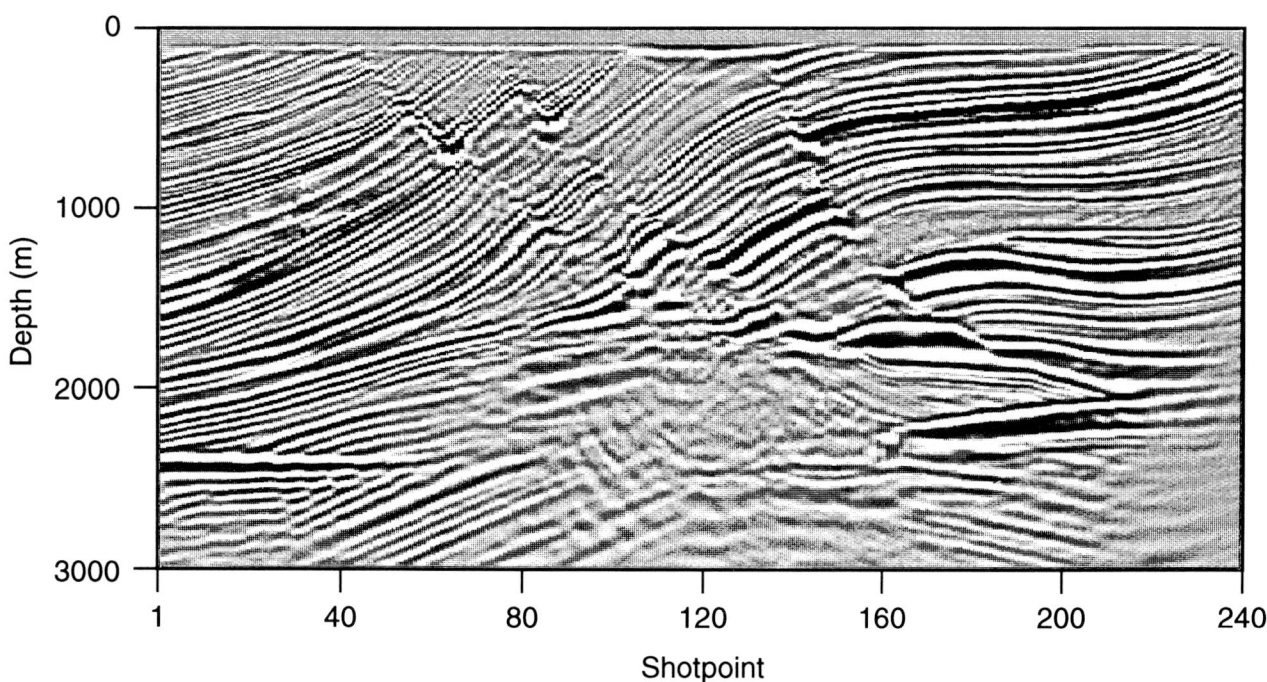

FIG. 4. Marmousi migration: stack of individual migrated images. Kirchhoff migration with traveltimes computed by the eikonal equation solver described in the text.

nature of the scheme. Finally, we smooth the propagated traveltimes (with precomputed coefficients) before interpolating them onto the migration grid. However, the unsmoothed traveltimes are used for the extrapolation to the next depth.

The 3-D generalization of this scheme involves replacing the 2-D eikonal equation in polar coordinates with the 3-D eikonal equation in spherical polar coordinates,

$$\left(\frac{\partial \tau}{\partial r}\right)^2 + \frac{1}{r^2}\left(\frac{\partial \tau}{\partial \theta}\right)^2 + \frac{1}{(r \sin \theta)^2}\left(\frac{\partial \tau}{\partial \phi}\right)^2 = \frac{1}{c^2}. \quad (3)$$

Angular derivatives are now approximated by finite differences in both θ and ϕ, and the difference approximations are selected to maintain an upwind scheme.

Like all other gridded eikonal solvers, ours breaks down when it computes traveltimes whose gradient (a vector in the same direction as a raypath) coincides with the expanding front, here a circle centered at the source point. This happens when the magnitude of the computed angular derivative equals the product of slowness and the radius; then, by the eikonal equation, the value of the radial derivative will be zero. Our fix for this problem is very simple, and in the same spirit as that of Aldridge and Oldenburg (1992). When a propagation direction is within a specified angle of turning parallel to a circle centered at the source point, we arbitrarily reset it normal to the circle. This effectively overwrites a certain branch of the traveltime field (the one expressing propagation parallel to the circular front) with a new branch. This biases the solver toward computing traveltimes by integrating slowness along straight raypaths, a method known to be stable though not necessarily of good accuracy. It also resolves in a straightforward manner any conflicts that occur when two distinct branches of the traveltime field intersect.

All the operations described are vectorizable, and they are also suitable for computation on a parallel machine, either single instruction multiple data (SIMD) or multiple instruction multiple data (MIMD). For the 2-D Marmousi data set, this method for generating traveltimes was significantly faster, by approximately a factor of two, than ray tracing/interpolation.

CONCLUSIONS

We have compared three Kirchhoff migrations of the Marmousi data set, each using a different method of generating traveltimes, with each other and with a benchmark image obtained by F-X migration. We have also discussed a variety of reasons why present alternatives for generating traveltimes in 3-D are not entirely satisfactory, and we have proposed a gridded eikonal solver that compromises some accuracy for a large gain in speed. Encouraged by the migrated result using the new method, we recognize the need for further testing and optimization in three dimensions, both of the gridded eikonal solver and of ray tracing/interpolation. We speculate that optimizing the eikonal solver will lead to greater accuracy or to more efficient implementations of methods like Vidale's (1988) or Moser's (1991), and optimizing the ray tracer will exploit certain aspects of ray tracing such as the use of Fermat's principle (Carter and Frazer, 1984; Gray, 1988) which are not applicable to eikonal solvers.

ACKNOWLEDGMENTS

We thank Dan Whitmore for producing the F-X image used in the migration comparison. We also thank John Vidale for making available his eikonal equation solver, and we thank Tijmen-Jan Moser for providing his shortest-path traveltime solver and for extensive conversations about traveltime solvers in general. Also, we are grateful to the reviewers, especially Paul Fowler, for many constructive comments.

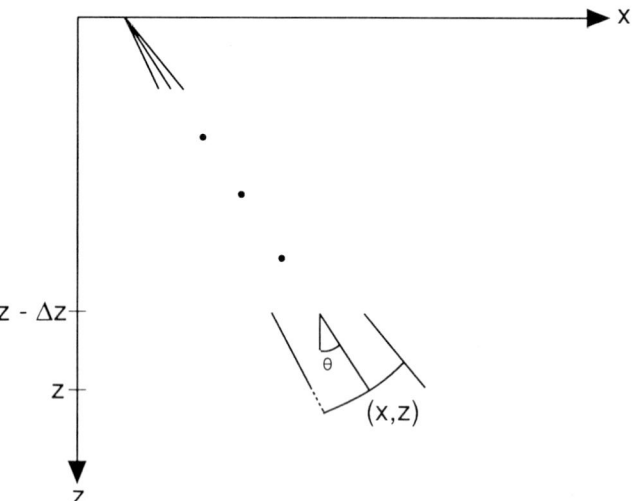

FIG. 5. To compute forward and backward difference approximations to the angular derivative of traveltime at a depth point (x, z), traveltime at neighboring points on depth level z must be interpolated or extrapolated to approximate traveltime at neighboring points on the circle centered at the source point.

REFERENCES

Aldridge, D. F., and Oldenburg, D. W., 1992, Refractor imaging using an automated wavefront reconstruction method: Geophysics, **57**, 378–385.

Berkhout, A. J., 1984, Seismic migration: Imaging of acoustic energy by wavefield extrapolation B. Practical aspects, Elsevier Science Publ.

Berryhill, J. R., 1979, Wave-equation datuming: Geophysics, **44**, 1329–1344.

——— 1984, Wave-equation datuming before stack: Geophysics, **49**, 2064–2066.

Bleistein, N., 1986, Two-and-one-half dimensional in-plane wave propagation: Geophys. Prosp., **34**, 686–703.

Bleistein, N., Cohen, J. K., and Hagin, F. G., 1987, Two and one-half dimensional Born inversion with an arbitrary reference: Geophysics, **52**, 26–36.

Carter, J. A., and Frazer, L. N., 1984, Accommodating lateral velocity changes in Kirchhoff migration by means of Fermat's principle: Geophysics, **49**, 46–53.

Geoltrain, S., and Brac, J., 1993, Can we image complex structures with first-arrival traveltimes?: Geophysics, **58**, 564–575.

Gray, S. H., 1986, Efficient traveltime calculations for Kirchhoff migration: Geophysics, **51**, 1685–1688.

―――― 1988, Computational inverse scattering in multi-dimensions: Inverse Problems, **4**, 87–101.

Moser, T. J., 1991, Shortest-path calculation of seismic rays: Geophysics, **56**, 59–67.

Reshef, M., and Kosloff, D., 1986, Migration of common shot gathers: Geophysics, **51**, 324–331.

van Trier, J., and Symes, W. W., 1991, Upwind finite-difference calculation of traveltimes: Geophysics, **56**, 812–821.

Versteeg, R., and Grau, G., 1991, The Marmousi experience: Proc. 1990 EAEG workshop on practical aspects of seismic data inversion, Eur. Assoc. Expl. Geophys.

Vidale, J., 1988, Finite-difference calculation of traveltimes: Bull. Seis. Soc. Am., **78**, 2062–2076.

Whitmore, N. D., and Garing, J. D., 1990, Interval velocity estimation using iterative pre-stack depth migration in the constant angle domain: Proc. of the 1990 EAEG/SEG Research Workshop on the Estimation and Practical Use of Seismic Velocities, Eur. Assoc. Expl. Geophys.

APPENDIX
F-X MIGRATION AND KIRCHHOFF MIGRATION

F-X and Kirchhoff migrations are both wave-equation based, but they are implemented differently. *F-X* migration operates on temporally Fourier transformed data, while Kirchhoff migration operates on data in the time domain. The *F-X* migration used here downward continues reflection data recursively, from depth to depth. The imaging is performed by a weighted crosscorrelation of downward continued source and receiver wavefields followed by a sum over frequencies. The bulk of the work is performing the downward continuation, which can be expressed in the following nested loop structure:

For each frequency
  For each depth
    For each lateral location
      Combine the wavefield from an aperture at the previous depth level
    Next lateral location
  Next depth
Next frequency.

The inner loop, or combination, is a weighted sum that expresses the downward extrapolation of a wavefield from one depth to the next using a 2-D Green's function. The inner two loops are expressed schematically in Figure A-1 [essentially the same as Figure III-7b(a) in Berkhout (1984)], which shows that seismic energy can travel many paths in propagating from any one of the source or receiver points to a possible reflection point within the earth. All of these paths are used by the migration method in constructing the wavefield at a depth. Kirchhoff migration can also be performed recursively, as a variant of Berryhill's (1979, 1984) time-domain downward-continuation method. In that scheme, a temporal convolution inner loop replaces the frequency outer loop in the above loop structure (the temporal convolution is actually a transformation of an integral over the third space dimension *y*), and the imaging step consists in picking off the convolved wavefield at the appropriate time for each depth point. However, it is much more efficient to perform Kirchhoff migration nonrecursively, by using an asymptotic approximation to the downward continuation operator. Here, assumed invariance of the geologic structure and the 2-D recorded wavefield in the *y*-direction (orthogonal to the line direction) allows a stationary phase evaluation of the integral over *y* (Bleistein et al., 1987). This produces a *nonrecursive* scheme for downward-continuing source and receiver wavefields directly from the recording surface to all depth points independently. As before, imaging is performed by picking off the wavefield at the imaging time which, for a given source, receiver, and depth point, is the total traveltime from the source to the depth point to the receiver. The following loop structure holds for nonrecursive Kirchhoff migration:

For each depth
  For each lateral location
    Combine the wavefield from an aperture at the upper surface
  Next lateral location
Next depth.

This structure contains one loop less than the one for *F-X* migration, reflecting the fact that nonrecursive Kirchhoff migration involves one integral less than *F-X* migration. In terms of the computing operation count, although the inner loop of this structure involves a larger aperture than for *F-X* migration at moderate and great depths, the absence of the outer loop over frequency tends to compensate completely for this fact, resulting in a considerable speedup of nonrecursive Kirchhoff migration over *F-X* migration. In the text, when we refer to Kirchhoff migration, we mean nonrecursive Kirchhoff migration.

Figure A-2 displays schematically the inner loop for Kirchhoff migration. The downward continuation of the upper-surface wavefield to a depth point is performed using a so-called 2.5-D Green's function, which means that the propagation paths are raypaths and that amplitudes along the

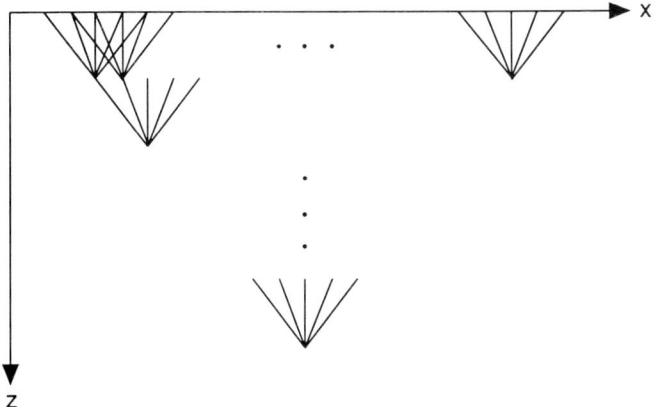

FIG. A-1. Propagation paths from the upper surface to a depth point for *F-X* downward continuation. There are many propagation paths from any upper surface point to the depth point, and all these paths are used to contribute energy to the wavefield at depth.

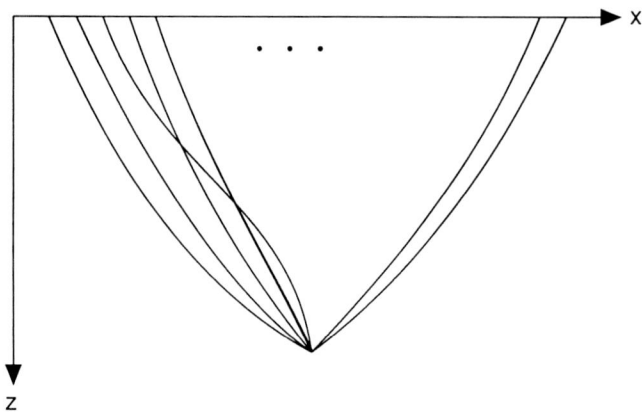

FIG. A-2. Propagation paths from the upper surface to a depth point for Kirchhoff downward continuation. Although a single propagation path connecting each upper surface point to the depth point is indicated, according to ray theory there might be no raypaths (a shadow zone) or more than one (a caustic).

raypaths account for 3-D geometrical spreading in a 2-D model by assuming invariance in the y-direction (Bleistein, 1986).

F-X migration automatically handles—in fact, forces— energy multipathing from upper surface points to depth points, while Kirchhoff migration allows a few paths at most to connect an upper surface point with a depth point. From an imaging perspective, this is the most important difference between F-X migration and Kirchhoff migration. F-X migration can guarantee that a portion of a wavefield will propagate from the upper surface to a depth point, while Kirchhoff migration, by choosing at most a very few (one, in our examples) of the many possible propagation paths and times, cannot make this guarantee. This fact explains in part the difference in imaging quality at the Marmousi objective formation: the complete image at the objective is built up from multipath contributions, and some of the minimum-time arrivals do not carry all the energy. Also, within the context of one-way (downward) wavefield extrapolation, F-X migration guarantees accurate 2-D amplitudes while Kirchhoff migration requires detailed dynamic ray calculations to handle complicated propagation (e.g., through caustics) properly. However, this issue is less important, because neither correct 2-D amplitudes nor incorrect 2.5-D amplitudes are completely desirable.

Seismic imaging: Use the right tool for the job

SAMUEL H. GRAY, *Amoco EPTG and Amoco Canada, Calgary, Canada*

Seismic imaging problems come in all shapes and sizes, from stratigraphic plays, where the objective is to map subtle features in flat geology by focusing the seismic wavelet, to structural plays, where the aim is to see clearly below geology with significant structural variations. Likewise, seismic imaging methods come in all shapes and sizes, from NMO/stack to multicomponent anisotropic prestack depth migration. For a given play, the most fundamental seismic imaging challenge is to match the play with the appropriate imaging technology. Is there a simple scheme for doing this? Building a catalog of imaging methods useful for the various play types is impossible because of ongoing technical developments and shifting economics. Still, it is possible to discuss how the available imaging tools can be applied to today's exploration and development plays and, indeed, whether some of the imaging tools should be applied at all. For example, do 2-D seismic lines have any value in a 3-D earth? Does time migration have any value, or should we always apply depth migration? When can we get away with NMO/DMO/stack/migration? In answering questions like these, I use the following principle of parsimony: Use the most advanced imaging technology that the play needs technically and can support economically.

A hierarchy of seismic imaging methods. The most advanced imaging technology available today is 3-D multicomponent anisotropic prestack depth migration. Most plays need considerably less than that, but do any plays actually need such technology? Perhaps surprisingly, the answer is yes, technically at least. In "Effects of salt-related mode conversions on subsalt prospecting" (GEOPHYSICS 1996), Ogilvie and Purnell show high-amplitude converted-wave reflections from the base of salt in several Gulf of Mexico surveys. Such events, considered to be noise on P-wave surveys, can be properly acquired, processed, and analyzed using multicomponent receivers. Also, in "Estimation of anisotrophy and anisotropic 3-D prestack depth migration, offshore Zaire" (GEOPHYSICS 1995), Ball estimated P-wave anisotropy in order to perform prestack depth migration. The value of 3-D multicomponent surveys, with their additional leverage in anisotropy estimation as well as their attendant demands on (especially depth) imaging algorithms, is gradually becoming recognized. Today, however, even though some plays can justify such effort technically, very few can justify the effort economically. Can other plays benefit from the application of this technology? Yes, all plays can to some degree, simply because 3-D multicomponent anisotropic prestack depth migration makes fewer incorrect assumptions about wave propagation in the earth than any other imaging method. But if most high-risk, high-reward subsalt plays can not economically support 3-D multicomponent surveys (presently, at least), most other plays can not, either.

The next most advanced imaging tool is single-component 3-D prestack depth migration. This tool, almost

Editor's note: A version of this paper was presented at the Geophysical Society of Tulsa 1997 Spring Symposium.

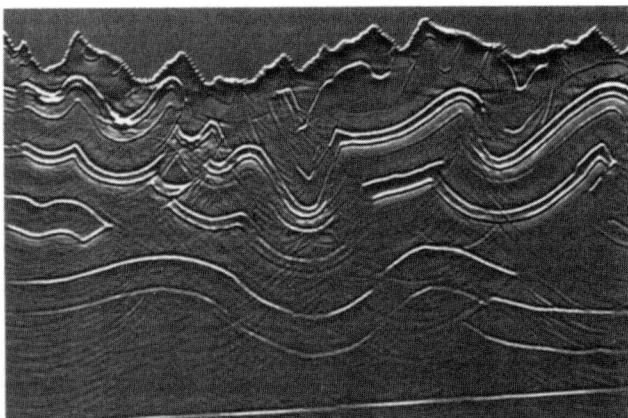

Figure 1. Prestack depth migration of a structural synthetic data set with significant velocity variations and extreme topographic variations. The velocities used for the migration are the correct interval velocities, so this is a correct image of the subsurface. From "Migration from topography: improving the near-surface image" by Gray and Marfurt (*CSEG Journal*,1995).

completely unavailable (i.e., almost infinitely expensive) even five years ago, is now being commonly applied to marine data. In "Geophysical imaging of subsalt geology" (*TLE*, 1997), Ratcliff and Weber discuss a gigantic 3-D prestack depth migration project, covering several subsalt plays in the Gulf of Mexico. Oil companies, large and small, are applying 3-D prestack depth migration in basins around the world, even where there are only moderately complicated geologic structures. Why this rush to 3-D prestack migration? Why not simply use 3-D poststack migration, in time or depth? For plays with plenty of complicated geologic structure, the answer is easy: 3-D poststack migration doesn't work because of the harm done to the data by the process of NMO/DMO/stack. Even though prestack migration demands much more of the interpreter's time in determining the velocity model and much more of the computer's time in imaging the data, the ability to see geology that was previously obscured by the approximations made in the stacking process justifies the extra effort. In other words, subsalt plays need, and can economically support, 3-D prestack depth migration. For other plays, less complicated structurally, we find that we still see images more clearly with prestack migration than with poststack migration, and that the extra interpretational cost is recovered by the increased interpretability of the image. Can all plays benefit from 3-D prestack depth migration? Yes, because 3-D prestack depth migration makes fewer incorrect assumptions about wave propagation in the earth than any other imaging method (except, of course, 3-D multicomponent anisotropic prestack depth migration). Can all plays that need 3-D prestack depth migration justify it economically? No, in land areas such as the Canadian foothills, sufficiently high-fold acquisition for 3-D prestack depth imaging, including velocity analysis and the resolution of statics problems, can be prohibitively expensive.

Figure 2. NMO/DMO/stack applied to the synthetic data, showing good event continuity despite severe topographic variations.

Figure 3. Depth migration of the stack in Figure 2, using the correct velocities. The image is much poorer than the image in Figure 1, indicating the failure of the standard flow for this structural play.

For geologic reasons, most structurally complex plays are three-dimensional, and because of the failure of various assumptions inherent in NMO/DMO/stack, nearly all structurally complex plays need prestack migration. So is there any need at all for 2-D or poststack imaging in plays with structural complexity? Yes, but not for the final result. 3-D poststack migration, in time or depth, is an extremely useful shortcut in building initial velocity models for 3-D prestack depth migration. Often, 2-D prestack depth migration is useful for the same purpose. Also, 2-D prestack time or depth migration is useful in imaging reflectors in structurally complicated land areas. Here, though, using only 2-D migration makes it difficult to distinguish in-plane reflectors from out-of-plane reflectors, adding a huge amount of risk to deciding well locations based on even the best possible 2-D imaging. Unfortunately, in structurally complicated land areas where the play economics do not permit the acquisition of high-fold 3-D data, 2-D imaging is usually used to decide well placement. While shallow wells can often be correctly placed using 2-D images, experience tells us that deeper wells can not. Using 2-D imaging, even prestack depth migration, for the final result in structurally complicated areas is not usually valid technically, and might even lead eventually to the loss of economic support for structural plays on land.

In structural plays, then, 2-D and poststack imaging are useful only as intermediate results. Are they useful as final results in stratigraphic plays? In the absence of complicated geologic structure, we can often think of a single direction that is dip to the possible stratigraphic target. If there is no well-defined dip direction (e.g., for surveys shot over a meandering channel system), it makes no more sense to do 2-D imaging for a stratigraphic play than it does for a structural play. But if there is a well-defined dip direction, 2-D prestack migration can be as effective as 3-D prestack migration. But is prestack migration needed, or even migration? Yes! Even poststack migration is known to improve the resolution of lateral truncations and sequence variations of seismic data. And with improvements in computing power, 2-D prestack migration is no longer significantly more expensive than DMO followed by poststack migration. So a stratigraphic play that justifies exploration or development justifies migration, and a 2-D stratigraphic play that justifies exploration or development justifies prestack migration (either time or depth), at least economically. Typically, 3-D stratigraphic plays can not yet economically justify prestack migration, except as quality control for the poststack migration, because 3-D prestack migration is still significantly more expensive than current implementations of 3-D DMO followed by poststack migration. Within a few years, however, prestack migration will be applied routinely on these plays, with poststack migration providing intermediate results. Do stratigraphic plays need prestack migration for technical reasons? Often, purely stratigraphic plays are investigated before stack, using tools such as amplitude variation with offset (AVO) analysis. As Mosher et al. show in "The impact of migration on AVO" (GEOPHYSICS, 1996), these tools are more appropriately applied after migration, which undoes the wave-propagation-induced distortions introduced into the unmigrated records.

If it's true, as I've argued, that we should migrate practically all data before stack, is the "standard" processing flow NMO/DMO/stack/migration ever useful for more than intermediate results? With well costs greatly exceeding the costs of processing and interpreting seismic data, the "most correct" image, i.e., the one most faithful to the requirements of actual wave propagation in the earth for the play at hand, should be used to decide drilling risks and well placement. The standard flow can contribute to producing the most correct image for both structural and stratigraphic plays, but seldom can it produce the most correct image itself.

The interpretive aspect of prestack seismic imaging: velocity, velocity, velocity. Who will do all this prestack migration, and how should it be done? Typically, the processor runs the standard flow to produce a poststack time migration with very little input from the interpreter. For the standard flow, picking velocities is the processor's job; the choice of velocity helps to improve the image, but does not otherwise help with the interpretation. For prestack depth migration, the situation is different. Here, the choice of velocity directly affects the image quality *and* the interpretation; in fact, some will argue that the primary goal of imaging is to determine the interval velocity field that positions the reflectors both at their correct lateral locations and in depth. The interpreter might do none, a small amount, or most, of the prestack depth migration processing, but he or she alone is responsible for determining the velocities. This responsibility adds to the interpreter's duties, and sometimes to the interpreter's woes (even though interval velocity analysis tools have improved significantly in recent years). As a result of the interpreters' distress, prestack time migration has become a popular, though not always appropriate, alternative to prestack depth migration. Here, as with the standard flow, the choice of velocity is used only to improve the image, not to help with the interpretation. Since interpreters cannot usually pick imaging (or stacking) velocities as well as processors can, interpreters are often willing to yield the imaging responsibility to the processors. This is both good and bad. Prestack time migration is, in theory and practice, a better imaging tool than the standard flow; it makes fewer incorrect assumptions about wave propagation and is capable of producing better, and more correct, images than the standard flow. In fact, for stratigraphic plays with insignificant velocity variations, prestack time migration is equivalent to prestack depth migration. On the other hand, prestack time migration is, in theory and usually in practice, an inferior structural imaging tool to prestack depth migration. Although it is capable of producing an interpretable structural image at least as quickly and easily as prestack depth migration, it is seldom capable of producing a more correct image, especially when it is performed without the benefit of the interpreter's detailed knowledge of the geology and the interval velocities. For structural plays, prestack time migration can serve the same purpose, at a more advanced level, as the standard flow, namely to produce well-imaged intermediate results that are understood not to be completely correct in a structural sense, without using too much of the interpreter's time.

The following example illustrates the interpretive nature of seismic imaging on a structural play. Although the example is somewhat unrealistic because it is two-dimensional, it still shows the limitations in imaging quality that can be obtained by various imaging methods. The synthetic data set was obtained by finite-difference modeling of a 2-D cross-section that is based on geology of the foothills of the Canadian Rockies. As is typical of overthrust areas, the geology is very complicated, and seismic velocities, which range between 3500 m/s and 6000 m/s, vary rapidly. Figure 1 shows the result of a prestack depth migration using the correct velocities. This is a correct image of the subsurface, and it serves as a basis for comparison. This section measures 22500 m across by 10 000 m deep. Figure 2 shows the result of the standard flow NMO/DMO/stack applied to the data, and Figure 3 shows a depth migration of the stack in Figure 2, again using the correct velocities for the migration. Although an expert processor applied the standard flow, producing a very appealing stack for the migration, the migrated image falls far short of the ideal image shown in Figure 1. Since the algorithms for the prestack and the poststack migrations were identical, we conclude that applying the standard flow to this data set, with substantial velocity variations and extreme topographic variations, has resulted in inadequate imaging for all but the shallowest drilling targets on this line. (On field data, the stack never looks as good as the stack in Figure 2, and the migration often does not look much worse than the one in Figure 3. On synthetic data, inaccuracies introduced in the standard flow lead to noise in the final migrated image, but although these inaccuracies are also present on a field data stack, their effects on the migrated image tend to be masked by the effects of random static errors, out-of-plane contributions, etc.) Figure 4 shows a prestack time migration of the data set, using velocities derived from the correct velocity field. The prestack migrated image is much clearer than the poststack migrated image, and the prestack migration has the added advantage that the imaging velocities can be esti-

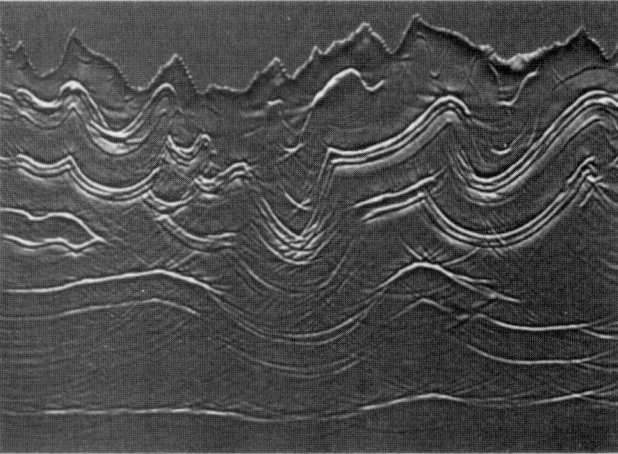

Figure 4. Prestack time migration of the synthetic data, using velocities derived from the correct velocities. The image is a vast improvement over the poststack migration in Figure 3, but still shows some structural uncertainties.

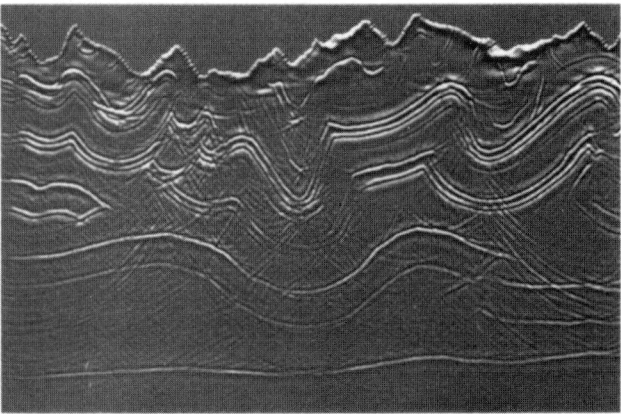

Figure 5. Prestack depth migration, using velocities estimated from a modern velocity analysis tool. The structural image is more correct than those in Figures 3 and 4.

mated by standard velocity analysis techniques. Still, several structural features of the prestack time migration, such as the variations in basement structure, are incorrect, reducing our confidence in this image. Finally, Figure 5 shows a prestack depth migration, using velocities estimated by a modern depth migration velocity analysis technique. This was the last of a few iterations of depth migration. Although there are errors in this image, especially in depth conversion, the structural errors are less severe than for either the poststack migration or the prestack time migration, both of which used the correct velocities. So for this structural example, prestack time migration performed significantly better than poststack depth migration, and prestack depth migration, applied by a pair of skilled processors but unskilled interpreters, gave a more correct (and, in this case, more appealing) structural image than prestack time migration.

To summarize the interpreter's seismic imaging responsibilities: Marine structural plays require 3-D imaging of 3-D data. Among the imaging tools applied will be noise reduction techniques, such as multiple removal, for which the interpreter is ultimately responsible. The interpreter is

Conclusions. We should migrate all seismic data, and we should (although we don't) always rely on prestack migration to provide the final image. The standard flow, NMO/DMO/stack/migration, by now largely obsolete as a final product, is useful only to provide intermediate results for structural imaging. The increasing emphasis on prestack migration, especially prestack depth migration, adds velocity analysis to the interpreter's responsibilities. This, in turn, challenges the developers of imaging technology to provide meaningful but easy-to-use velocity analysis tools for the interpreters.

Suggestions for further reading. "Estimation of aniso-tropy and anisotropic 3-D prestack depth migration, offshore Zaire" by G. Ball (GEOPHYSICS, 1995). "In the Foothills, prestack depth migration IS interpretive processing" by S. H. Gray and G. Maclean (*CSEG Recorder*, 1996). "Migration from topography: improving the near-surface image" by S. H. Gray and K. J. Marfurt (*Canadian Journal of Exploration Geophysics*, 1995). "Effects of salt-related mode conversions on subsalt prospecting" by J. S. Ogilvie and G. W. Parnell (GEOPHYSICS, 1996). " The impact of migration on AVO" by C. C. Mosher, T. H. Keho, A. B. Weglein, and D. J. Foster (GEOPHYSICS, 1996). "Subsalt imaging via target-oriented 3-D prestack depth migration" by D. W. Ratcliff, C. A. Jacewitz, and S. H. Gray (*TLE*, 1994). "Geophysical imaging of subsalt geology" by D. W. Ratcliff and D. J. Weber (*TLE*, 1997). **Ŀ**

Acknowledgments: I thank Gary Maclean for providing the synthetic data set, Randy Cameron for providing the processing expertise leading through the standard flow, and Gary Murphy for providing (1) most of the unskilled prestack depth migration interpretation, and (2) the interpretation tool.

Corresponding author: Sam Gray, 403-233-1304

also responsible for analyzing interval velocities for prestack depth migration. Leading up to the final migrations will be an application of 2-D prestack depth migration to several lines to gain an initial estimate for the velocities, and an application of the standard flow, leading to one or more iterations of 3-D poststack depth migration to refine that estimate. Using 2-D prestack and 3-D poststack migration adds to the preliminary work, but reduces the number of iterations of the more expensive 3-D prestack migration.

Land structural plays should, but for economic reasons often don't, involve 3-D imaging of 3-D data. Unfortunately, 3-D data acquired on land over structural areas are seldom of high enough fold to resolve completely the problems of statics and velocity, and 2-D data fail to provide adequate leverage against out-of-plane reflectors. Although the data volumes are smaller than for marine plays, they are often noisier, partly as a result of inadequate solutions of statics and velocity problems. As a preliminary step, prestack time migration is preferable to the standard flow, bringing out reflectors that prestack depth migration will eventually place in the "correct" locations.

Marine stratigraphic plays usually involve 3-D data. The imaging problems are not as complicated as they are for structural plays, and the standard flow is not needed to produce intermediate images. After an application of prestack noise removal, prestack time migration is usually the ultimate imaging tool required. Unfortunately, present-day play economics often restrict the imaging to the standard flow.

Land stratigraphic plays can be imaged much as marine stratigraphic plays can, as long as we remember that the noise problems are different from those encountered on marine data.

SIMPLE PRESTACK MIGRATION OF CROSSWELL SEISMIC DATA

C.L. LINER[1] and L.R. LINES[2*]

[1] Dept. of Geosciences, University of Tulsa, 600 S. College Ave., Tulsa, OK 74104-3189, U.S.A.
[2] Amoco Production Research, P.O. Box 3385, Tulsa, OK 74102-3385, U.S.A.

* Present address: Dept. of Earth Sciences, Memorial Univ. of Newfoundland, St. John's, Newfoundland, Canada A1B 3X5.

(Received January 7, 1994; accepted March 3, 1994)

ABSTRACT

Liner, C.L. and Lines, L.R., 1994. Simple prestack migration of crosswell seismic data. *Journal of Seismic Exploration*, 3: 101-112.

Crosswell prestack migration can be accomplished by using a beautifully simple method. The method assumes constant velocity and combines the use of elliptical impulse responses and wavefront summation. For given source and receiver borehole positions and a seismic velocity, there is an ellipse of possible locations for scatterers (an aplanatic surface) for each trace sample. Trace amplitudes for the entire trace form a series of concentric ellipses. We show that if these surfaces are generated for a large number of traces at various source and receiver positions and summed together, the position of the scatterers can be successfully imaged. The method is tested on a synthetic model and reflectors are correctly imaged if the velocity of the medium is valid. This crosswell method also produces encouraging results for a real data case.

KEY WORDS: seismic, crosswell, migration, imaging, borehole seismology, prestack.

INTRODUCTION

Crosswell imaging has seen increased use in recent years as a reservoir characterization tool. The advent of energetic non-destructive borehole sources has led to an increase in the number of high-resolution data sets which can be used to characterize variations in seismic velocities of the reservoir rock. The methods for imaging such data have generally included seismic traveltime tomography and reflection imaging. As mentioned by Gray and Lines (1992),

0963-0651/94/$5.00   © 1994 Geophysical Press Ltd.

traveltime tomography can be used to estimate seismic velocities, while depth migration methods can utilize these velocities in the imaging of reflected arrivals. Methods for accomplishing prestack or poststack migration have been described by Hu et al. (1988) and Zhu and McMechan (1988). Prestack migration of field data has also been reported by Findlay et al. (1991). Another approach is to generate a stacked section rather than a migrated section (Stewart and Marchisio, 1990; Lazaratos, et al., 1992).

Depth migration requires a detailed interval velocity model and typically involves many user-defined parameters to control aspects of migration such as aliasing and aperture effects. However, by assuming constant velocity, our implementation of prestack migration involves a minimum number of parameters.

MIGRATION CONCEPTS AND ALGORITHM

The constant velocity impulse response (IR) for prestack migration of surface seismic data is an ellipse with source and receiver at the foci (Claerbout, 1985, p. 171).

The equation for this impulse response is

$$(x - x_c)^2 / (vt/2)^2 + (z - z_c)^2 / [(vt/2)^2 - h^2] = 1 \qquad (1)$$

where

(x, z) = coordinates in migrated image
x_c = x-center of IR (midpoint)
z_c = z-center of IR (zero)
v = migration velocity (constant)
t = reflection time of pulse
h = half-offset.

An ever-present concern in migration is operator aliasing which can occur on the steep limbs of the IR if sampling is inadequate as the operator turns toward vertical. A simple, but effective, method of avoiding spatial aliasing is to represent the ellipse parametrically. To show the concept, write the Cartesian ellipse equation (1) in the general form

$$(x^2 / a^2) + z^2 / (a^2 - h^2) = 1 \qquad (2)$$

and introduce an angular parameter, θ, by

$$x \equiv a \cos\theta$$
$$z \equiv \sqrt{(a^2 - h^2)} \sin\theta \quad , \qquad (3)$$

where the range on θ is $0 \leq \theta \leq 2\pi$. Direct substitution of equation (3) into equation (2) reduces to the identity $\sin^2\theta + \cos^2\theta = 1$, showing that equation (3) is a valid substitution. With this parametric representation, the avoidance of spatial aliasing reduces to a question of how finely θ is sampled. In our crosswell migration implementation, θ is sampled 500-1000 times on the interval $0 \to 2\pi$. The sines and cosines can be precomputed (tabled) to enhance speed of the algorithm.

Equation (1) assumes a special situation, namely, that the source and receiver are at the same z-level, $z = 0$. Fig. 1(a) shows the surface recording IR, while (b)-(d) show other possible configurations. Equation (1) must be generalized to allow migration of downhole recording geometries. In particular, the idea of a single half-offset must be extended to include a horizontal half-offset, h_x, vertical half-offset, h_z, and net half-offset, h.

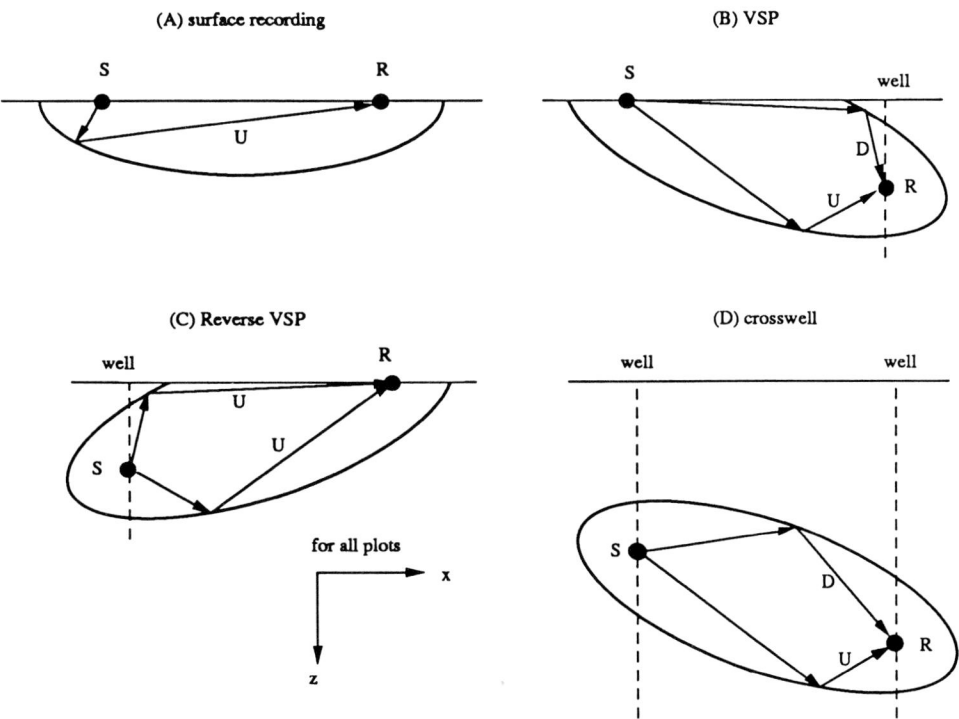

Fig. 1 - Elliptical impulse response geometry of prestack migration for various acquisition methods. The source location is signified by S and the receiver by R. Each labeled ray, D and U, denote downgoing and upgoing waves, respectively. Note that only VSP and crosswell acquisition include downgoing primary reflections. These are rare in the VSP case (requiring shallow receivers and steep dips), but common in crosswell data.

If the source and receiver coordinates are (x_s, z_s) and (x_g, z_g), respectively, then the half-offsets are given by

$$h_x = (x_g - x_s)/2$$
$$h_z = (z_g - z_s)/2 \qquad (4)$$
$$h = \sqrt{(h_x^2 - h_z^2)} \quad .$$

The geometry is shown in Fig. 2.

With these definitions, the formulas describing our migration algorithm are parametric equations of an ellipse with given arbitrary foci,

$$x = [(x_1 - x_c)h_x/h] - [(z_1 - z_c)h_z/h] + x_c$$
$$z = [(x_1 - x_c)h_z/h] + [(z_1 - z_c)h_x/h] + z_c \quad ,$$

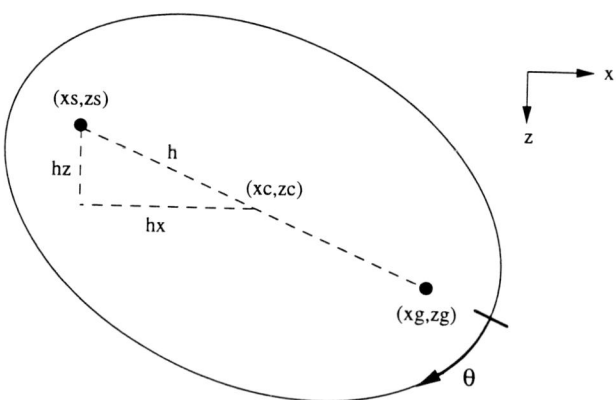

Fig. 2. Geometry of an ellipse with arbitrary foci at (x_s, z_s) and (x_g, z_g). The center of the ellipse is at (x_c, z_c), the various half-offsets are shown and the parametric angle, θ, is indicated.

PRESTACK MIGRATION

where

$$x_1 = a \cos\theta + x_s + h_x$$

$$z_1 = (a^2 - h^2)^{1/2} \sin\theta + z_s + h_z$$

$$a = vt/2 > h$$

$$x_c = x_s + h_x$$

$$z_c = z_s + h_z$$

and (h_x, h_z, h) are given by equation (4).

One subtlety of implementation involves a sign change between upgoing and downgoing primary reflections (Findlay, et al., 1991). This has been incorporated into our method by allowing the operator to change sign according to

$$\text{sign} = +1 \text{ for } 0 \le \theta \le \pi$$

$$\text{sign} = -1 \text{ for } \pi < \theta < 2\pi \ .$$

This sign convention is strictly correct for horizontal reflectors and approximately correct for dipping ones.

This formulation is general enough to allow migration of crosswell data as well as surface (pre- or post-stack), VSP, reverse VSP and salt proximity data (downhole with vertical offset only). This flexibility is a consequence of aliasing treatment and using the actual source/receiver (x, z) coordinates rather than assuming any special acquisition symmetry. The methodology also extends to migration of data acquired in curved 2-D boreholes.

Of primary importance in this algorithm is the simultaneous migration of upgoing, downgoing and horizontal traveling reflection energy. Therefore, the common practice of wavefield separation before migration is not required.

Another appealing feature of this algorithm is natural avoidance of direct arrival energy. Mathematically, the direct arrival corresponds to vt/2 = h, which is a non-physical ellipse. Only times beyond 2h/v will be processed as possible reflection events. This constraint can be further strengthened by defining migration time, t, to be t > 2h/v + delay to allow for a broad direct pulse. Thus, direct arrival muting is not required as a pre-process. The migrated images can have an arbitrary grid spacing (dx, dz), but the number of θ-steps should reflect the fineness of the grid. If θ is sampled too coarsely, grid points will be skipped as larger ellipses are placed into the image plane.

To summarize, we have developed a non-aliased, efficient algorithm which is completely general in terms of 2-D acquisition geometry, that simultaneously process all reflection events, naturally avoids migration of the direct arrival, and involves few user-supplied parameters.

Finally, since the method is analytic it has a straight forward extension to 3-D. This extension is currently being developed.

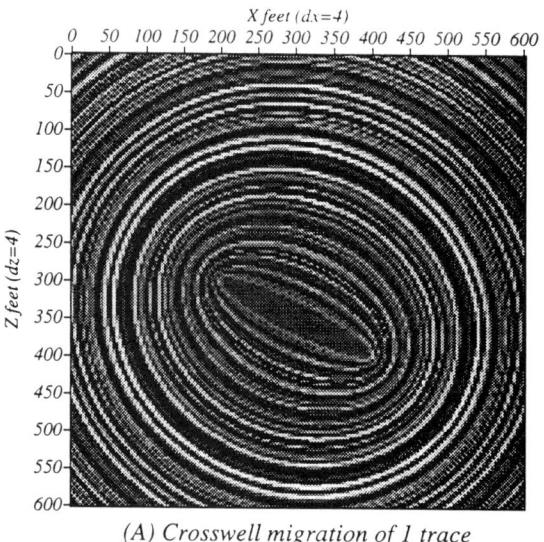

(A) Crosswell migration of 1 trace

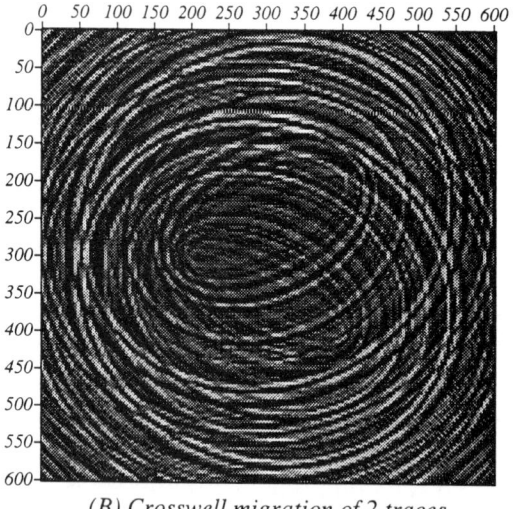

(B) Crosswell migration of 2 traces

Fig. 3. Impulse response of crosswell migration. Source well is at x = 200 and receiver well at x = 400. (a) Crosswell migration of one trace. (b) Crosswell migration of two traces.

IMPULSE RESPONSE

Fig. 3 shows impulse responses of the crosswell migration algorithm. For each plot the source well and receiver well are at x = 200 and 400, respectively.

In the upper plot, Fig. 3(a), a single trace has been migrated. The long axis of each ellipse is along a line connecting the source and receiver. The blank zone at the center is unmigrated because this area is related to direct arrival energy.

Fig. 3(b) shows the result of migrating two traces. Note that only the receiver location has changed between traces. For any number of traces, the final migrated image is a linear superposition of individual trace migrations.

MODEL RESULTS

In order to illustrate the validity and restrictions of this migration technique, we use a three-layer model of flat beds in which a low velocity layer (v = 14000 ft/s or 4267 m/s) is sandwiched between two high velocity layers (v = 21000 ft/s or 6400 m/s). The top of the low velocity layer has a depth of 100 ft (30.5 m) and the bottom of the low velocity layer has a depth of 150 ft (45.7 m). Finite-difference synthetic seismograms were computed for a crosswell simulation in which the wells were 100 ft (30.5 m) apart. The modeled recording geometry was such that the sources were located at 415 ft (126.5 m) and 465 ft (141.7 m) depths and data were recorded by 50 receivers located between depths of 10 and 500 ft, at 10 ft (3 m) intervals. In this case, upgoing waves from sources at depth were reflected from the base of the two layers back down into receivers below the reflecting layers. The model data were then processed to isolate reflections by f-k filtering to remove the upgoing energy, including direct arrivals. Since the sources were near the bottom of the hole, most of the direct arrival energy was upgoing. The next step involved surgical muting to remove downgoing multiples in order to leave only the primary reflected arrivals as shown in Fig. 4a. It turns out that both of these steps are not overly important if one directly runs the migration program with the option to process only arrivals which are slightly delayed after the first arrival. If one leaves the multiples in, these are migrated to a position which is mainly outside the zone of interest.

The migration of the processed data in Fig. 4a produces a very good image of the two layer model when using the velocity of the bottom layer, 21000 ft/s (6400 m/s). The bottom interface is correctly positioned while the upper interface is too shallow by about 50 ft (15 m). In order to correctly position the top reflector, we would need to migrate with a lower velocity. In cases where velocity changes are more gradual, as is usually the case, these depth errors would be less pronounced. One solution to the variable velocity problem is to migrate with a series of constant velocities and examine the images at various depths. In essence, a focusing analysis of images themselves could be used in velocity estimation.

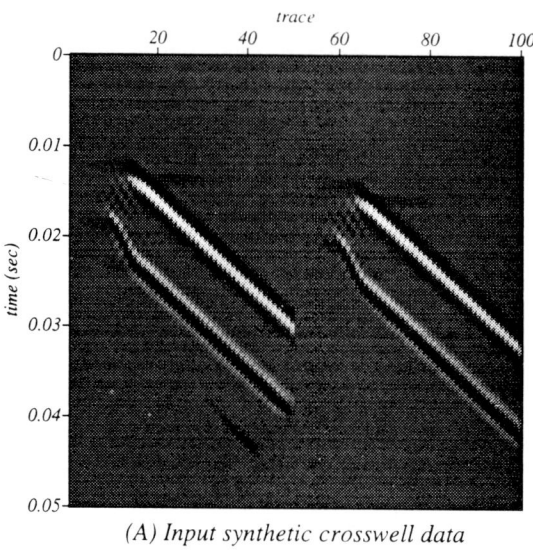

(A) Input synthetic crosswell data

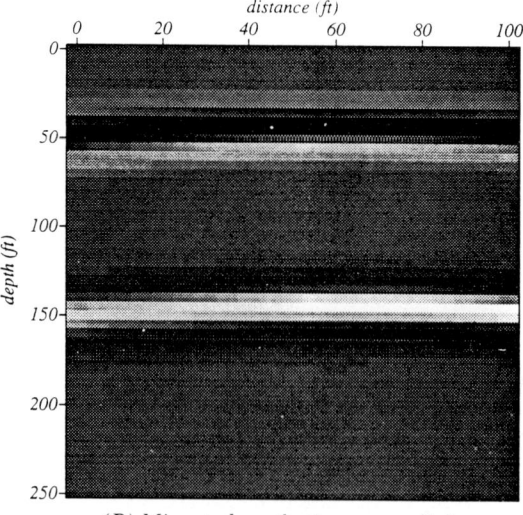

(B) Migrated synthetic crosswell data

Fig. 4. Synthetic model test. Source well is at x = 0 and receiver well is at x = 100. (a) Input synthetic crosswell data. (b) Migrated synthetic crosswell data.

MIGRATION OF REAL DATA

Our final test for the new crosswell migration method was to process real data from a West Texas survey described by Lines et al. (1993). The data were recorded in two wells that are approximately 720 ft (220 m) apart. It generally involves imaging layers of siltstone and dolomite in the Grayburg formation where dips are typically less than 5°. The zone of interest is a siltstone unit at about 4370 ft (1332 m). Fig. 5 shows a record after f-k filtering was applied to remove upgoing energy including direct arrivals.

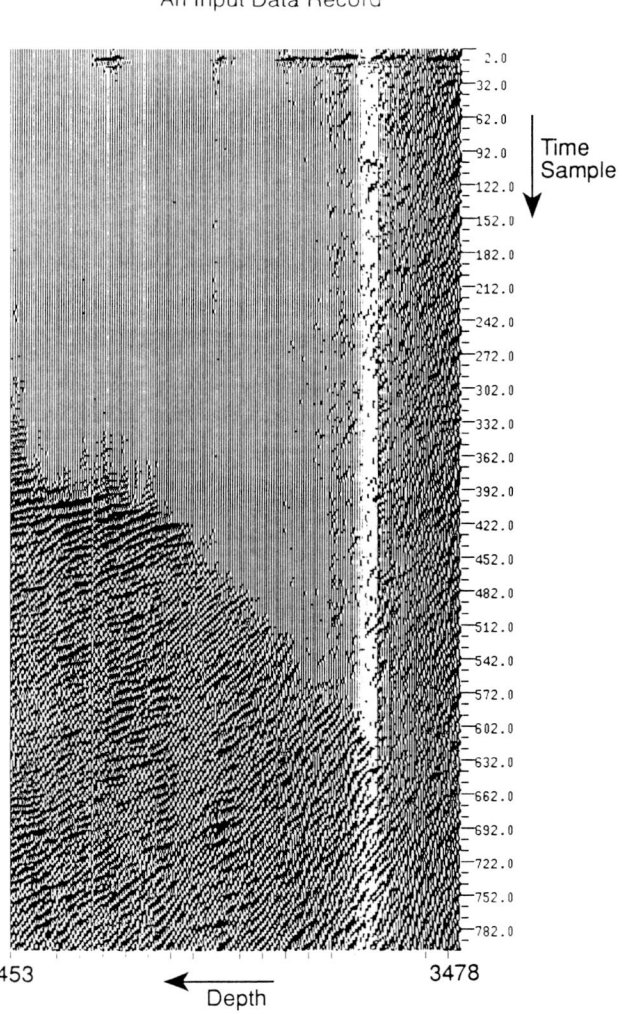

Fig. 5. An input common receiver record of 196 traces for sources between depths of 3478 and 4453 ft (sample interval = 0.2 ms). The receiver depth is at 4440 ft.

For this record, the receiver was at 4440 ft (1353 m) and the sources ranged in depth from 4453 ft (1357 m) to 3478 ft (1060 m) at 5 ft (1.5 m) intervals. The reflections for the deeper sources are apparent on the input data. The time sample interval for this data was 0.2 ms.

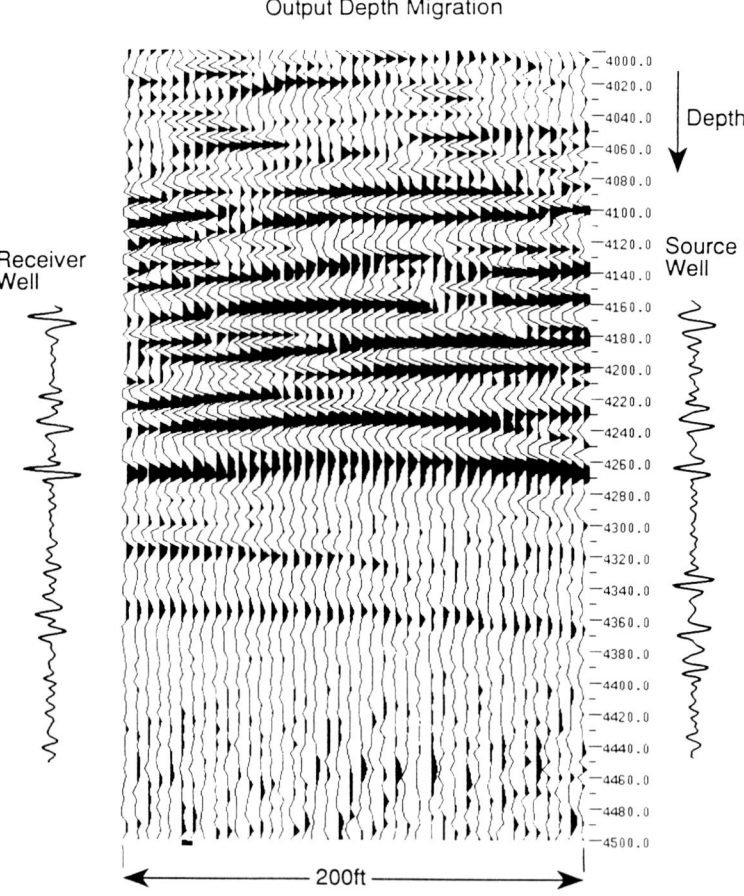

Fig. 6. Output migration with nearby source and receiver wells synthetics for 200 ft illumination zone between wells.

Our migration with a velocity of 19000 ft/s (5791 m/s) produced a coherent image for reflectors in the range 4000-4400 ft (1219-1341 m). Fig. 6 shows the depth image in a 200 ft (60 m) horizontal span in the highly illuminated zone between wells. The reflections compare reasonably well with the sonic logs shown for the source and receiver wells. An overlay of the velocity tomogram on the depth migration in Fig. 7. generally shows that the highest reflection image amplitudes correlate well with the largest velocity gradients. The dips on the migrated image are consistent with known information. We believe the migration of crosswell reflections provides a high resolution image which compliments the velocity tomogram. The optimum velocity for migration is determined for the depths of interest and is basically within the range of velocities for these depths (15500-21000 ft/s or 4730-6400 m/s). It is encouraging that, due to the high frequency content in our data (400-1600 hz), the wave lengths are only 15-20 ft (4.5-6 m), providing a high-resolution image.

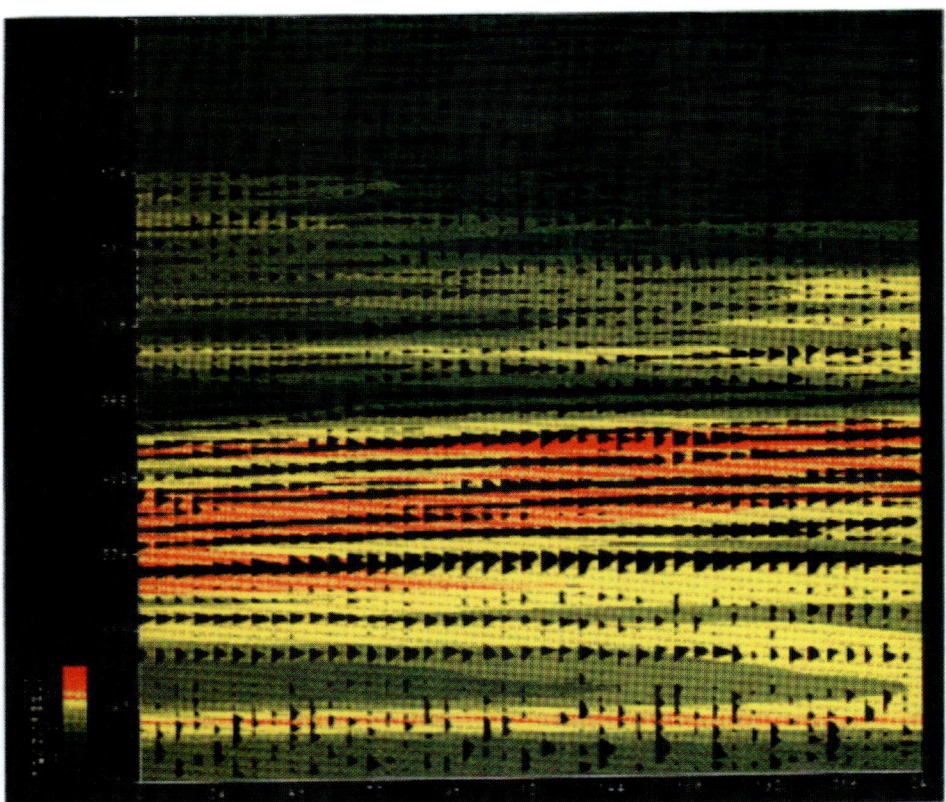

Fig.7. An overlay of a velocity tomogram on the depths migration obtained by the method. Depth range is from 3700-4470 ft (1128-1362 m) over the central illumination zone.

CONCLUSIONS

We have described a simple crosswell migration method which is based on the principle of elliptical impulse responses and wavefront summations for reflected crosswell arrivals. Despite the restriction of constant velocity, this method has shown encouraging results for synthetic and real data.

REFERENCES

Claerbout, J., 1985. Imaging the Earth's Interior. Blackwell Scientific Publications, Oxford.
Dillon, P.B., 1990. A comparison between Kirchhoff and GRT migration on VSP data. Geophys. Prosp., 38: 757-778.
Hu, L., McMechan, G.A. and Harris, J.M., 1988. Acoustic prestack migration of cross-hole data. Geophysics, 53: 1015-1023.
Findlay, M.J., Goulty, N.R. and Kragh, J.E., 1991. The crosshole seismic reflection method in opencast coal exploration. First Break, 9: 509-514.
Gray, S.H. and Lines, L.R., 1992. Crosswell tomographic migration. Jrnl. of Seismic Explor., 1: 315-324.
Lazaratos, S.K., Harris, J.M., Rector, J.W. and Van Schaak, M., 1992. High- resolution cross-well imaging of a West Texas carbonate reservoir, Pt. 4: Reflection imaging. 62nd Ann. Internat. SEG Mtg., New Orleans, Expanded Abstr.: 49-51.
Lines, L.R., Miller, M., Tan, H., Chambers, R.L. and Treitel, S., 1993. Integrated interpretation of borehole and crosswell data from a West Texas field. The Leading Edge, Jan.: 13-16.
Stewart, R.R., and Marchisio, G., 1990. Cross-well seismic imaging using reflections; Proceedings of the First SEGJ Internat. Symp. on Geotomography. SEGJ Publication, Tokyo.
Zhu, X. and McMechan, G.A., 1988. Acoustic modeling and migration of stacked cross-hole data. Geophysics, 53: 492-500.

Speed and accuracy of seismic migration methods

Samuel H. Gray*

*Amoco Exploration and Production Technology, P.O. Box 240, Calgary, AB T2P 2H8 Canada

Summary

Seismic migration is the culmination of a series of processing steps applied to seismic data. In the oil industry, migration provides the image of the Earth's subsurface for geophysicists and geologists to decide on prospective locations for hydrocarbon traps. Because of the high costs of drilling wells, it is crucial that the seismic imaging be performed accurately; because of the high costs of acquiring and processing seismic data, it is important (though not crucial) that imaging be performed cheaply. Here, I present and discuss several of the most popular seismic migration methods, investigating their accuracy and speed. The acoustic wave equation with variable seismic velocity provides the model for seismic wave propagation, and the various methods solve the wave equation in various domains: two in the time domain (Kirchhoff and reverse-time migration) and two in the frequency domain (an explicit finite-difference method and phase-shift migration with its extension to handle lateral velocity variations, phase-shift plus interpolation). Comparing their implementations shows advantages and disadvantages of all the methods; although Kirchhoff migration emerges as a workhorse method, none can be said to be the best.

Introduction

Seismic imaging problems come in all shapes and sizes: small and big, shallow and deep, structural and stratigraphic, and on and on. Historically, the solutions to these problems have been constrained to some degree by the economics of acquiring and processing seismic data, and of drilling wells. These constraints have always resulted in a final product (the final seismic image and the ability to interpret it) that was almost acceptable. The earliest digital seismic data was almost well-enough acquired and processed to see the relatively shallow, easy targets that were the hydrocarbon traps discovered twenty to thirty years ago, and present seismic data, with one or two orders of magnitude greater redundancy than 1970's data, provides almost enough information about the Earth's subsurface to yield reliable images of deeper, better hidden, and generally more subtle traps.

Among the many seismic acquisition and processing solutions whose quality has increased steadily over the years (just enough to maintain a fairly constant frustration level among seismic interpreters in their attempts to map the subsurface), a certain class of seismic imaging techniques, known as seismic migration, has improved about to the point of acceptibility. That is, a range of seismic migration techniques exists that can adequately solve most imaging problems. This is not to imply that the appropriate migration tool is always used, but rather that for most imaging problems, the appropriate migration tool has been developed.

These seismic migration techniques will be the subject of this paper. Seismic migration methods form a subset of a broader class of seismic imaging methods. Migration attempts to form images of subsurface reflectors by moving, or "migrating" (a term with a great deal of historical baggage) reflection events from one place on a seismic record to their correct spatial locations. Other imaging techniques, such as the application of normal-moveout corrections followed by common-midpoint stacking (NMO/stack) attempt, less ambitiously, to simulate a "zero-offset" seismic cross section, where many seismic records each having many source-receiver offsets have been replaced by a single section with a source-receiver pair located at each point along the Earth's surface. After this process, it is still necessary to migrate the data to move the reflection events to their correct spatial locations and collapse diffractions. Often, a process called dip-moveout (DMO) is inserted between NMO and stack in order to preserve the reflection character of steeply dipping reflection events, but the resulting stack still needs to be migrated. Another subset of imaging methods attempts to remove the damaging effects of multiple reflections from seismic records. Confusingly, the terms "imaging" and "migration" are often used interchangeably.

Seismic data result from reflections of energy sent into the Earth from specific source locations. The reflections occur at interfaces between different types of rocks from energy, and are consequently localized in time. Then seismic data are recorded in space and time, and it is natural to expect that some methods will process data in space and time. It is also natural to expect seismic researchers to take advantage of many powerful digital signal processing advances by transforming data into the temporal frequency domain. It is perhaps less natural to go one step further and investigate seismic data in the *spatial* frequency domain, but this domain turns out to give perhaps the best all-around migration technique. All combinations of these domains are useful for various types of seismic processing, and we often transform data from one to another. For migration, however, the most useful domains appear to be space-time (x-t), space-frequency (x-ω), and wavenumber-frequency (k-ω). In this paper, I shall describe two space-time methods (Kirchhoff and reverse time), one space-frequency method (explicit finite-difference), and one wavenumber-frequency method (phase-shift). I shall also describe a powerful generalization of phase-shift migration, phase-shift plus interpolation (PSPI) which shuttles back and forth between the wavenumber domain and the space domain. I shall discuss these methods with an eye on their accuracy and speed. The bulk of this discussion will take place in the context of two-dimensional (2-D) poststack migration, but I shall indicate how the methods are enhanced and/or limited when they are considered as prestack migrations.

In limiting the discussion to these migration methods, I am necessarily excluding several interesting techniques. Most of these techniques are variations of the ones presented here, although at least one (Gaussian beam migration - Hill, 1990) is an extremely powerful hybrid that seems to improve on Kirchhoff migration. Also, I am completely ignoring the vitally important problem of providing the velocity field for migration. Depth migration requires a detailed interval velocity model - an actual propagation velocity at each point in the subsurface. It turns out that estimating the velocities to be used for depth migration is much harder than the problem of migration, since the major tool available for estimating those velocities is to migrate the data!

What is poststack migration?

Ideally, all the seismic traces acquired will be migrated; otherwise, why would they be acquired? Historically, most of the traces from a 2-D or 3-D seismic survey were not migrated; even today, the vast majority of migrations are performed on data sets that have been culled from the original shot records. The reason for this was, and is, economics. Although computing power has advanced to the point where prestack migration (all the input traces are migrated before the images are stacked to form the final image) is performed on perhaps ten percent of all data volumes in both two and three dimensions, still poststack migration (all the input traces are stacked into a greatly reduced data volume before migration) is far more common. The process that reduces the original data set to a simulated zero-offset volume is called the common-midpoint (CMP) stack. Yilmaz (1987) provides an excellent discussion of this procedure, as well as many other seismic processing techniques. Briefly, the seismic traces are sorted from common shot records to CMP records, i.e., records where all the traces have the same source-receiver midpoint. Then, assuming that all reflectors are flat and that the seismic velocity to each flat reflector is constant (although perhaps a different constant from the reflector above or below the one being considered - an absurd assumption!), the Pythagorean time-distance relationship is used to replace the hyperbolic moveout of the reflection event (i.e., the relationship between source-receiver offset and reflection time) with a linear moveout. This process is called normal moveout (NMO) correction. After all events on all traces in a CMP gather have been NMO corrected, the traces are summed together, or stacked, into a single trace that simulates the zero-offset trace. Performing NMO/stack on all the CMP gathers in a survey yields a single data volume that is ready to be migrated.

As bad as the flat-reflector assumption is, this process works remarkably well in producing a data volume that can be migrated. However, the CMP stack prevents one of the major benefits of prestack migration, namely estimating actual earth velocities. This velocity estimation step is crucial in seismic interpretation where the geology is structurally complicated. The combination of this ability to estimate velocity with the added signal-to-noise improvement of prestack migration over poststack migration makes prestack migration a far more powerful imaging tool than poststack migration.

After NMO/stack, the data are ready for migration. It turns out, though, that migration is not quite ready for the data yet. We'd like to use the scalar wave equation

$$\frac{\partial^2 P}{\partial x^2} + \frac{\partial^2 P}{\partial z^2} - \frac{1}{v^2(x,z)}\frac{\partial^2 P}{\partial t^2} = 0 \tag{1}$$

as if the section to be migrated is a single wavefield $P(x,z=0,t)$, but we've performed the CMP stacking process to simulate a section of independent zero-offset experiments, that is, portions of many wavefields. Each of these zero-offset experiments will require its own migration (an enormous computer expense) unless we somehow visualize the section differently. We do this by imagining that all the diffractors in the Earth exploded at time $t = 0$, and that the energy from these explosions propagated to the Earth's surface with one-half the actual propagation velocity. In that case, we've recorded essentially the same data, now in the form of the single wavefield that we wanted, and we won't have to migrate each zero-offset experiment independently. This concept of exploding reflectors (Loewenthal et al., 1976) as an expression of the stacked section is hardly less realistic than the approximation of a zero-offset trace by a stacked trace.

Now the data are ready to be migrated and migration is ready for the data. In fact, the exploding reflector model gives us an explicit goal for migration: to transform the recorded data *P(x,z=0,t)* into a picture of the subsurface reflectors at the instant of explosion, i.e., to recover *P(x,z,t=0)*. In describing the migration methods, it will be convenient to express this process in two steps, the first being a downward continuation of the wavefield, expressed by *P(x,z=0,t)* -> *P(x,z,t)*, followed by an application of the imaging condition *P(x,z,t)* -> *P(x,z,t=0)*. Because the exploding reflector model has replaced the seismic velocity with the half-velocity, the velocity *v* in the wave equation is now understood to be the half-velocity.

What does migration do?

The acoustic wave equation is linear, and migration is linear too. The wavefield *P(x,z=0,t)* is a linear sum of its components, and the action of migration on the individual components can be summed to obtain the action of migration on the entire wavefield. This fact allows us to consider an *unmigrated* seismic section as being the superposition of individual spikes, and to consider the action of migration on any one of those spikes. Alternatively, we may build up the migrated section diffractor by diffractor, considering the *migrated* seismic section as being the superposition of diffractors. Although we haven't derived it yet, the constant-velocity Kirchhoff migration formula allows us to view the action of migration either upon an isolated spike on the input section (migration impulse response) or onto an isolated diffractor location on the output section (diffraction stack). This formula is

$$P(x, z, 0) = \int W(x, z, x') \left(\frac{\partial}{\partial t}\right)^{1/2} P(x', 0, r/v) \, dx' \tag{2}$$

where $r = \sqrt{(x-x')^2 + z^2}$ is the distance between an output location (x,z) (where an input sample is about to accumulate) and the input location $(x',0)$ (where the input sample was collected), and W is a weight applied to the input data. We shall discuss W in detail in the Appendix. The half-derivative operator $\left(\frac{\partial}{\partial t}\right)^{1/2}$ is the temporal expression of the filter $\sqrt{i\omega}$ applied in the frequency domain; it restores the correct phase to the data after migration without greatly altering the locality of the input samples. If we consider an input section consisting entirely of zeroes except for a single nonzero sample, then the migration formula will smear that sample out onto a semicircle (Fig. 1), which is the set of possible reflector locations for that sample. Alternatively, if we want to compute the migrated output at a single point (x,z), Eq. (2) tells us to sum the input data collected along a hyperbolic curve in the unmigrated section.

The depth migration methods in two dimensions

This section relies heavily on the presentation of Whitmore et al. (1988). We use the scalar wave equation (1) to derive migration methods for the acoustic pressure field P. Our objective is to transform the surface measurements $P(x,z=0,t)$ to the subsurface wavefield at the instant of explosion $P(x,z,t=0)$. The unmigrated sections are considered to have N_x traces, each with N_t time samples, and the migrated sections also have N_x traces with N_z depth samples. When we express the traces in the frequency domain, we recognize that the number of frequencies carrying meaningful data, N_ω, might be considerably less than N_t.

v(z) Phase-shift migration

We start with the method that is the most elegant and powerful in its simplicity, and at the same time the most economical. Introduced by Gazdag in 1978, phase-shift migration is blindingly fast and extremely accurate, able to image dips up to 90 degrees (and beyond, in an extension). The only limitation of this method is its restriction to lateral velocity invariance. That is, as long as the background velocity is laterally constant, phase-shift migration can image very steep dips, making it ideal for imaging, say, the flanks of salt domes that intrude in the otherwise fairly flat Gulf of Mexico sedimentary basin.

Phase-shift migration begins with Eq. (1), with the understanding that v is really one-half the propagation velocity and is independent of x. Then we express $P(x,z,t)$ as an inverse spatial and temporal Fourier transform and substitute this expression into Eq. (1). Interchanging the order of integration and differentiation and performing the differentiations gives

$$\frac{d^2 p}{dz^2} + k_z^2 p = 0 \tag{3}$$

where $p(k_x, z, \omega)$ is the spatial/temporal Fourier transform of $P(x,z,t)$ and

$$k_z = \pm \sqrt{\frac{\omega^2}{v^2} - k_x^2} \tag{4}$$

Assuming that $v(z)$ is constant within strips of depth Δz leads to exact solutions of Eq. (3) of the form

$$p(k_x, z + \Delta z, \omega) = p(k_x, z, \omega) \exp(i k_z \Delta z) \tag{5}$$

We extrapolate the wavefield downward, to increasing depths, and backwards in time, forcing k_z

and ω to have the same sign and specifying the branch of the square root to be used in Eq. (4). Eq. (5) is the downward continuation formula for phase-shift migration. It is an expression for taking the wavefield through a depth layer of thickness Δz. It is unconditionally stable, and it can be made exact by including the (k_x, ω)-dependent transmission coefficients at the assumed velocity discontinuities at the layer boundaries.

To do phase-shift migration, we first perform the double Fourier transform $P(x,z,t)$ -> $p(k_x,z,\omega)$, and then we implement Eq. (5) for all ω, z, and k_x. (Actually, wavenumbers $k_x > \omega/v$ in magnitude refer to evanescent modes; if included in the computations, these modes must be decayed in the downward continuation to preserve stability.) Finally, we apply the imaging condition, recovering $P(x,z,t=0)$ by performing an inverse spatial Fourier transform and an integral over frequency. Thus the total amount of work for phase-shift migration is: initial 2-D Fourier transform of P into p, downward continuation of p, and final 1-D Fourier transform and integral over frequency. The 2-D Fourier transforms require $O(N_x N_t \log(N_x N_t))$ operations. Downward continuing a single Fourier component p through all depths requires N_z complex multiplications. The number of these Fourier components is $N_\omega N_x$, so the total number of complex multiplications is $O(N_z N_\omega N_x)$, where the O symbol implies that only a certain percentage of horizontal wavenumbers k_x are included for each frequency. The imaging step consists of summation over frequencies followed by inverse spatial Fourier transform. The first of these must be done for all depths and wavenumbers, requiring $N_z N_x N_\omega$ additions, and the second requires N_z Fourier transforms, or $O(N_z N_x \log(N_x))$ operations. This total is dominated by the $O(N_z N_\omega N_x)$ complex multiplications of the downward continuation. For poststack migration, it makes little sense to try to improve on this kind of performance.

v(z) Kirchhoff migration

Now that we have the expression for phase-shift migration, we can use it to derive other migration methods, starting with Kirchhoff migration. Schneider (1978) has done this in two dimensions by performing a 2-D inverse Fourier transform of Eq. (5) to obtain the downward continuation formula in space and time, then setting $t = 0$ in that expression. Schneider's final formula for 2-D Kirchhoff migration is somewhat different (and more time-consuming to implement) from the one shown in Eq. (2), which can be obtained only after an asymptotic approximation of the integral over time (Jakubowicz and Levin, 1983). When Eq. (5) is Fourier transformed, the product becomes a convolution, and the wonderful locality of the phase-shift downward continuation expression is broken. Expressing the time integral (asymptotically) as a single number speeds up Kirchhoff migration considerably, but not to the point where it is competitive with phase-shift migration. On the other hand, Kirchhoff migration is much more versatile than phase-shift migration, or any other kind of migration because it allows us to migrate input data (either all of it or any subset of it) onto a subset of the actual output section.

Eq. (2) is an expression for constant-velocity migration; the expression for $v(z)$ migration replaces r/v (constant-velocity traveltime) with two-way traveltime along the raypath joining the source-receiver point with the image point (x,z). According to this modification of Eq. (2), then, the total amount of work for Kirchhoff migration consists of half-differentiating the input traces, computing traveltimes between all upper-surface points and image points, and performing the weighted sum to obtain the output at each of the image points. The half-derivative is implemented by Fourier transforming the input traces from time to frequency, multiplying by the filter $\sqrt{i\omega}$, and inverse Fourier transforming the traces back to time, at a computational cost of $O(N_x N_t \log(N_t))$. The traveltimes can be computed by raytracing, and since the velocity is laterally invariant, the raytracing needs to be done from only one source-receiver point to all image points. There are many ways to do this, and the simplest is to shoot a fan of rays downward into the Earth, note the location and time along each raypath at each depth step, and interpolate the traveltimes along the raypaths at each depth onto the regular x-grid of migration image points. Assuming that each incremental ray calculation from depth z to depth Δz requires $O(1)$ operations, and that N_{ray} rays are traced, the traveltime calculations cost $O(N_{ray} N_z)$ operations plus $O(N_x N_z)$ operations for the interpolation. As shown in the Appendix, computing the migration weights is a smaller problem than computing the traveltimes. The Kirchhoff summation usually involves the bulk of the operations in Kirchhoff migration (although a very careful operation count reveals that applying the half-derivative filter is a significant percentage of the total number of operations). If the output section consists of all the N_x trace locations as the input section, then there are $N_x N_z$ output points, each of them requiring a summation over the input traces to form the image. This summation is over a migration aperture of $N_{x'}$ traces, where for long seismic lines, $N_{x'}$ is usually less than N_x. So the work of Kirchhoff migration is dominated by the $O(N_x N_z N_{x'})$ summation cost. When we investigate $v(x,z)$ migration, we'll need to work a little harder to keep the cost of Kirchhoff migration from being dominated by the cost of raytracing.

Because of the asymptotic collapse of the time integral, Kirchhoff migration is not exact with the error tending to be confined to the shallow part of the section. Nevertheless, $v(z)$ Kirchhoff migration can be very accurate, depending on the care taken in the raytracing. Like phase-shift migration, Kirchhoff migration can be used to image very steep dips (even greater than 90 degrees). However, Kirchhoff migration suffers from one shortcoming unknown to any of the frequency-domain methods. That is, when it sweeps data out to the steeply dipping part of the semi-circular migration impulse response curve, the high frequencies tend to be handled incorrectly, causing aliased noise to appear on the migrated section. Anti-aliasing the migration operator (Gray, 1992; Lumley et al., 1994) is a relatively recent improvement of Kirchhoff migration, bringing its accuracy very close to the level of phase-shift migration. Still, Kirchhoff migration is a bit slower than phase-shift migration for the same accuracy, and is not as easy to code up.

v(x,z) Kirchhoff migration

When the velocity of the waves varies laterally, the migration problem becomes fundamentally more difficult for all methods except for Kirchhoff migration, and even for Kirchhoff migration, one must take care in order to avoid increasing runtime by an order of magnitude. The Kirchhoff summation for *v(x,z)* migration remains as it is for *v(z)* migration. However, the traveltime calculations are considerably more complicated. We still need traveltimes along raypaths joining each source-receiver point along the upper surface, but we can't get away with calculating these traveltimes from a single upper-surface point and using lateral velocity invariance. In fact, in the worst case we can envision having to connect N_x upper-surface points with $N_x N_z$ image points with raypaths. Assuming that each raypath crosses $O(N_z)$ points where the velocity might vary, and that each incremental ray calculation involves $O(1)$ operations, we are faced with $O(N_x N_x N_z N_z)$ operations, which is more than we use for the Kirchhoff summation. However, we can do much better than this worst case. Instead of tracing a new raypath from each surface location to each image point, we can much more efficiently shoot a fan of rays downward and interpolate the traveltimes onto the grid of image points (Gray, 1986). This reduces the operation count for the traveltime calculations to $O(N_x N_{ray} N_z)$ for the raytracing plus $O(N_x N_x N_z)$ for the interpolation. Although this count is similar to the one for the Kirchhoff summation, the traveltime calculations usually take less time than the summation.

Another way to compute traveltimes for Kirchhoff migration is provided by directly solving the eikonal equation, with a source placed at every surface location (Vidale, 1988; van Trier and Symes, 1991). This method solves for the field of traveltimes from a surface location. Its operation count is similar to that for raytracing plus interpolation, and it doesn't appear to be more accurate than raytracing (Gray and May, 1994).

Some notable recent attempts to make the traveltime calculations more accurate are wavefront construction (Vinje et al., 1993), the use of traveltimes associated with maximum wavefield energy instead of minimum time (Audebert et al., 1997), and a method that alternates migration (down to a certain depth) and downward continuation of the wavefield to the same depth for subsequent migration (Bevc, 1997).

v(x,z) Phase shift plus interpolation migration

The beauty of phase-shift migration lies in its simplicity and economy. Unfortunately, that simplicity and economy are due entirely to the lateral invariance of velocity. The generalization of phase-shift migration to handle lateral velocity variations (phase-shift plus interpolation, or PSPI - Gazdag and Sguazzero, 1984) is neither simple nor economical, but it is very accurate. Like phase-shift migration, PSPI is a frequency-domain method; unlike phase-shift migration, PSPI shuttles back and forth between space and wavenumber, and that shuttling back and forth at each depth for each frequency slows it down considerably.

In its simplest form, we assume that, between depth z and depth Δz there are N_v discrete regions where the velocity takes on different (constant) values. PSPI begins with the wavefield at depth z in space and frequency, denoted by $\tilde{P}(x, z, \omega)$, Fourier transforms the field to

$p(k_x, z, \omega)$, and then downward continues p to depth $z + \Delta z$ using Eq. (5) N_v times, once for each velocity. Now there are N_v wavefields $p(k_x, z + \Delta z, \omega)$, and each of these is inverse transformed to a corresponding $\tilde{P}(x, z + \Delta x, \omega)$. Then these fields are combined to a single field by choosing at each x the field $\tilde{P}$ corresponding to the velocity at the point (x,z). This produces the wavefield $\tilde{P}(x, z + \Delta z, \omega)$ to be downward continued to the next depth.

~~Operationally, this requires computing $\tilde{P}$ at each depth and N_v intermediate wavefields p~~ and $\tilde{P}$ that are discarded immediately after their use. It eliminates the initial and final spatial Fourier transforms, but it replaces them with a forward and N_v inverse spatial Fourier transforms at each depth. All these extra Fourier transforms add $O(N_\omega N_v N_z N_x \log(N_x))$ operations, making PSPI much more expensive than $v(x,z)$ Kirchhoff migration.

v(x,z) Explicit finite-difference migration

PSPI is very accurate but, because it shuttles back and forth between the space and wavenumber domains, it is very expensive. Why not just stay in space? This is a good idea, yielding a popular migration method that is somewhat cheaper and about as accurate as PSPI except at the steepest dips, but one that has problems with stability.

When we stay in space ($\tilde{P}(x, z, \omega)$ instead of $p(k_x, z, \omega)$), we change the multiplication in Eq. (5) to a convolution. Thus for the extrapolation step, $\tilde{P}(x, z + \Delta z, \omega)$ is computed as the spatial convolution of $\tilde{P}(x, z, \omega)$ and the inverse Fourier transform of $\exp(ik_z \Delta z)$. This inverse Fourier transform is expressed in terms of the Hankel function $H_1^{(1)}$ (Jakubowicz and Levin, 1983), so using carefully tabulated values for the Hankel function in the convolutions spanning the entire N_x traces will yield results equivalent to PSPI. However, the amount of work is prohibitive in this naive approach, since a quadrature of length N_x for each of $N_\omega N_x N_z$ points replaces a small number of spatial Fourier transforms at each of $N_\omega N_z$ output depths, yielding even more operations than PSPI. To overcome this problem, we reduce the length of the convolution to an aperture much smaller than N_x, but this causes two new problems to arise. First, we limit the dip of the downward continuation operator and second, we introduce instabilities into the downward continuation. The first problem is not extremely serious as long as we don't need to image dips near 90 degrees, and it can be solved by making the operator uncomfortably large (about twenty to forty traces for typical seismic data). The second problem is very serious. Truncating the convolution is effectively the same as truncating the inverse spatial Fourier transform of the phase-shift operator $\exp(ik_z\Delta z)$, but doing that introduces a Gibbs phenomenon into the phase-shift operator itself (Nautiyal et al., 1993), so that some of the wavenumber components are modified to have value greater than unity. These components will be amplified exponentially upon the recursive application of the truncated operator over several depth levels, with disastrous consequences. Holberg (1988) and Hale (1991a), among others, have shown how to overcome this problem with little loss of accuracy.

Having overcome the stability problem, we now have a fairly simple algorithm. For each frequency, we downward continue the wavefield from depth to depth, at each depth point $(z, z+\Delta z)$ computing the wavefield from its values at several (the twenty to forty mentioned above) nearby x-locations at the previous depth. The velocity used to obtain the extrapolated wavefield at $(x, z+\Delta z)$ is the velocity at that point, so that the downward continuation operator needs to be precomputed for every value that the seismic velocity takes on. If there are N_v velocity values, precomputing the operators requires $O(N_\omega N_v N_x \log(N_x))$ operations. Then performing the downward continuation requires, for each frequency, a convolution at each output point, or $O(N_\omega N_x N_z N_{x'})$ complex operations, where $N_{x'}$ is the convolution length. Finally, the downward continued wavefield is evaluated at zero time by summing over frequencies. All this work is typically cheaper than PSPI.

v(x,z) Reverse-time migration

As opposed to the migration methods studied so far, reverse-time migration is a brute-force method. Instead of seeking a domain where the wave equation can be expressed elegantly or solved efficiently, reverse-time migration simply solves a discretized version of the wave equation. It does this ingeniously, though, providing an extremely accurate migration method that can image all dips. In discretizing the wave equation, the time derivative is expressed as a forward difference, and at time t_j the value of the discretized wavefield is expressed in terms of its values at future times. Unusual as this sounds, remember that migration always runs the wave equation backwards in time, obtaining $P(x,z,t=0)$ from $P(x,z=0,t)$. "Initially" (more correctly, "finally"), P and its time derivative within the subsurface are set to zero, and the observed values of the wavefield, and finite-differenced values for its derivative, at the greatest time t_{max} are introduced at $z = 0$. At the next time step, those boundary values will have propagated a short distance into the subsurface, and more of the recorded wavefield and its derivative (its values at the next-greatest time $t_{max} - \Delta t$) are injected at $z = 0$. Continuing, at successive times from t_{max} to zero, values for P and its time derivative appear at $z = 0$ to be propagated into the subsurface.

The earliest implementations of reverse-time migration (McMechan, 1983; Whitmore, 1983) used second-order difference approximations for all the derivatives. These approximations require a small number of operations for each grid point but, to avoid numerical grid dispersion, they require a computational grid considerably finer than the migration grid so that the wavefield can be sampled at about ten points per wavelength. Higher-order differences require more operations per grid point but need to sample the wavefield at fewer (approaching two) points per wavelength. Presently, it appears that using tenth-order (or so) differences minimize the total number of operations for acoustic finite-difference modeling or reverse-time migration in two dimensions.

It is easy to write a reverse-time migration program, and to count its operations. At each time step, the finite-difference stencil is applied at all points in space: this adds up to $O(N_t N_x N_z)$ operations, where the constant implied by the O symbol depends on the size of the finite-difference stencil, the ratios of the migration grid size to the finite-difference grid size, and for reasons of numerical stability, the lowest velocity and the highest frequency. Although this estimate appears to be similar to the cost estimate for Kirchhoff migration, reverse-time migration is usually more expensive than even PSPI; the O symbol seems to be hiding a very large number. To reduce the amount of work by nearly half, we can begin the calculations at t_{max}, by computing only at the shallowest depths of the migration grid, because the wavefield can't propagate past these depths in a small number of time steps. As time decreases from t_{max}, the range of z-values included in the calculations must increase to allow the final data values to propagate down towards the greatest depths.

A disadvantage of discretizing Eq. (1) for reverse-time migration is the tendency of the method to generate artificial multiple reflections at model boundaries. Such reflections are considered to be artificial because (1) if the model boundaries are correctly located, multiple reflections from them will have been attenuated somewhat by the stacking process and (2) if the model boundaries are not correctly located, the reflections create noise that obscures the actual reflection events on the migrated image. A simple way around this problem is to smooth the slowness (reciprocal velocity) field before migration.

Migration cost versus temporal frequency

The more temporal frequencies, both high and low, are present in the seismic data, the greater the resolving power of any processing technique, and migration is no exception. But this added resolution carries a price. We've seen that the cost of the frequency-domain migration methods depends in part on the number of frequencies processed. Less obvious is the full effect of frequencies on migration cost.

For all migration methods, the depth step size depends on the frequencies present: The sharper the seismic wavelet, the smaller the depth steps need to be to sample the wavelet in space, making N_z proportional to the value of the highest frequency present in the seismic data. So even Kirchhoff migration's cost depends linearly on the number of frequencies present. Since we can apply the same argument to the frequency-domain methods we've investigated, and the costs of those methods already contains a factor proportional to the number of frequencies present, we see that the cost of frequency-domain migration increases *quadratically* with the number of frequencies (although the use of multigrid techniques, with larger depth step sizes and larger sampling increments for the spatial convolutions at lower frequencies, can reduce this dependence to linear). And for reverse-time migration, the cost is proportional to the step size in x, z, and t, and stability requires that these step sizes all depend on the highest frequency present. So reverse-time migration's cost increases *cubically* with the highest frequency.

A word about 3-D poststack migration

3-D poststack migration can be carried out just as 2-D poststack migration is, with the addition of an extra spatial dimension. This extra dimension actually adds *two* more orders of N to the amount of work for Kirchhoff migration, since the amount of work is proportional to the number of input traces times the number of output traces, and each of these numbers increases by an order of N. On the other hand, the operation count for reverse-time migration appears to increase only by a single factor of N (which tends to be larger than either of the extra N's for Kirchhoff migration). So the 3-D poststack operation count disparity between reverse-time migration and Kirchhoff migration appears to be less than the 2-D disparity.

For explicit finite-difference migration in three dimensions, Hale (1991b) showed how to speed up the process dramatically, to the point where this method provides completely acceptable accuracy at a cost comparable to (and often lower than) 3-D Kirchhoff migration.

Prestack migration

In short this section, I discuss migration before stack, concentrating on a single computational aspect that gives Kirchhoff migration a significant speed advantage over the other methods.

Migration before stack puts off the stacking process until the very end, at which point stacking is a summing process, not an imaging process. Instead of performing a CMP stack to simulate an exploding reflection, thereby reducing the number of traces input to the migration, prestack migration migrates all the input traces, and then sums the migrated sections together.

Recall that poststack migration consists conceptually of two steps, downward continuation of the wavefield followed by the application of an imaging condition. The same holds true for prestack migration. But the concept of exploding reflectors has no analogue in prestack migration, so we must formulate a new imaging condition to go along with the process of downward continuation. This is done by recognizing that there are now *two* wavefields, one associated with the seismic source and the other associated with the recorded data. These are both downward continued into the Earth (already twice as much work as for poststack migration). The wavefield from the source is downward continued forward in time (advancing from the source initiation time, which is considered to be zero), and the wavefield from the receivers is downward continued backward in time (retarding from the time of recording). Since reflection at a subsurface point has occurred sometime between time zero and the recorded time for the reflection, the two wavefields are thus allowed to meet at the subsurface point at the same time, which is the actual time of reflection or scattering from that point. So the imaging condition at an image point could be as easy as selecting from the downward continued reflected wavefield the sample for which the source wavefield time equals the recorded wavefield time. In practice, it is better to normalize this value, dividing the recorded wavefield evaluated at the imaging time by source wavefield evaluated at the imaging time. Doing this gives the acoustic reflection coefficient at the reflector.

The imaging condition for prestack migration puts different computational requirements on the various migration methods. For example, as mentioned above, there are now potentially two downward continuation operations, one for the source wavefield and one for the recorded wavefield. To make matters worse, the source wavefield, originally concentrated in space, spreads out to cover all space. So the computational grid for its downward continuation is the entire subsurface ($N_x N_z$ image points), or as much of it as can be fit into computer memory (usually some aperture $N_{x'} N_z$ points), originally filled entirely with zeroes except for a single trace located at the source position. For all but Kirchhoff migration, there is no convenient way around computing with all these zeroes during the downward continuation process until the wavefield finally fills the grid. Similarly, the recorded wavefield, usually a recording spread from a common-shot record, needs to be padded with many zero traces to accommodate downward continued energy that comes from dipping subsurface reflectors not directly under the spread. These zero traces add meaningless work (addition and multiplication of zeroes) until the downward continuation finally moves energy into the padded traces.

Prestack Kirchhoff migration is a wave-equation process and is therefore subject in principle to these requirements imposed by the imaging condition. And, indeed, Kirchhoff migration can be formulated so as to be at the same disadvantage as the other methods. But that isn't necessary. The same asymptotic evaluation of integrals that speeds up poststack Kirchhoff migration can be applied to prestack Kirchhoff migration (Bleistein et al., 1987), with the result that Eq. (2), with r/v replaced by the total source-to-reflector-to-receiver traveltime and with a different weight W, still provides the migration formula. In fact, this formulation collapses the two downward continuations into one, which can be expressed as the exact same raytracing procedure used in poststack migration; this raytracing now becomes a fairly insignificant overhead operation when compared with the extra work of migrating many more input traces. Also, as with poststack migration, prestack Kirchhoff migration allows the smearing of a subset of the input traces onto a subset of the output section, where now Fig. 1 becomes the ellipse of possible reflector locations to which a single sample on a trace with nonzero source-receiver offset can migrate. Thus, the cost of prestack Kirchhoff migration, again proportional to the number of input traces times the number of output traces, is usually at least one order of magnitude less than the cost of the most popular (and more accurate) competitor, explicit finite-difference migration.

The great flexibility of prestack Kirchhoff migration yields a further advantage. That is, it can be used to migrate data from records that are not the result of a single wave-equation experiment. For example, Kirchhoff migration can migrate data from common-offset records, where each trace has a different source and a different receiver. This is a great advantage over the rest of the other methods, which require the unmigrated record to be the result of a single experiment (a common-shot record or, by acoustic reciprocity, a common-receiver record, or even a plane-wave synthesized record).

Migration examples

Two examples will suffice to show the difference in imaging quality among the migration methods. Both examples use synthetic data, and both use the actual seismic velocities in the migrations, so that we are testing only the speed and accuracy of the migration methods.

The first example (O'Brien and Gray, 1996) tests the ability of seismic migration to image beneath high-velocity salt bodies. The extreme lateral velocity variations around salt distorts the recorded wavefield considerably, requiring very accurate seismic migration. Fig. 2 shows two migrations of the exploding-reflector data set shown in Fig. 2(a); Fig. 2(b) shows an explicit finite-difference migration, and Fig. 2(c) shows a Kirchhoff migration. Clearly, the explicit finite-difference migration has imaged the subsalt reflection events with greater clarity and less noise than the Kirchhoff migration. Events labelled D, E, and F show where even explicit finite-difference migration has failed to produce a perfect image. Neither PSPI nor reverse-time migration could improve appreciably on the image in Fig. 2(b).

The second example is from the Marmousi synthetic model data set (Versteeg and Grau, 1991), which is realistic and difficult enough to test the accuracy of good-quality migration methods and to cause poor-quality migration methods to fail completely. Fig. 3, from Gray and May (1994) shows the final stacks from two prestack migrations, explicit finite-difference migration in Fig. 3(a) and Kirchhoff migration in Fig. 3(b). The differences between the images are not great, but the explicit finite-difference migration shows a slightly better-focussed image throughout the section, most notably near the target horizon at about Shotpoint #140 at depth 2600 m. While the image from Kirchhoff migration is not the best that can be obtained from that algorithm, both images are from well-tuned production codes, and the runtime for the Kirchhoff migration was approximately twenty times less than that for the finite-difference migration.

These comparisons are typical. For poststack $v(x,z)$ migration, Kirchhoff migration usually produces images that are not quite as good as PSPI, explicit finite-difference, or reverse-time migrations. For prestack migration, the differences in imaging quality tends to decrease as more migration noise is cancelled by the stacking process, and Kirchhoff migration usually produces acceptable images at a small fraction of the cost of the other migration methods. There are cases, however, when "nearly as good" is not good enough. In these cases, usually involving extreme lateral variations in propagation velocity (such as imaging around salt in the Gulf of Mexico), Kirchhoff migration is often unable to to produce a completely satisfactory image while the other methods perform quite well.

Conclusions

Migration is the final step of seismic data processing, the culmination of a long flow that typically includes editting, deconvolution and other filtering, and scaling steps, among others. If it is performed correctly, migration shows where the reflectors are, allowing seismic interpreters to pick drilling locations. Depending on the quality of the input seismic data and the accuracy of the velocity model, depth migration is capable of imaging reflectors with great accuracy, often placing hydrocarbon-bearing structures within tens of meters of their actual locations.

As shown in the migration comparison, the methods described in this paper are all accurate in geologic regimes of "typical" structural complexity. Usually, the difference in image quality among the migration methods is less than our uncertainty in estimating the velocity model, so that for most cases, using a more expensive method to gain a small amount of accuracy is not justified economically. Because of its good accuracy and relative cheapness, then, Kirchhoff migration has become the depth migration workhorse of the industry. As the examples show, however, Kirchhoff migration is not always accurate enough for all imaging problems.

In discussing several popular seismic migration methods, I have completely ignored the fundamentally important problem of estimating the velocity model for migration. If we had no ability to estimate seismic velocities for migration, then extremely accurate migration methods wouldn't be helpful and we wouldn't need to develop them; the very fact that the methods described here have been developed indicates that the velocity estimation problem can be solved, at least to some degree. As it turns out, the most useful tool in estimating migration velocities is (prestack) migration itself. Because of this, the cursory treatment of the prestack migration methods presented here needs to be extended and completed, especially in three dimensions.

References

Audebert, F., Nichols, D., Rekdal, T., Biondi, B., Lumley, D., and Urdaneta, H., 1997, Imaging complex structure with single-arrival Kirchhoff prestack depth migration: Geophysics, **62**, 1533-1543.

Bevc, D., 1997, Imaging complex structures with semi-recursive Kirchhoff migration: Geophysics, **62**, 577-588.

Bleistein, N., Cohen, J. K., and Hagin, F. G., 1987, Two and one-half dimensional Born inversion with an arbitrary reference: Geophysics, **52**, 26-36.

Gazdag, J., 1978, Wave equation migration with the phase-shift method: Geophysics, **43**, 176-185.

Gazdag, J., and Sguazzero, P., 1984, Migration of seismic data by phase shift plus interpolation: Geophysics, **49**, 124-131.

Gray, S. H., 1986, Efficient traveltime calculations for Kirchhoff migration: Geophysics, **51**, 1685-1688.

Gray, S. H., 1992, Frequency-selective design of the Kirchhoff migration operator: Geophysical Prospecting, **40**, 656-571.

Gray, S. H., and May, W. P., 1994, Kirchhoff migration using eikonal equation traveltimes: Geophysics, **59**, 810-817.

Hale, D., 1991a, Stable explicit depth extrapolation of seismic wavefields: Geophysics, **56**, 1770-1777.

Hale, D., 1991b, 3-D depth migration by McClellan transformations: Geophysics, **56**, 1778-1785.

Hill, N. R., Gaussian beam migration: Geophysics, **55**, 1416-1428.

Holberg, O., 1988, Towards optimum one-way wave propagation: Geophys. Prosp., **36**, 99-114.

Jakubowicz, H., and Levin, S., 1983, A simple exact method of 3-D migration - theory: Geophysical Prospecting, **31**, 1-33.

Loewenthal, D., Lu, L., Robertson, R., and Sherwood, J., 1976, The wave equation applied to migration: Geophysical Prospecting, **24**, 380-399.

Lumley, D., Claerbout, J., and Bevc, D., 1994, Anti-aliased Kirchhoff 3-D migration: 64th Ann. Internat. Mtg., Soc. Expl. Geophys., Expanded Abstracts, 1282-1285.

McMechan, G. A., 1983, Migration by extrapolation of time-dependent boundary values: Geophysical Prospecting, **31**, 413-420.

Nautiyal, A., Gray, S. H., Whitmore, N. D., 1993, Stability versus accuracy for an explicit wavefield extrapolation operator: Geophysics, **58**, 277-283.

O'Brien, M., and Gray, S., 1996, Can we image beneath salt?: The Leading Edge, **15**, 17-24.

Schneider, W. A., 1978, Integral formulation for migration in two and three dimensions: Geophysics, **43**, 49-76.

van Trier, J., and Symes, W. W., 1991, Upwind finite-difference calculation of traveltimes: Geophysics, **56**, 812-821.

Versteeg, R., and Grau, G., 1991, The Marmousi ecperience: Proc. 1990 EAEG workshop on practical aspects of seismic data inversion, Eur. Assoc. Expl. Geophys.

Vidale, J., 1988, Finite-difference calculation of traveltimes: Bull. Seis. Soc. Am., **78**, 2062-2076.

Vinje, V., Iversen, E., Gjoystdal, H., 1993, Traveltime and amplitude calculation using wavefront construction: Geophysics, **58**, 564-575.

Whitmore, N. D., 1983, Iterative depth migration by backward time propagation: Presented at the 53rd Ann. Internat. Mtg., Soc. Expl. Geophys.

Whitmore, N. D., Gray, S. H., and Gersztenkorn, A., 1988, Two-dimensional poststack depth migration: a survey of methods: First Break, **6**, 189-197.

Yilmaz, O., 1987, Seismic data processing, Society of Exploration Geophysicists.

Appendix: Collapsing the weight calculations for Kirchhoff migration

In this Appendix, I present some efficiencies that can be realized from looking closely at the Kirchhoff migration weight W. These efficiencies, resulting in a factor of three or so speedup for Kirchhoff migration, are much more modest than the order-of-magnitude runtime differences among the various migration methods, but they serve to illustrate a pragmatic philosophy towards scientific computing that can provide a particular implementation of an algorithm with a speed advantage over its competitors in the industry.

Eq. (2) expresses all the steps of poststack Kirchhoff migration in two dimensions. The half-derivative is a type of filter to be applied to the input traces. Since the traces usually need some frequency bandpass filtering as well, the most convenient domain for the combination of half-differentiation and bandpass filtering is the frequency domain. After this combination of operations, the data are returned to the time domain in order to contribute to the final summation, which occurs in the innermost loop of the migration program. The weight W multiplies an input sample as it accumulates into the final image, and its presence requires at least one extra multiplication in the innermost loop. (The presence of W is important. If W is ignored, then the amplitudes along the migration impulse response shown in Fig. (1) are uniform; this will result in incorrect cancelling of amplitudes when migrating flat reflection events and the generation of significant migration artifacts.) If, in addition, W must be computed in the innermost loop, all the operations of calculating W then dominate the work performed in the migration program. With some care, however, all the work of calculating W can be taken out of the innermost loop and placed in outer loops. Then, instead of dominating the work performed by the program, the effort spent calculating W becomes negligible. Not only can we eliminate most of the work of calculating W, though, we can remove the very presence of W from the innermost loop, eliminating even that extra inner-loop multiplication. The approximations we make in reducing the work of the Kirchhoff summation usually have no adverse effect on the final image.

For constant-velocity migration, this procedure is trivial, and exact. Here, W is proportional to $z/\sqrt{r}$, where z is the depth of the image point and r is the distance from the source-receiver pair on the Earth's surface to the image point (Bleistein et al., 1987). Since the velocity is constant, r is proportional to the reflection time t of the input trace sample that is being accumulated into the image. So if, upon reading the input trace into memory, we divide each sample by the square root of its time index, and upon writing the final migrated trace onto disk, we multiply each sample by its depth index, we've accomplished the multiplication by W at a computational cost of $N_x(N_t + N_z)$ operations.

For variable-velocity poststack migration, W has a more complicated expression, involving the in-plane and out-of-plane spreading of ray tubes joining each source-receiver pair with each image point. If the velocity varies laterally, it is possible for the raypaths to pass through caustics, causing the downward-continued wavefield to undergo phase changes. However, experience has shown that using the procedure of the previous paragraph gives images that are nearly indistinguishable from the images obtained when the raytube spreading is "correctly" taken into account, even on synthetic zero-offset data where the effects of the phase shifts due to caustics have not been averaged out by the CMP stacking process. Without being able to explain why this so, I attribute it very roughly to the tremendous healing power of the Kirchhoff summation.

For prestack migration, reducing the calculation of W, though not trivial, is elementary. I illustrate the procedure on common-offset migration, whose expression is given in Bleistein et al. (1987). A careful application of the true-amplitude migration formula given in that references requires computing several raypath quantities, namely integral of the velocity along a ray, ray angle, in-plane geometrical spreading loss along a ray, and ray takeoff angle from the upper surface. These terms all appear in the innermost loop of the program (the Kirchhoff summation); if they have been computed during the (preliminary) raytracing to find traveltimes, their values can be looked up and combined to provide W. When this is done, the migration runs at least three times slower than a migration where the terms combining to form W are not computed at all (so that W is taken to be unity). A much better approximation to W than unity can be obtained at essentially no expense as follows. First, assume, as we did for poststack migration, that velocity is constant. Then we can use the amplitude given in Eq. (63) in Bleistein et al. (1987):

$$W = z\sqrt{1 + \frac{(x_s - x)\cdot(x_r - x)}{|x_s - x||x_r - x|}} \sqrt{\frac{|x_s - x| + |x_r - x|}{|x_s - x||x_r - x|}} \left[\frac{|x_r - x|}{|x_s - x|} + \frac{|x_s - x|}{|x_r - x|}\right]$$

where $x_s = (x_s, 0)$ is the source location, $x_r = (x_r, 0)$ is the receiver location, and $x = (x,z)$ is the image point location and, for convenience, I ignore here and later constant factors involving powers of 2, π, and v. If the velocity is constant, then all distances are proportional to times, so that we can write, for example, $|x_s - x|$ as vt_s. The argument of the first square root is $1 + \cos\theta$ for the angle θ between the raypaths from x_s and x_r to x; using the double-angle formula for $1 + \cos\theta$ and the law of cosines yields a value for the square root proportional to

$$\frac{t_s + t_r}{\sqrt{t_s t_r}} \sqrt{1 - \frac{(2h)^2}{(vt_s + vt_r)^2}}$$

where $2h$ is the source-receiver offset of the input trace that is being migrated. Then the expression for W simplifies to

$$W = z\left[\frac{1}{t_s^2} + \frac{1}{t_r^2}\right](t_s + t_r)^{3/2}\sqrt{1 - \frac{(2h)^2}{(vt_s + vt_r)^2}}$$

Now we note that $(t_s + t_r) = t$, the total source-to-image point-to-receiver traveltime, which is just the time of a sample on an input trace that is about to be accumulated into an image point. So we can account for the final two terms in this expression by multiplying the input trace (as it is being read into computer memory) by the product of $t^{3/2}$ and a square-root that depends on the source-receiver offset and the upper-surface velocity. Also, we can account for the first term z by a simple multiplication of the output trace as it is being written out to disk. This leaves only the term $1/t_s^2 + 1/t_r^2$, which must be somehow kept away from the Kirchhoff summation. This term is approximately equal to $8/t^2$, as we indicate next. We take the sum and convert the times to distances along (straight) raypaths from the source and the receiver to an image point. Then we express those distances in terms of the source-receiver midpoint x_m and half-offset h, i.e.,

$$\frac{1}{t_s^2} + \frac{1}{t_r^2} = \frac{v^2}{z^2 + (x - x_m + h)^2} + \frac{v^2}{z^2 + (x - x_m - h)^2}$$

To second order in

$$\frac{h(x - x_m)}{z^2 + (x - x_m)^2 + h^2}$$

the sum on the right side of the last equation is

$$\frac{1}{t_s^2} + \frac{1}{t_r^2} \approx \frac{2v^2}{z^2 + (x - x_m)^2 + h^2}$$

By expanding out $t^2 = (t_s + t_r)^2$ in terms of the midpoint and half-offset and performing a somewhat more detailed calculation, we can show that, to the same order,

$$t^2 \approx \frac{4}{v^2}\left(z^2 + (x - x_m)^2 + h^2\right)$$

so that

$$\frac{1}{t_s^2} + \frac{1}{t_r^2} \approx \frac{8}{t^2}$$

as desired. This allows us to evaluate the term $\frac{1}{t_s^2} + \frac{1}{t_r^2}$ as a simple division by t^2 as an input trace is read into memory. Because of the approximations made in deriving the last expression, it loses validity as both h and $|x - x_m|$ become large and z becomes small, referring to large-offset data being migrated to steep dips at shallow depths. Luckily, such data are usually muted, or set to zero, during preprocessing because they often contain refracted, nonreflected events inappropriate for migration.

So all the terms that make up W can be evaluated as we read the input traces or write the output traces. By doing this, we can save significantly on the cost of a Kirchhoff migration. The Kirchhoff migrations shown in the examples made use of these speedups.

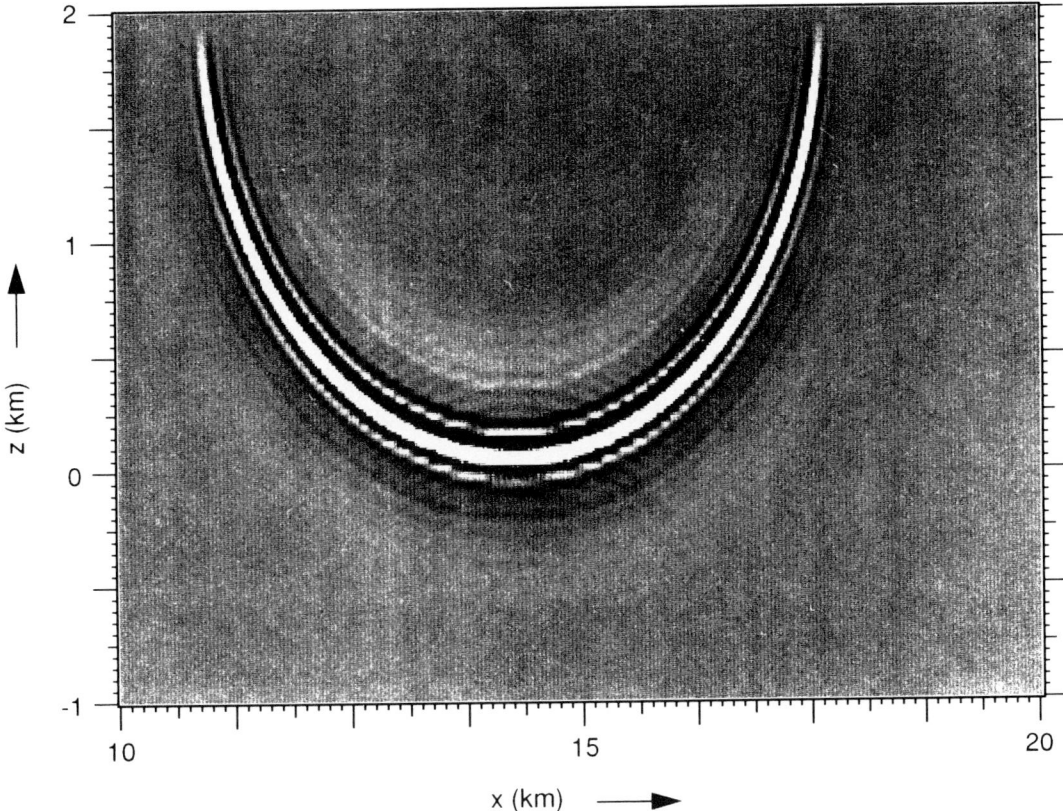

Figure 1. A migration impulse response, the action of migration, including bandpass filtering, on an unmigrated section consisting of zero data except for a single nonzero sample. Note the amplitude decay at the steepest dips.

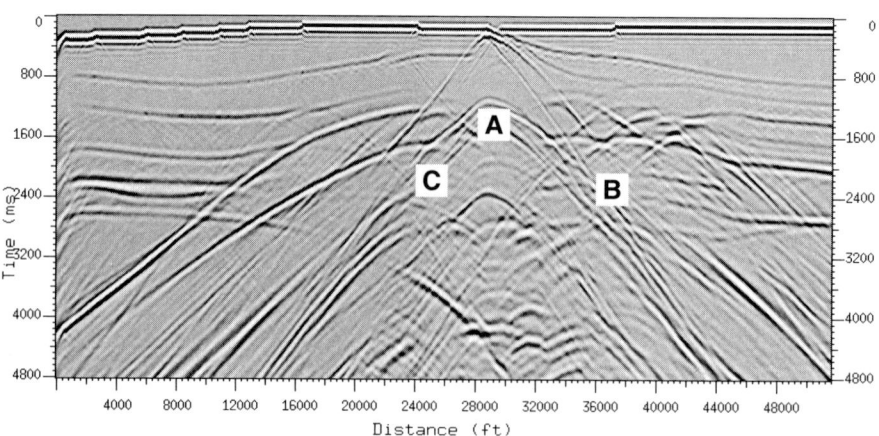

Figure 2(a). An unmigrated exploding-reflector synthetic section from a salt body with significant lateral velocity variations. From O'Brien and Gray (1996).

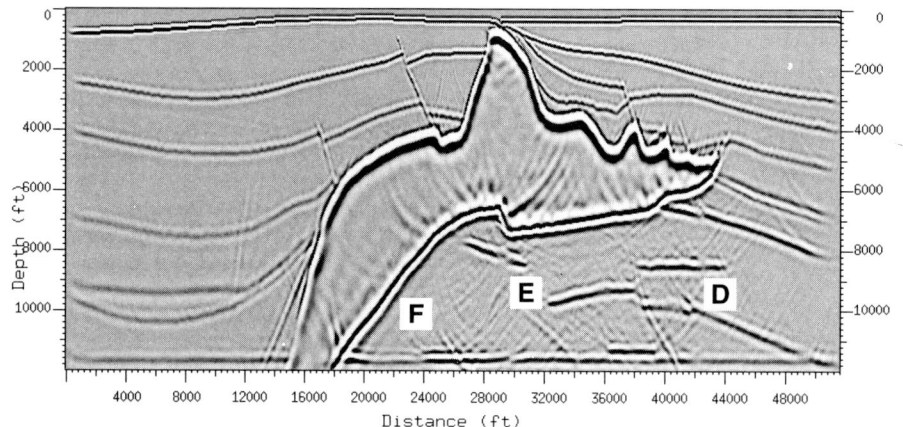

Figure 2(b). Explicit finite-difference migration of the section in Figure 2(a).

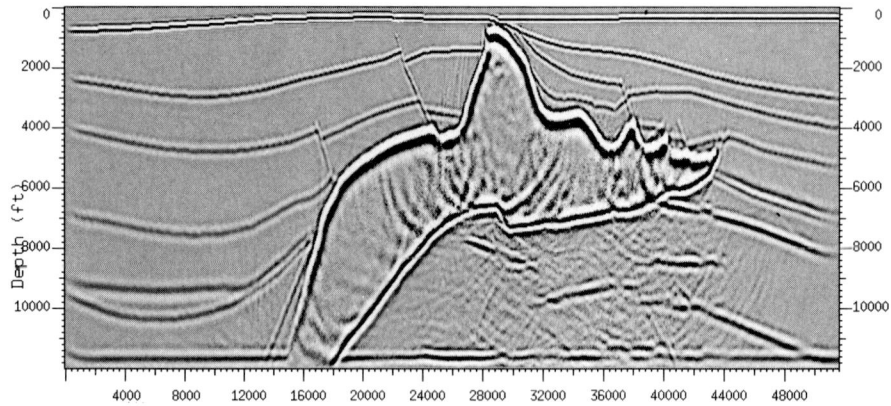

Figure 2(c). Kirchhoff migration of the section in Figure 2(a).

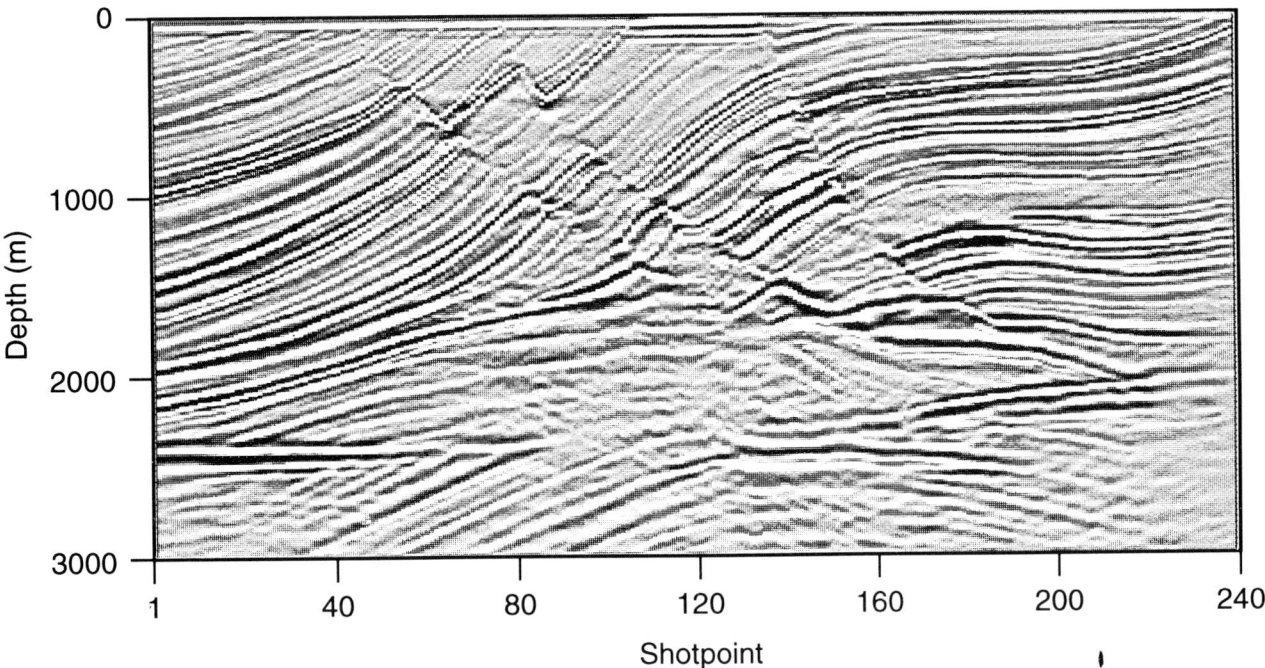

Figure 3(a). Explicit finite-difference prestack migration/stack of the Marmousi data. From Gray and May (1994).

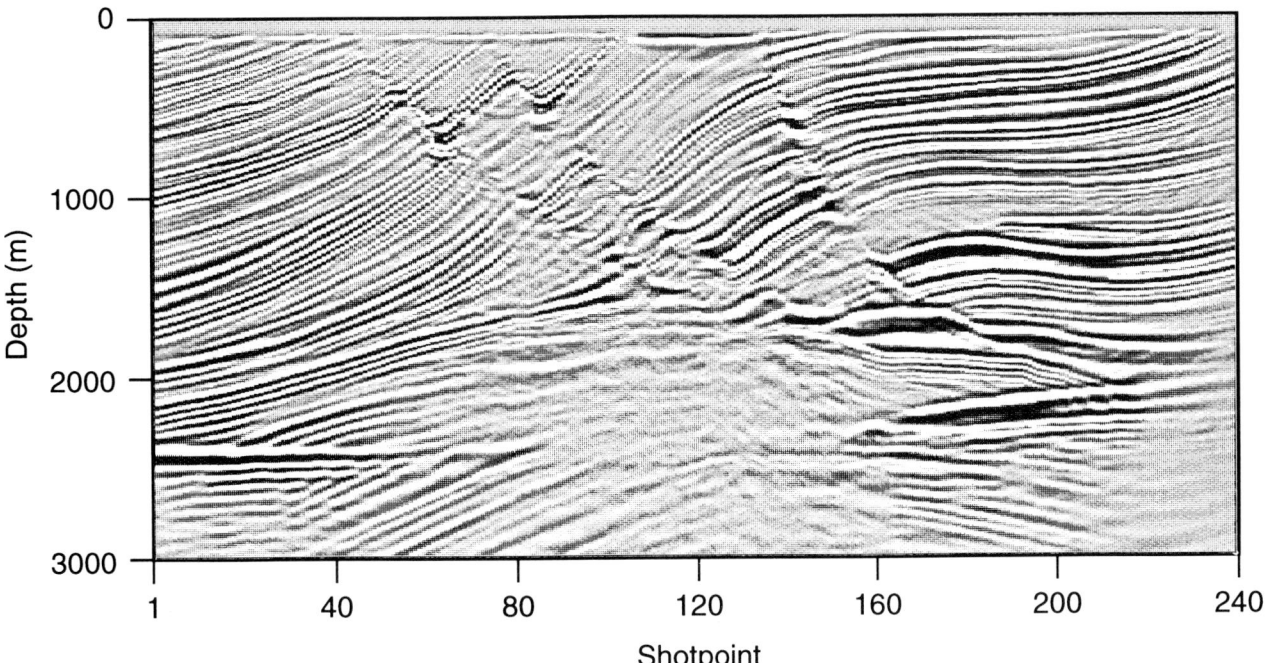

Figure 3(b). Kirchhoff prestack migration/stack of the Marmousi data set.

Tutorial

Smiles and frowns in migration/velocity analysis

Jinming Zhu[*], Larry Lines[‡], and Sam Gray[**]

ABSTRACT

Reliable seismic depth migrations require an accurate input velocity model. Inaccurate velocity estimates will distort point diffractors into smiles or frowns on a depth section. For both poststack and prestack migrated sections, high velocities cause deep smiles while low velocities cause shallow frowns on migrated gathers. However, for prestack images in the offset domain, high velocities cause deep frowns while low velocities cause shallow smiles. If the velocity is correct, there will be no variation in the depth migration as a function of offset and no smiles or frowns in the offset domain. We explain migration responses both mathematically and graphically and thereby provide the basis for depth migration velocity analysis.

INTRODUCTION

Depth migration is the process of positioning reflected seismic arrivals at their proper subsurface locations. The accurate depth migration of seismic reflectors requires accurate velocity estimations. Inaccurate velocity estimates will cause moveout artifacts such as smiles and frowns to appear on depth-migrated images. The elimination of these moveout features by adjusting seismic velocities allows depth migration to be used as a powerful velocity analysis tool. It can be argued on the basis of model studies (Lines et al., 1993) and with real data examples (Whitmore and Garing, 1993) that iterative prestack depth migration provides a very general velocity analysis method for structurally complex media.

In this paper, we examine migration moveout for both poststack and prestack depth migration of point diffractors, both mathematically and geometrically. Point diffractors are used for their simplicity and because reflected wavefields can be considered a superposition of point diffractor arrivals, according to Huygens's principle. We show the effects of velocity on the depth migration of point diffraction arrivals so that one can establish criteria for velocity analysis. We do this for both zero-offset (poststack) and nonzero-offset (prestack) recording configurations.

SMILES AND FROWNS IN POSTSTACK MIGRATION

In understanding the case of poststack migration, consider two point diffractors in the middle of a uniform medium with velocity $v = 4000$ m/s. The depths of the diffractors are 600 and 800 m, respectively. We now examine the migration of a point diffraction seismogram from this model recorded by coincident source-receiver positions. The zero-offset section is shown in Figure 1. The images obtained by migrating the input diffraction hyperbola in Figure 1 are compared for the cases where the velocity is too low (Figure 2a), exactly correct (Figure 2b), and too high (Figure 2c). If the velocity is too low, the poststack migration images are shallow frowns. If the velocity is correct, the images are concentrated blobs at the proper depth. If the migration velocity is high, the images are deep smiles. These are described by data examples in Yilmaz (1987). The shallow frowns are described as undermigration, while the deeper smiles are described as overmigration.

The phenomenon of undermigrations and overmigrations for poststack data can be described by considering the case of coincident source-receiver positions. For simplicity, we consider the case of a point diffractor for a coincident source-receiver (or zero-offset) recording configuration. This recording configuration, as shown in Figure 3, is the situation which we simulate with stacked data.

Suppose there is a point diffractor P on the vertical axis of a Cartesian system (x, z). The diffractor at depth z is vertically below the origin. Consider a coincident source-receiver pair S/R on the surface of the earth with lateral offset x from the

---

Published on Geophysics Online May 6, 1998. Manuscript received by the Editor November 26, 1997; revised manuscript received November 26, 1997.
[*]Formerly Memorial University, Department of Earth Sciences, St. John's, Newfoundland, Canada A1B 3X5; presently GX Technology Corporation, 5847 San Felipe, Suite 3500, Houston, Texas 77057. E-mail: jzhu@gxt.com.
[‡]Formerly Memorial University, Department of Earth Sciences, St. John's, Newfoundland, Canada A1B 3X5; presently The University of Calgary, Department of Geology and Geophysics, 2500 University Drive, N.W., Calgary, Alberta, Canada T2N 1N4. E-mail: lines@geo.ucalgary.ca.
[**]Amoco Canada Petroleum Company, P.O. Box 200, Station M, Calgary, Alberta, Canada T2P 2H8. E-mail: shgray@amoco.com.
© 1998 Society of Exploration Geophysicists. All rights reserved.

origin. Let the one-way vertical traveltime from the surface to P be t_0; let the total traveltime from S/R to P be t; and let the medium velocity be v. The one-way traveltime for an arrival traveling from S/R to P is given by using the Pythagorean theorem and the distance relationship between the sides of the right triangle in Figure 3. That is,

$$x^2 + v^2 t_0^2 = v^2 t^2 \qquad (1)$$

or

$$t = \sqrt{t_0^2 + \frac{x^2}{v^2}}. \qquad (2)$$

If we wish to use the two-way reflection times ($2t$ and $2t_0$), which are the arrival times for a reflection experiment, we can consider the same distance relationship by using the medium's half-velocity, $v/2$. Then the model in Figure 3 is essentially the exploding reflector model of Loewenthal et al. (1976). In this model, reflection seismograms containing arrivals with two-way propagation are described by one-way propagation of explosions that propagate to the surface with half the velocity of the medium. Except for a few pathological cases (Claerbout, 1985; Yilmaz, 1987), this exploding reflector model can be considered a suitable model for poststack data. Figure 4 is a record of such an explosion experiment from the diffractor model.

Let us consider the migration of a diffraction arrival of a trace at offset x from the origin. Suppose the arrival time is t. To migrate this arrival, we use the principle of aplanatic surfaces (Sheriff, 1991). An aplanatic surface defines the locus of possible reflection depth points that could exist for a given one-way traveltime, t, and a migration velocity, v_m. For a coincident source-receiver position and a constant migration velocity, the aplanatic surface is defined by the following equation of a circle giving all possible locations of reflection points:

$$(x - x_m)^2 + z_m^2 = v_m^2 t^2, \qquad (3)$$

where (x_m, z_m) defines the migrated domain. If we substitute the expression for t from equation (2), we obtain

$$z_m^2 = v_m^2 \left(t_0^2 + \frac{x^2}{v^2} \right) - (x_m - x)^2. \qquad (4)$$

Note that the observed reflection time in equation (2) is expressed in terms of the actual velocity, v, which we generally do not know but hope to determine, whereas the estimated

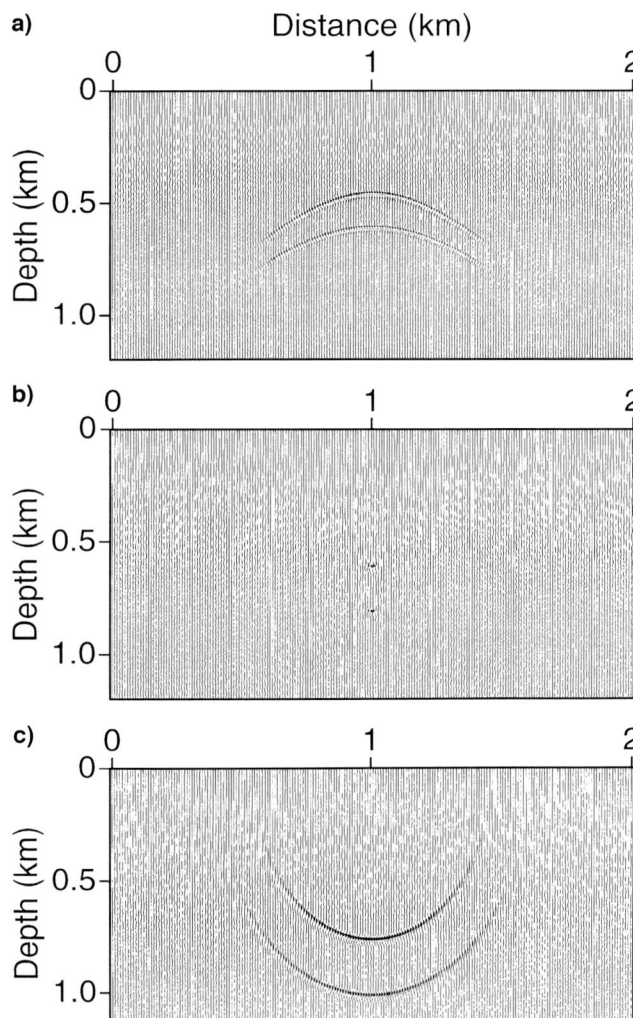

FIG. 2. Poststack migration of a model with two point diffractors. (a) A smaller migration velocity produces shallow frowns. (b) The true velocity collapses the hyperbolae to concentrated blobs. (c) A larger migration velocity results in deep smiles.

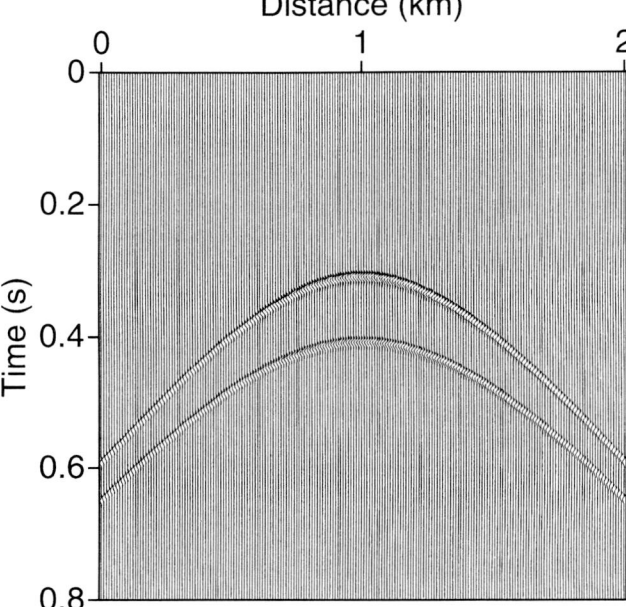

FIG. 1. Synthetic zero-offset seismograms of two point diffractors. The diffractors are at depths of 600 and 800 m, respectively, in the middle of the model.

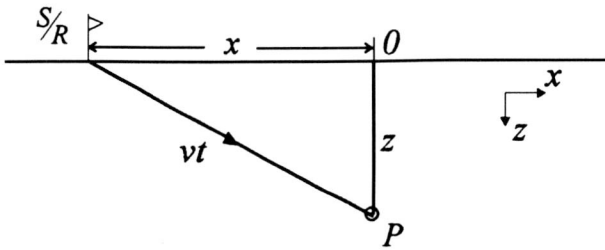

FIG. 3. The zero-offset (poststack) geometry for a point diffractor at P.

velocity used in migration is v_m. Migration of the record in Figure 4 is essentially the superposition of all these aplanatic surfaces, one for each source-receiver pair. We hope to find geometric criteria corresponding to cases where $v_m < v$, $v_m = v$, $v_m > v$, which will allow us to find cases where the velocity is correct.

In migration, we are dealing with the superposition of wavefield amplitudes that are distributed along aplanatic surfaces. For the poststack situation, this is the wavefront superposition method described by Robinson and Treitel (1980). The migrated image represents a summation of those amplitudes that are in phase such that they will interfere constructively. Mathematically, this constructive interference can be described by the method of stationary phase (Scales, 1995). The basic idea of stationary phase is that highly oscillatory time sequences tend to cancel upon migration except where the phase function has a stationary point. This stationary point occurs where the first derivative of the phase function equals zero. In migration, the phase function is the phase difference between the migration curve and the diffraction signatures of the data; locations where the migration curve is tangent to diffraction or reflection events in the data give the stationary phase contributions to the migration.

In migration, an alternative kinematic description of the amplitude summation along aplanatic surfaces is given by the envelope curves for the aplanatic surfaces. If a set of aplanatic curves is described by $F(x, z, t) = 0$, then its envelope is defined by curves satisfying $F(x, z, t) = 0$ and $dF/dx = 0$.

For doing this, we need to find the envelope of the aplanatic curves defined by equation (4), as shown by Maeland (1989). Essentially the envelope is the solution of the system consisting of equation (4) and its tangent curve (Sneddon, 1957). The tangent curve of equation (4) is given by differentiating equation (4) with respect to the source-receiver position x,

$$\frac{v_m^2}{v^2} x - (x - x_m) = 0. \quad (5)$$

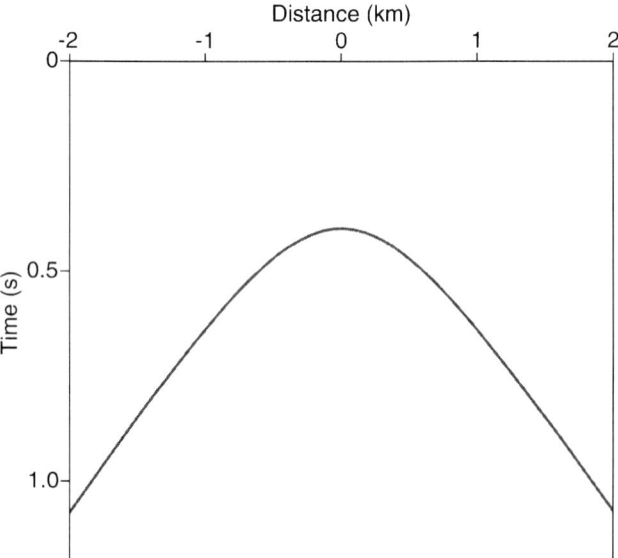

FIG. 4. Geometrical expression of a zero-offset section for a point diffractor.

Equivalently, if we define $\beta = v_m/v$,

$$(1 - \beta^2)x = x_m. \quad (6)$$

Now let's consider the case of $v_m \neq v$. For this case, we can have

$$x = \frac{x_m}{1 - \beta^2}. \quad (7)$$

Substituting equation (7) into equation (4) leads to

$$\frac{z_m^2}{v_m^2 t_0^2} - \frac{x_m^2}{(v^2 - v_m^2)t_0^2} = 1. \quad (8)$$

Now let's consider three specific cases.

1) Migration velocity smaller than the medium velocity, i.e., $v_m < v$.—In this case, equation (8) represents a hyperbola. Its vertex is on the depth axis with coordinate of $v_m t_0$ below the origin. The apex is thus above the diffractor position as $v_m t_0 < v t_0$. Obviously, it opens downward because the center of the hyperbola is just on the coordinate origin. Thus, the poststack migration of the diffraction curve will be a shallow hyperbolic frown when the migration velocity is too small, so that undermigration partially collapses the original hyperbola into a second, better focused hyperbola. This observation forms the basis of residual migration and cascaded migration.

In fact, the formation of such shallow frowns can also be well illustrated geometrically. Figure 5 illustrates this migration case. The cyan circle is the diffractor point. The red curve is the original record we simulated for such an explosion. The blue curves are the migration aplanatics that finally superpose to form the envelope of another hyperbola (in green) that is laterally much narrowed. This essentially indicates undermigration.

2) Migration velocity greater than the medium velocity, i.e., $v_m > v$.—In this case, equation (8) can effectively be reformulated as

$$\frac{z_m^2}{v_m^2 t_0^2} + \frac{x_m^2}{(v_m^2 - v^2)t_0^2} = 1. \quad (9)$$

This is the equation of a semiellipse with the center at the coordinate origin. The vertex on the depth z-axis is still $v_m t_0$ ($>v t_0$), which is now above the diffractor point P. The mouth of the ellipse is toward the negative axis of depth, as we are only interested in the positive z-values. Therefore, the migration of the diffraction curve in Figure 4 will be deep elliptic smiles on the migrated section when the migration velocity is too large.

Such a migration procedure is also geometrically represented in Figure 6. Most of the curves in Figure 6 are just the same as in Figure 5. Figure 6 is an excellent example showing that when the migration velocity is too high, the superposition of the individual aplanatics forms an elliptic smile (in green) in the migrated section.

3) Migration velocity equal to the medium velocity, i.e., $v_m = v$.—In this case, we have to start from equations (4) and (6) because equation (8) is no longer valid. When $v_m = v$, equation (6) gives $x_m = 0$. Substituting this value of x_m into equation (4), we obtain $z_m = v t_0 = z$. This implies that when the

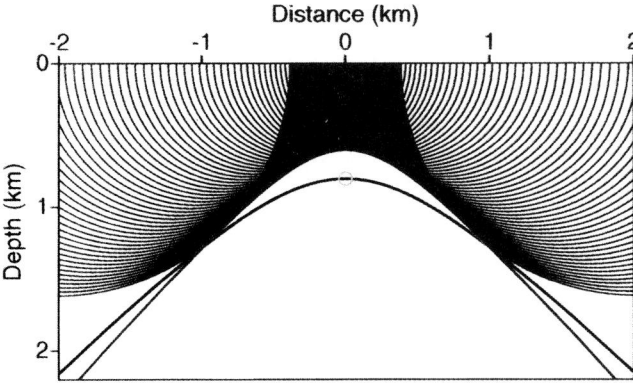

FIG. 5. Poststack migration of the diffractor model when a velocity smaller (3 km/s) than the true velocity (4 km/s) is used. The cyan circle represents the diffractor position. The red is the scaled recorded hyperbolic diffraction curve. The superposition of all the wavefronts in blue results in a shrunken hyperbola in green in the final migration section.

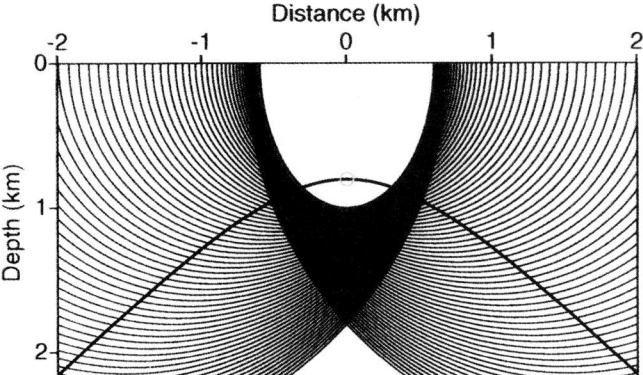

FIG. 6. Poststack migration of the diffractor model when a velocity larger (5 km/s) than the true velocity (4 km/s) is used. The superposition of all the wavefronts in blue results in an elliptic smile in green in the final migration section.

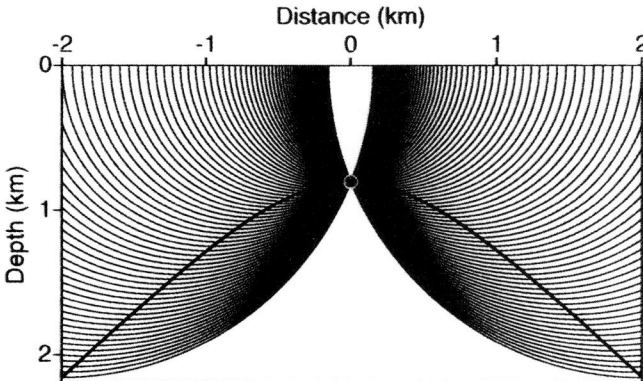

FIG. 7. Poststack migration of the diffractor model when the true velocity (4 km/s) is used for migration. The green cross is the result of the constructive superposition of all the wavefronts in blue. This indicates a perfect recovery of the diffractor point in the migration.

migration velocity is correct, the superposition of the migration aplanatics finally collapses the recorded diffractions to the correct spatial position, $(x_m, z_m) = (0, z)$. Parallel to such mathematical development, Figure 7 expresses geometrically the procedure of reconstructing the true diffraction point by simple superposition of aplanatic curves.

SMILES AND FROWNS IN THE PRESTACK MIGRATED SECTION

In this section, we will show that the smiles and frowns discussed above are common to prestack migrated stacked sections. They can also be verified easily, both mathematically and geometrically. Since the mathematical development is very similar to that in the poststack migration case, we will focus on the geometrical aspects.

Let us first consider the same point diffractor P on the vertical axis of a Cartesian system. Now assume there is a source at S and a receiver at R on the earth's surface, with x-coordinates of x_s and x_r, respectively (see Figure 8). If the medium velocity is v, the total traveltime from the source S to the diffractor P and back up to the receiver R can thus be given by the so-called double square root relationship (Claerbout, 1985),

$$t = \frac{1}{v}\left(\sqrt{x_s^2 + z^2} + \sqrt{x_r^2 + z^2}\right). \quad (10)$$

Migration of a single diffraction arrival at R attributable to the source at S can still be described by the concept of aplanatic surfaces. The aplanatic surface for the source-receiver pair shown in Figure 8, analogous to equation (3), can be represented by

$$\sqrt{(x_s - x_m)^2 + z_m^2} + \sqrt{(x_r - x_m)^2 + z_m^2} = v_m t. \quad (11)$$

Substituting equation (10) into the above, we have

$$F(x_m, z_m; x_s, x_r) = \beta\left[\sqrt{x_s^2 + z^2} + \sqrt{x_r^2 + z^2}\right]$$
$$-\left[\sqrt{(x_s - x_m)^2 + z_m^2} + \sqrt{(x_r - x_m)^2 + z_m^2}\right] = 0. \quad (12)$$

This is essentially an ellipse in the migrated space (x_m, z_m) for this special case of constant velocity.

The final migration of all these arrivals recorded by different receivers from many sources is the envelope of the individual ellipses. The envelope of these ellipses is essentially the solution of equation (12) and its derivative equations (Sneddon, 1957),

$$F_{x_s}(x_m, z_m; x_s, x_r) = 0 \quad (13)$$

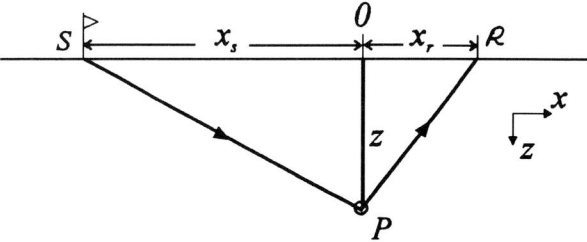

FIG. 8. Prestack geometry for a point diffractor at P.

and

$$F_{x_r}(x_m, z_m; x_s, x_r) = 0. \quad (14)$$

Following the same procedures as in the last section, we can develop the same conclusions as those in the poststack migration, although the mathematical derivations will be much more complicated. Figure 9 schematically shows four shot gathers from the diffraction model in Figure 8. Figure 10 geometrically summarizes the migration procedure when a velocity that is too small is used for prestack migration. Just as expected, the superposition of all individual migration ellipses results in a shrunken hyperbola (in green). Notice, however, that not all of the ellipses are tangent to the envelope. Figure 11 illustrates that the correct velocity allows the migration to reconstruct the point diffractor model almost perfectly as long as the recording coverage is sufficiently wide and dense. In contrast to Figure 11, Figure 12 demonstrates that when a velocity that is too large is used for prestack migration ($v_m > v$), the superposition of individual elliptical aplanatics (in blue) results in another ellipse (in green) in the final migrated section.

Figure 13 shows a numerical example of prestack depth migrations that correspond to cases of velocity lower than, equal to, and higher than the true medium velocity. A low migration velocity ($v_m = 3000$ m/s) results in frowning migrated image (a), caused by an insufficient collapse of diffractions. In contrast, using a velocity that is too high ($v_m = 5000$ m/s) in migration results in smiling images (b). In either of these two cases, the migrated images of the diffractors are not properly focused. A smaller velocity results in image shallowing, while a larger velocity leads to image deepening. Only when the true velocity is used will the diffractions completely collapse to their true positions (c). Thus, the final migrated section exhibits smile and frown patterns whether the migration is performed on

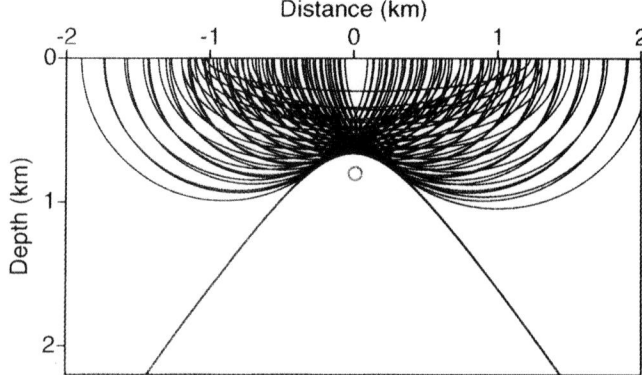

FIG. 10. Prestack migration of the diffractor model when a velocity smaller (3.3 km/s) than the true velocity (4 km/s) is used. The cyan circle represents the diffractor position. The superposition of all the wavefronts in blue results in a shrunken hyperbola in green in the final migration section.

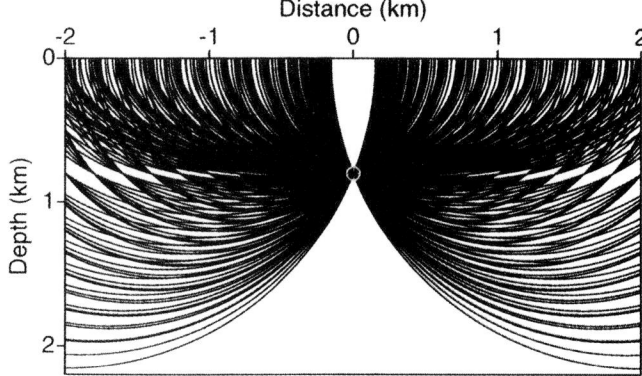

FIG. 11. Prestack migration of the diffractor model when the true velocity (4 km/s) is used for migration. The green cross is the result of the constructive superposition of all the wavefronts in blue. This indicates a perfect recovery of the diffractor point in the migration.

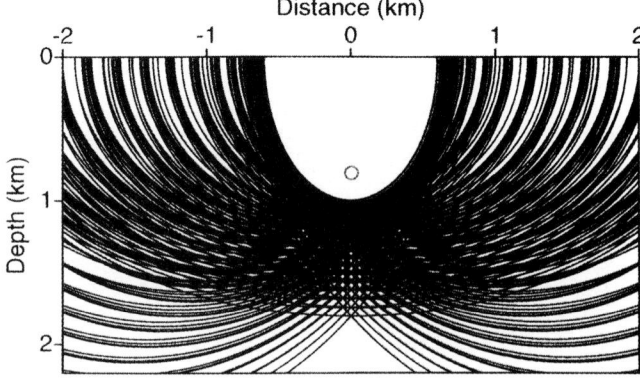

FIG. 12. Prestack migration of the diffractor model when a velocity larger (5 km/s) than the true velocity (4 km/s) is used. The superposition of all the wavefronts in blue results in an elliptic smile in green in the final migration section.

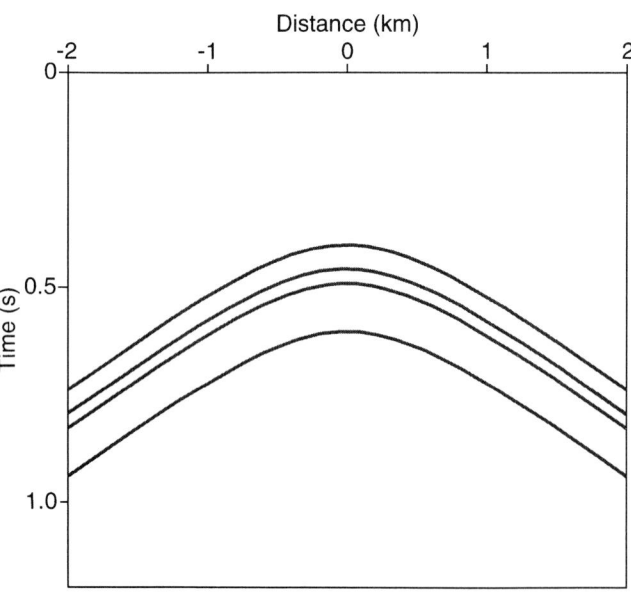

FIG. 9. Some representative prestack recording events from a diffractor point model. Each event corresponds to a source position. The diffractor lies at 0.8 km in depth in the middle of the model.

poststack or prestack seismic data whenever errors exist in the migration velocity.

These undermigration and overmigration features are generally observed in both poststack and prestack migration sections. However, smiles and frowns are often very difficult to observe on the migration sections because individual diffractors are often just an element of the reflecting interface. Generally, the migration moveout effects on common image gathers (CIGs) from prestack migrations are quite pronounced and allow for effective velocity analysis. Interestingly enough, the moveout characteristics on CIGs are very different from those in the migrated stacked sections.

PRESTACK DEPTH MIGRATION MOVEOUT

Many recent studies demonstrate that prestack depth migration can accurately image reflections and diffractions without dip restriction if a reasonable approximation of the velocity field is available (Versteeg, 1993; Lines et al., 1993). Nevertheless, the velocity model is the key component in these migrations. Theoretically, there exist several alternative methods for velocity analysis (Versteeg, 1993; Lines et al., 1993). Here we will analyze the prestack migration moveout features that are fundamental to the basic theories in interval velocity analysis

utilizing CIGs (Al-Yahya, 1989). Our analysis, however, will no longer depend on the assumption of a layered-earth model.

Consider the general subsurface structure and recording geometry as shown in Figure 14. P denotes the arbitrary scattering point in the earth's interior. S and R are a source-receiver pair illuminating P. D is the surface image of P. Assuming that the average velocity above P is $\bar{v}$ and that the diffraction received at R (because of a source wavelet from S and then diffracted at P) never travels beneath P, then its arrival time can be expressed as

$$t = \frac{1}{\bar{v}}(\overline{SP} + \overline{RP}). \tag{15}$$

When an incorrect average velocity $\bar{v}_m$ is used for migration, the diffraction signal received from P will be migrated to an incorrect point P'. P' generally has both vertical and lateral displacements from the true position P. We denote these displacements with $\Delta x = \overline{P'Q}$ and $\Delta z = \overline{QP}$. In this case, the traveltime will be

$$t = \frac{1}{\bar{v}_m}[\overline{SP'} + \overline{P'R}] = \frac{1}{\bar{v}_m}[(\overline{SP} - \overline{P_1P}) + (\overline{RP} - \overline{P_2P})]$$

$$= \frac{1}{\bar{v}_m}[(\overline{SP} - \overline{Q_1P}) + (\overline{RP} - \overline{Q_2P})]$$

$$+ \frac{1}{\bar{v}_m}(\overline{P_2Q_2} - \overline{P_1Q_1}). \tag{16}$$

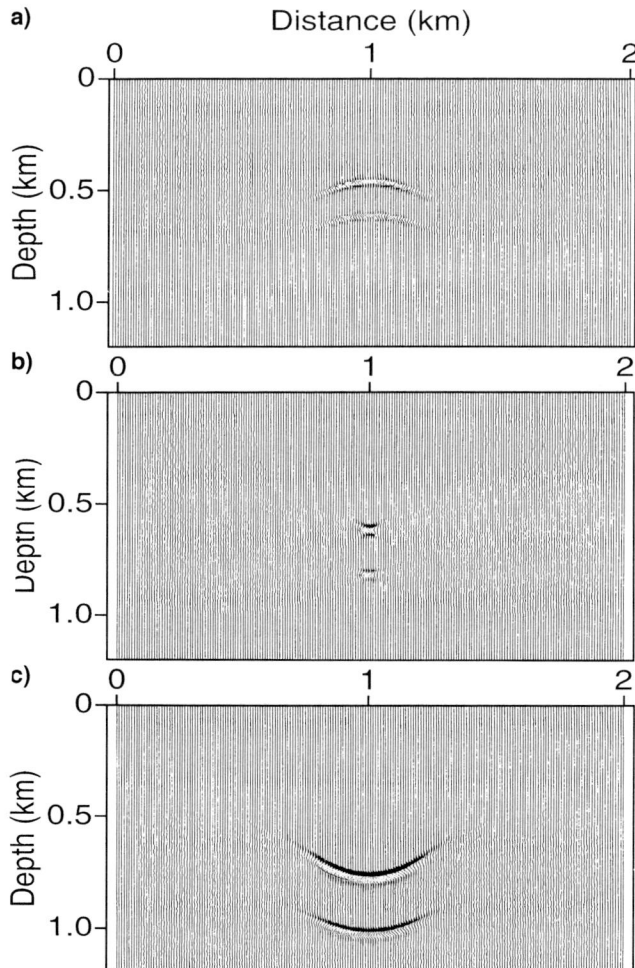

FIG. 13. Prestack depth migration of a model with two point diffractors. A migration velocity smaller than, equal to, and larger than the true velocity is used for images in (a), (b), and (c).

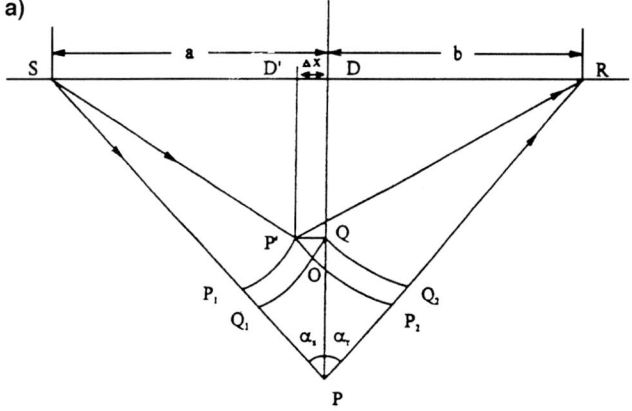

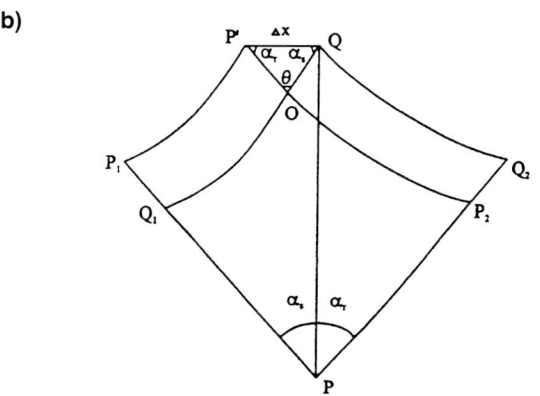

FIG. 14. Migration depth/velocity relationship diagram in a general subsurface structure. (a) A wrong velocity migrates the reflection to a position P', which has a lateral displacement Δx in addition to a vertical displacement Δz. (b) An enlarged view of the lower part of (a).

From Figure 14b, the following relationship holds in the triangle $\triangle P'QO$:

$$\frac{\sin \alpha_s}{\overline{P'O}} = \frac{\sin \alpha_r}{\overline{QO}} = \frac{\sin \theta}{\Delta x}. \quad (17)$$

Thus, we have

$$\begin{aligned}\overline{P_2Q_2} - \overline{P_1Q_1} &\simeq \overline{QO} - \overline{P'O} \\ &= \frac{\Delta x}{\sin \theta} \sin \alpha_r - \frac{\Delta x}{\sin \theta} \sin \alpha_s \\ &= \frac{\Delta x}{\sin \theta} (\sin \alpha_r - \sin \alpha_s) \\ &= \frac{\Delta x}{\cos(\theta/2)} \sin \frac{\alpha_r - \alpha_s}{2}. \end{aligned} \quad (18)$$

In the derivation of the last equation above, we have used the relationships $\alpha_r + \alpha_s + \theta = 180°$, $\sin \theta = 2\sin(\theta/2)\cos(\theta/2)$, and $\sin \alpha_r - \sin \alpha_s = 2\cos[(\alpha_r + \alpha_s)/2]\sin[(\alpha_r - \alpha_s)/2]$. Therefore, the distance part of the last term of equation (16) would be an order smaller than Δx as long as $|\sin(\alpha_s - \alpha_r)/2)| < 0.10$ and θ is not close to $180°$. The latter condition generally holds, as α_s and α_r would not be zero for most cases. The first condition is equivalent to $|\alpha_s - \alpha_r| < 12°$, i.e., the difference between the two illuminating angles being less than $12°$. Since $|\alpha_s - \alpha_r| = 2\alpha$, where α is the structural dip at P, the above inequality thus only holds for structures of gentle dip. In such cases, equation (16) can be properly approximated by

$$t = \frac{1}{\bar{v}_m}[(\overline{SP} - \overline{Q_1P}) + (\overline{RP} - \overline{Q_2P})]. \quad (19)$$

This equation is equivalent physically to the assumption that the lateral displacement Δx is negligible compared to the vertical one. Eliminating t from equations (15) and (19), we obtain

$$(1 - \beta)(\overline{SP} + \overline{RP}) = \overline{Q_1P} + \overline{Q_2P}, \quad (20)$$

where $\beta = \bar{v}_m/\bar{v}$. From Figure 14, the following general relations hold:

$$\overline{SP} = z/\cos \alpha_s; \qquad \overline{Q_1P} = -\Delta z \cos \alpha_s;$$
$$\overline{RP} = z/\cos \alpha_r; \qquad \overline{Q_2P} = -\Delta z \cos \alpha_r.$$

Substituting these relations into equation (20) leads to

$$\Delta z = (\beta - 1)\frac{z}{\cos \alpha_s \cos \alpha_r}. \quad (21)$$

In the case of a zero-offset source-receiver pair just at D, $\alpha_s = \alpha_r = 0$, we obtain

$$\Delta z = (\beta - 1)z. \quad (22)$$

This implies the migration depth z will be shallower than the true depth z if a velocity smaller than the true velocity ($\bar{v}_m < \bar{v}$) is used for migration, while it will be deeper if a higher velocity ($\bar{v}_m > \bar{v}$) is used. Only when $\beta = 1$, i.e., the true medium velocity is used for migration, will the diffractor be located properly. By denoting $\Delta z_0 = (\beta - 1)z$, equation (21) can be rewritten as

$$\Delta z(\alpha_s, \alpha_r) = \frac{\Delta z_0}{\cos \alpha_s \cos \alpha_r}. \quad (23)$$

Now let us consider the following three categories.

1) Migration velocity less than the true velocity ($\bar{v}_m < \bar{v}$).— In this case, $\beta < 1$, $\Delta z_0 < 0$, and $\Delta z(\alpha_s, \alpha_r) < \Delta z_0$. Generally the following relation,

$$\Delta z(\alpha_s + \epsilon_1, \alpha_r + \epsilon_2) < \Delta z(\alpha_s, \alpha_r), \quad (24)$$

also holds for any α_s, α_r and small nonnegative values of ϵ_1, ϵ_2. This relation indicates the migration image of P will form a smile that curves upward on a CIG, which is a display of migration traces versus the source-receiver offset corresponding to a fixed surface point.

2) Migration velocity greater than the true velocity ($\bar{v}_m > \bar{v}$).— In this case, $\beta > 1$, $\Delta z_0 > 0$, and $\Delta z(\alpha_s, \alpha_r) > \Delta z_0$. Similar to equation (24), we have

$$\Delta z(\alpha_s + \epsilon_1, \alpha_r + \epsilon_2) > \Delta z(\alpha_s, \alpha_r). \quad (25)$$

This relation indicates the migration image of P forms a frown that curves downward on a CIG.

3) Migration velocity equal to the true velocity ($\bar{v}_m = \bar{v}$).— In this special case, $\beta = 1$, $\Delta z_0 = 0$. Thus

$$\Delta z(\alpha_s, \alpha_r) = 0 \quad (26)$$

for any source-receiver pair. This simply means that when the true velocity is used for migration ($\bar{v}_m = \bar{v}$), the migration images of the diffractor point P will be at the exact depth, regardless of source-receiver offset. So, its images form a horizontal segment on the CIG displays.

To consider prestack migration velocity analysis in terms of offset and common midpoints (CMPs), consider again Figure 8 for a point diffractor at $(0, z)$. The midpoint can be denoted by $X = (x_r + x_s)/2$ and the offset denoted by $2h$ so that $x_s = X - h$ and $x_r = X + h$. Equation (10) gives the total traveltime for a particular point diffractor, but it can be embedded into an equation for a correct migration ellipse:

$$\begin{aligned}t_1 = t_1(X) &= \frac{1}{v}(\sqrt{(X - h - x)^2 + z^2} \\ &+ \sqrt{(X + h - x)^2 + z^2}). \end{aligned} \quad (27)$$

In terms of migration velocity and migration coordinates, we can also write the traveltime expression as

$$\begin{aligned}t_2 = t_2(X) &= \frac{1}{v_m}(\sqrt{(X - h - x_m)^2 + z_m^2} \\ &+ \sqrt{(X + h - x_m)^2 + z_m^2}). \end{aligned} \quad (28)$$

When $v_m = v$, the migration ellipse (28) is identical to the ellipse (27) for the true velocity. If the velocity analysis location x_m coincides with the diffractor location $x = 0$, we then have $z_m = z$ for all h. That is, the migration depth equals the correct depth of the diffractor regardless of the offset value when the velocity is correct. When the velocity is not correct ($v_m \neq v$), we need to evaluate the envelope of migration ellipses for all values of the midpoints X. That is, we need to set the derivatives of $t_1(X)$ equal to the derivatives of $t_2(X)$ for all X. If, again, we perform the velocity analysis at the actual diffractor location, then these derivatives give, respectively, the slopes of the correct and the incorrect migration ellipses at $x_m = x = 0$. These slopes are different unless both ellipses are flat at the analysis location, i.e., unless $X = 0$ also. So we set $X = x_m = x = 0$ into the expressions for $t_1(X)$ and $t_2(X)$ to obtain

$$\frac{h^2 + z^2}{v^2} = \frac{h^2 + z_m^2}{v_m^2} \tag{29}$$

or

$$z_m^2 = (\beta^2 - 1)x^2 + \beta^2 z^2, \tag{30}$$

where $\beta = v_m/v$.

In the case of $v_m < v$, we have $\beta < 1$. Equation (30) can be rewritten as

$$z_m^2 + (1 - \beta^2)x^2 = \beta^2 z^2, \tag{31}$$

which is an ellipse. So this is a special case of equation (24).

If $v_m > v$, then $\beta > 1$ and equation (30) is essentially the hyperbolic equation

$$z_m^2 - (\beta^2 - 1)x^2 = \beta^2 z^2. \tag{32}$$

This is a hyperbolic frown in the half-plane of positive depth.

If $v_m = v$, then $\beta = 1$ and equation (30) simplifies to $z_m = z$. This represents a horizontal line segment. Only in this special case of using the correct velocity for migration will the final stack of the migrated common reflecting point (CRP) gather reach its highest energy in the migration section.

The above procedure of migrating a CRP gather also can be represented geometrically. Figure 15 shows a CRP gather from the same diffractor model of Figure 8. This gather is essentially equivalent to that of CMP gather in this special case of a single diffractor point in a constant-velocity medium. Figure 16 shows the migrated CRP gathers for this diffractor model corresponding to different migration velocities. Because the migrated CRP gather is just the same as the so-called CIG gathers, Figure 16 equivalently illustrates the CIGs corresponding to different migration velocities. The recorded hyperbola in the CRP gather will appear as elliptical smiles when a smaller velocity is used for migration, while a hyperbolic frown appears when a larger velocity is used for migration. If the correct velocity is used, then the CIG gather will show a horizontally aligned segment.

Therefore, the velocity error in migration is very well expressed on CIGs. We can use the same point diffractor model to illustrate these theoretical predictions. Eighty-one shot profiles are theoretically simulated, with each record consisting of 60 traces. Figure 17 shows the CIGs for surface position $x = 1.0$ km when different velocities are used in migration. It clearly demonstrates that only when the true velocity is used

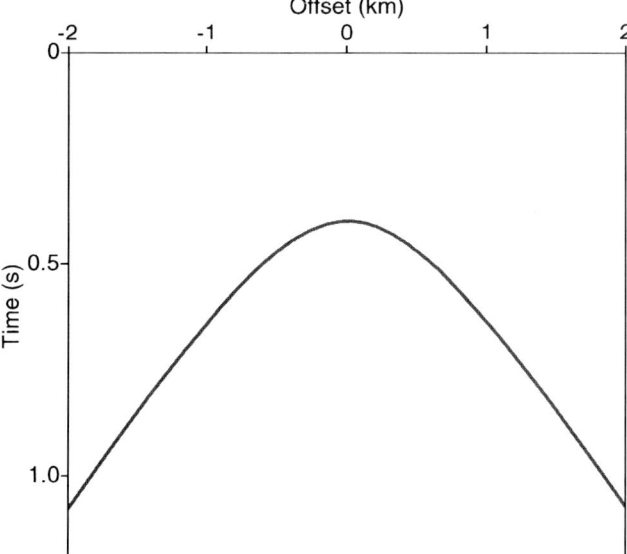

FIG. 15. CRP gather from a diffractor point model. The diffractor lies at 0.8 km depth in the middle of the model.

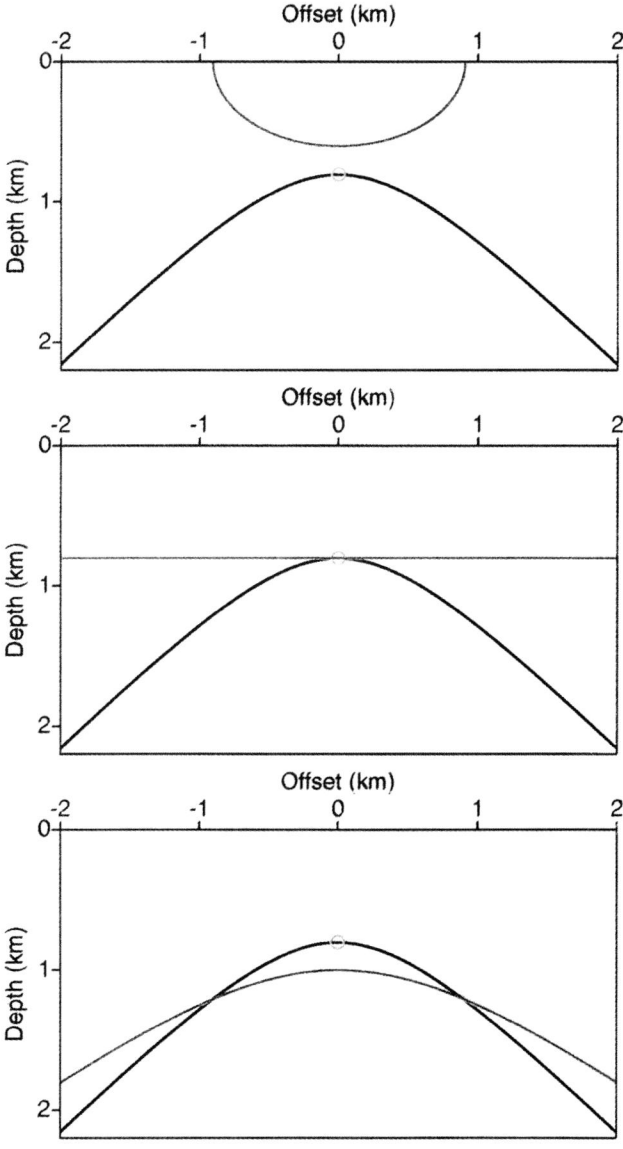

FIG. 16. CIGs when different migration velocities are used. Migration velocity is 3 (a), 4 (b), and 5 km/s (c), respectively. The circle in cyan represents the diffractor position. The red is the scaled CRP gather.

will the migration images of diffractors be independent of the source-receiver offset. When the velocity is lower than the true velocity ($v_m = 3000$ m/s), the diffractor images form smiles at a depth shallower than the true depth. This is total in agreement with equations (24) and (31). In contrast, when the velocity is higher than the true velocity ($v_m = 5000$ m/s), the diffractors are expressed as frowns on a CIG at a depth greater than the true depth. This is just what has been predicted by mathematical expressions (25) and (32).

Thus, the migration velocity error is documented well on both the final migration sections (Figure 13) and the CIGs (Figure 17). Interestingly, the CIGs show shallow smiles for low velocities and deep frowns for high velocities (Lines et al., 1993), whereas the final depth migration sections show shallow frowns for low velocities and deep smiles for high velocities (Yilmaz, 1987).

If we take a careful further look at Figures 13 and 17, we see that the curvatures of the shallow smiles/frowns are generally larger than the deep ones. This indicates that velocity errors are more pronounced on shallow reflections and thus are easier to correct. This also agrees with the general observation that sufficient offset/depth ratio should be kept to analyze properly the velocity errors (Lines, 1993). Luckily, we often have more constraints available on the shallow parts of the earth, such as well logs and geological exposures. We can also more effectively use techniques such as first-break tomography to constrain our near-surface velocity estimation. With respect to the deeper structure, we generally have to accept that it is coarsely defined and more ambiguous.

Though the diffractor model is oversimplified, it is of vital significance in migration and velocity analysis theory because any complicated structures can be considered as a continuum of diffractors. This is especially suitable for moveout analysis on a CIG that corresponds to a single surface point. The smile and frown patterns on CIGs can be effectively used to qualitatively and quantitatively analyze the migration velocity (Al-Yahya, 1989).

CONCLUSIONS

Depth migration is very sensitive to errors in the velocity model. Migration moveout features such as smiles and frowns have previously been reported (Yilmaz, 1987; Lines et al., 1993). We have shown both mathematically and geometrically these smiles and frowns on migration sections and CIGs. Using the simple diffractors model, we demonstrated that after migration, either using the stacked data or the prestack shot gathers, the diffractions migrate to shallow frowns when the migration velocity is too small and deep smiles when the velocity is too large. Only when the migration velocity is exactly the medium velocity will both the poststack and prestack migration produce concentrated blobs in the migration section. However, the CIGs produced in the prestack depth migration provide an effective migration velocity analysis domain. Our study starting from the very general subsurface structure showed that the migration moveout in the CIGs is interestingly different from the patterns in the migration sections. In CIGs, a lower velocity produces shallow smiles, while a higher velocity results in deeper frowns. When the migration velocity is correct, the CIGs present horizontal segments at the exact diffractor depths, which indicate that the migration image of the point diffractor is independent of source-receiver offset. These moveout patterns of smiles and frowns can serve as both qualitative and quantitative criteria for migration velocity analysis.

ACKNOWLEDGMENTS

We thank Jianhua Pan for his technical assistance. The authors also thank Sven Treitel for providing us a copy of the Maeland paper.

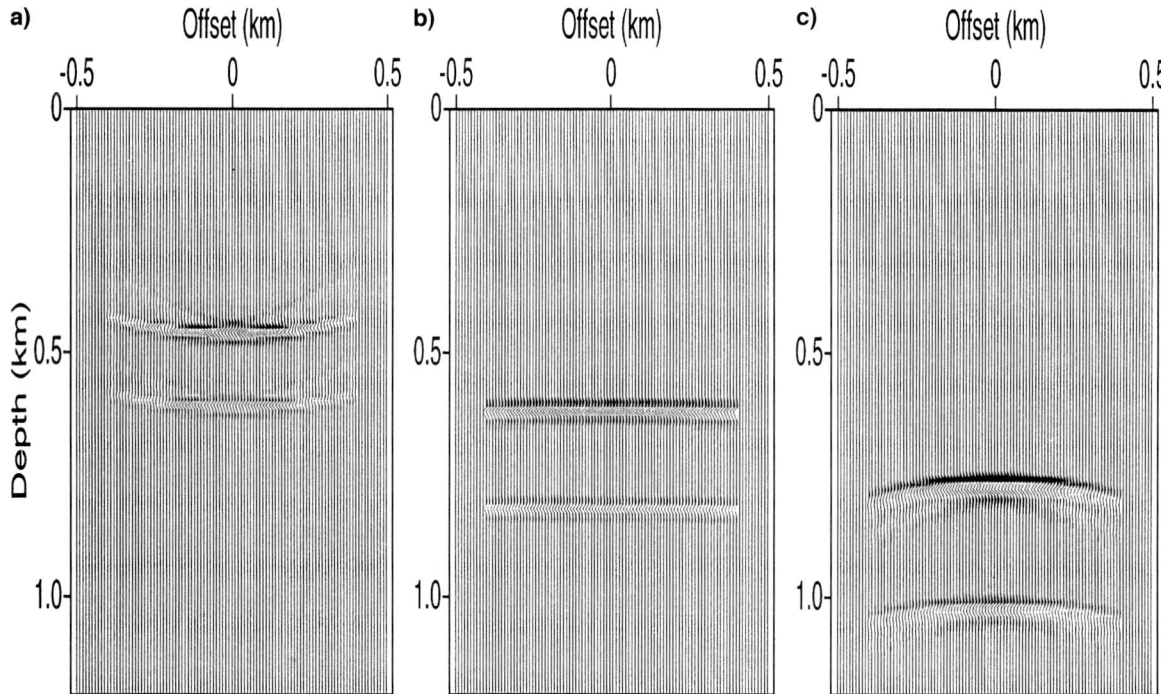

Fig. 17. CIGs for a model with two point diffractors.

REFERENCES

Al-Yahya, K., 1989, Velocity analysis by iterative profile migration: Geophysics, **54**, 718–729.

Claerbout, J. F., 1985, Imaging the earth's interior: Blackwell Scientific Publications, Inc.

Lines, L. R., 1993, Ambiguity in analysis of velocity and depth: Geophysics, **58**, 596–597.

Lines, L. R., Rahimian, F., and Kelly, K. R., 1993, A model-based comparison of modern velocity analysis methods: The Leading Edge, **12**, 750–754.

Loewenthal, D., Lu, L., Roberson, R., and Sherwood, J., 1976, The wave equation applied to migration: Geophys. Prosp., **24**, 380–399.

Maeland, E., 1989, Focusing aspects of zero-offset migration: Geophys. Trans., **35**, No. 3, 145–156.

Robinson, E. A., and Treitel, S., 1980, Geophysical signal analysis: Prentice-Hall, Inc.

Scales, J. A., 1995, Theory of seismic imaging: Springer-Verlag, New York, Inc.

Sheriff, Robert E., 1991, Encyclopedic dictionary of exploration geophysics: Soc. Expl. Geophys.

Sneddon, I. N., 1957, Elements of partial differential equations: McGraw-Hill Book Co.

Versteeg, R. J., 1993, Sensitivity of prestack depth migration to the velocity model: Geophysics, **58**, 873–882.

Whitmore, N. D., and Garing, J. D., 1993, Interval velocity estimation using iterative prestack depth migration in the constant angle domain: The Leading Edge, **12**, 757–762.

Yilmaz, O., 1987, Seismic data processing: Soc. Expl. Geophys.

Comparison of Kirchhoff and reverse-time migration methods with applications to prestack depth imaging of complex structures

Jinming Zhu* and Larry R. Lines[‡]

ABSTRACT

The performance of two popular migration methods—the Kirchhoff integral and reverse-time migrations—is evaluated through applications to imaging complex structures using prestack shot records. The migration results from the Marmousi model data demonstrate that reverse-time migration is more accurate than Kirchhoff migration in imaging the steeply dipping faults. However, the improved accuracy of reverse-time migration requires higher computational costs. In the application to the Alberta Foothills data where a good estimate of the velocity model is available, however, both the Kirchhoff and the reverse-time migration methods produce almost identical results. This implies that in the real world of exploration seismology, it will be relatively difficult to identify which method performs better because we never know the exact answer of the subsurface.

INTRODUCTION

Migration is the processing step of constructing the true subsurface structure from the recorded seismic data. Because of its significance in interpretation, many advanced migration methods have been proposed in the past couple of decades. Migration can now be performed in different domains, such as space-time, space-frequency, wavenumber-time, or wavenumber-frequency. Kirchhoff integral migration (Schneider, 1971, 1978; French, 1975; Berkhout, 1982) and reverse-time migration (Hemon, 1978; McMechan, 1983; Whitmore, 1983) are often implemented in the x-t domain. Both methods are based soundly on the wave equation, the mathematical description of seismic wave phenomena. Theoretically, they both are capable of migrating steep dip reflections.

Kirchhoff integral migration is the first digital migration scheme ever developed for imaging purposes in the form of hyperbolic summation (Schneider, 1971). Its early success in poststack seismic migration was a result of its accuracy and its low cost. However, it suffers greatly from the difficulty of calculating the Green's function in media with both vertical and lateral variations, where depth migration is a necessity. In such cases, the two-point ray-tracing procedure becomes very time consuming and often fails. Tracing rays from the top down (Gray, 1986) greatly reduces the computational requirement. Gray's idea of top-down ray tracing has encouraged the successful application of Kirchhoff integral method to 2-D and 3-D data from the Gulf of Mexico (Ratcliff et al., 1992). The recent progress in the finite-difference calculation of traveltimes through the eikonal equation and Fermat's principle has pushed the Kirchhoff method further to being more accurate and practically applicable to prestack seismic data imaging (Gray and May, 1994).

In contrast to the Kirchhoff integral method, reverse-time migration is a relatively new seismic migration technique. Similar to finite-difference simulation, reverse-time migration uses the finite-difference solution of the wave equation to drive the wavefield. The difference is that extrapolation is done backward in time in migration, but is done forward in time during simulation. However, since wave equation computations are insensitive to the directionality of time, both forward modeling and reverse-time migration essentially use the same finite-difference scheme in wavefield extrapolation. Reverse-time migration can be accurate because it is based directly on the seismic wave equation. Improved accuracy can be achieved by using higher order finite-differences to approximate the differentials (Dablain, 1986). Nonetheless, the solution of the wave equation by finite-differences can be costly. It requires a large amount of central memory and CPU time, and it can often exceed computer resources. However, thanks to the

---

Published on Geophysics Online, May 6, 1998. Manuscript received by the Editor November 25, 1996; revised manuscript received October 20, 1997.
*Formerly Memorial University Seismic Imaging Consortium, Department of Earth Sciences, St. John's, Newfoundland, Canada A1B 3X5; presently GX Technology Corporation, 5847 San Felipe, Ste. 3500, Houston, Texas 77057. E-mail: jzhu@gxt.com.
[‡]Formerly Memorial University Seismic Imaging Consortium, Department of Earth Sciences, St. John's, Newfoundland, Canada A1B 3X5; presently The University of Calgary, Department of Geology and Geophysics, 2500 University Drive, N.W., Calgary, Alberta, Canada T2N 1N4. E-mail: lines@geo.ucalgary.ca.
© 1998 Society of Exploration Geophysicists. All rights reserved.

recent enormous advance in computer sciences, reverse-time migration is now applicable to real data (Rajasekaran and McMechan, 1995; Lines et al., 1996). Mufti et al. (1996) have further extended its application to 3-D data by using relatively large horizontal grid steps to save significantly on memory requirement and computations. Their treatment is based on the observation that waves travel nearly vertically in stacked sections, so that dispersion is a mild problem horizontally due to large horizontal wavelengths.

Thus far, both Kirchhoff integral and reverse-time migrations have been applied to real seismic data, even in the case of 3-D with some degree of success. However, there are few publications on data sets where both Kirchhoff and reverse-time migration methods are applied to prestack depth imaging of complex geologies. Larner and Hatton (1990) give a very objective comparison of Kirchhoff integral and finite-difference migrations in the case of stacked data with the conclusion that both methods produce comparable migration accuracy, though their finite-difference migration is based on one-way wave equations. Whitmore et al. (1988) obtain similar conclusions in a comprehensive survey of depth migration methods on stacked data. Here, as we focus on prestack depth migration, however, we presume that the Kirchhoff integral and reverse-time migrations would somehow perform differently in some aspects.

In this paper, we provide a comparison of the Kirchhoff integral and reverse-time migration methods with applications to complex geological areas. We hope the examples given here can serve as a benchmark for future use of either method in the imaging of complicated areas.

THEORY

Both Kirchhoff integral and reverse-time migrations have been extensively studied and well developed. Our discussion will focus on algorithm evaluation, accuracy or migration effectiveness, and computational performances. Nonetheless, a brief summary of the theoretical basics is also included for self-completeness of the paper.

Kirchhoff integral migration

Both stacked and prestack Kirchhoff migrations are based on the integral solution of the wave equation. The solution is expressed as a surface integral over the known seismic observations. Based on the WKBJ approximation of the Green's function to the Kirchhoff integral solution of the acoustic wave equation (Aki and Richards, 1980, 415–419), the migration integral of a single shot generally can be expressed by the surface integral,

$$R(\mathbf{x}; \mathbf{x}_s) = \int_\Sigma \mathbf{n} \cdot \nabla \tau_r(\mathbf{x}_r; \mathbf{x}) A(\mathbf{x}_r; \mathbf{x}; \mathbf{x}_s) u^m(\mathbf{x}_r, \tau_s(\mathbf{x}; \mathbf{x}_s) + \tau_r(\mathbf{x}_r; \mathbf{x}); \mathbf{x}_s) \, d\mathbf{x}_r; \quad (1)$$

where Σ is the recording surface; τ_s and τ_r are the traveltimes from the source point $\mathbf{x}_s$ to the subsurface position $\mathbf{x}$, and from $\mathbf{x}$ to the receiver at $\mathbf{x}_r$, respectively; and $\mathbf{n}$ is the outward normal of the surface Σ. Here, u^m denotes the time derivative of the recorded traces. For a 2-D case, $m = 1/2$. The term $A(\mathbf{x}_r; \mathbf{x}; \mathbf{x}_s)$ is the geometrical spreading term that functions here as an amplitude modulator to the recording traces. Similar formulations can be found in the literature (Schneider, 1978; Berkhout, 1982; Keho and Beydoun, 1988; Docherty, 1991). Using a far-field approximation, migration using equation (1) is basically a weighted summation of the derivative traces along the presumed diffraction trajectory $t = \tau_s + \tau_r$. The weights are often approximated based on a constant velocity model. In such case, the weights can be analytically expressed as a function of velocity, traveled distance, and the obliquity of the emergence ray at the recording surface (Bleistein et al., 1987). Thus, the determination of traveltimes plays the key role in the calculation of the integral. These are traditionally accomplished by ray tracing. Ray tracing essentially will give all the information about τ_s and τ_r. It will simultaneously give us the directions of the rays. However, in this paper we employ a robust numerical solution to the eikonal equation for reasons given in Vidale (1988).

In the case of migrating prestack shot records, the problem becomes relatively simple and natural. Here, we use an improved Vidale algorithm to calculate both τ_s and τ_r (Vidale, 1988). The compensation of amplitudes because of the propagating distance is treated similarly as in Gray and May (1994). However, for the sake of economy, the obliquity of the emergency ray, $\cos\theta$, is not included in our implementation. This would not be a problem for the deeper part of the earth where rays will finally reach the surface at relatively small angles. The shallow steep reflectors, on the other hand, could suffer some accuracy deterioration. This deterioration is increased further by the far-field approximation.

In essence, the prestack Kirchhoff depth migration is performed here nonrecursively. It works simply on the data trace by trace. A single trace is migrated by distributing the recorded energy along aplanatic curves with the amplitudes modulated by some geometrical functions. These aplanatic curves are determined by solving the eikonal equation. Figure 1 shows the Kirchhoff migration impulses based on a faulted block geometrical model. In this migration result for a single trace consisting of six events, the geometrical distributions of the possible scatterers suggested by these events are imaged correctly. The migration amplitudes are close to those of the reverse-time migration except at the two zones where the refraction takes place. These refraction zones are imaged correctly with reverse-time migration but require special treatment with Kirchhoff

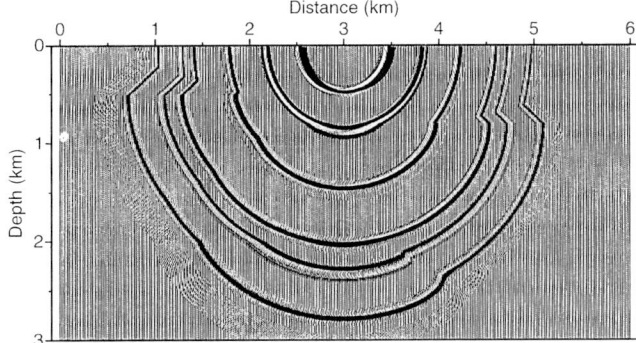

FIG. 1. Kirchhoff migration impulses. The migration is based on a blocked velocity model with a normal fault developed throughout the depth range. The velocity in each block linearly increases with depth.

migration if amplitudes are to be preserved. This treatment generally requires extra computations.

Reverse-time migration

Reverse-time migration emerged as a very powerful and general imaging tool for reflection seismic data in early 1980s (Hemon, 1978; Baysal et al., 1983; Loewenthal and Mufti, 1983; McMechan, 1983; Whitmore, 1983; Levin, 1984). It is based essentially on the symmetry of the acoustic wave equation in time. It uses the same finite-difference scheme to backward extrapolate the recorded wavefields as in forward modeling.

In this work, we use a finite-difference solution of the wave equation with accuracies of fourth order in space and second order in time. By treating the recorded traces as distributed sources, the finite-difference scheme generally can be expressed as

$$u_{i,j}^{k-1} = A + 2u_{i,j}^k - u_{i,j}^{k+1} + s_{i,j}^{k-1}, \qquad (2)$$

where u is the pressure wavefield, s is the surface recorded wavefield, A is the finite-difference approximation of the Laplacian operation on the wavefield, and i, j, k are indexes for $x, z,$ and t respectively. The effect of A is basically a 2-D spatial filtering on the present (t_k) wavefield. Mufti et al.'s (1996) general form of finite-difference formulation is very attractive, especially in three dimensions. In the case of two dimensions as described here, we still prefer the uniform square grid ($\Delta x = \Delta z$) to the rectangle grid ($\Delta x \neq \Delta z$). In terms of computations at each time step, there are only four multiplications per grid point when square grids are used. However, there will be eight multiplications for each grid when the grids are rectangles. Thus, for a given size of a physical model, even if the lateral step size of the rectangular grid is twice as large as its vertical one, the computation will still be about the same as that when square grids with the same vertical step size are used. In fact, the validity of using a large lateral mesh step will be challenged in the prestack case.

Thus, our implementation of reverse-time migration can be summarized by four steps: (1) determine the excitation-time imaging condition by solving the eikonal equation, (2) extrapolate the recorded wavefields, backward in time using equation (2), (3) apply the imaging condition, and (4) sum the individual migrated shot to produce a final stacked image. The first three steps are basically the same as described in Chang and McMechan (1986). Its essence is that of a shot record migration. Figure 2 shows the reverse-time migration impulses with the same input data as used by Figure 1. Compared to the Kirchhoff migration impulses, this result accurately recovers not only the geometrical shapes but the amplitudes as well. Specifically, the refracting zones are migrated with small amplitudes, which is to be expected. The results also indicate clearly that migration reflections will occur at the geological interfaces if the full wave equation is used and the velocity model is not properly smoothed (Loewenthal et al., 1987). Migration noise caused by interbed reflections also can be reduced significantly by use of the nonreflecting wave equations by introducing a density function inversely proportional to the velocity function (Baysal et al., 1984; Zhu and Lines, 1994). For the examples in this paper, generally we smooth the velocity model to reduce such migration noises (Versteeg, 1993).

Performance evaluations

Kirchhoff migration can be performed both recursively and nonrecursively. Our choice of the nonrecursive method is based largely on our ability to calculate traveltimes accurately. This choice eliminates (without much loss of accuracy) the need for extrapolating wavefields from depth to depth, and also significantly reduces the computations. The enhancement in computational efficiency is worth the loss of extrapolated snapshots. Multivalued arrivals are not handled by the Kirchhoff migrations, whereas they can be handled by prestack reverse-time migrations that invoke the full wave equation (Whitmore and Lines, 1986).

Generally, for a model of $N_x \times N_z$ grid points, migrating one shot with N traces, the Kirchhoff integral method will take $O(N_x' \cdot N_z \cdot N)$ operations (Table 1). In most cases, the migration aperture N_x' is much smaller than the model's lateral extent N_x. It is seen from this expression that the computation is dependent directly on the number of traces in the gather, so the computation for a gather of less traces will take less time. For example, the special case of migrating a single trace gather, as shown in Figure 1, only takes 24 s on our department's SPARC station 10/30. In fact, 17 s are a result of the overhead in preparation for the migration. Computing traveltimes takes about 40% of the total computations. This number is dependent somewhat on the complicated nature of the model that arises from the search of computing wavefronts in traveltime calculation. However, this number can be reduced significantly by preparing time tables prior to the integration procedure, if enough memory is available.

One of the most important attributes of the Kirchhoff method perhaps is that it can use selective shots and traces

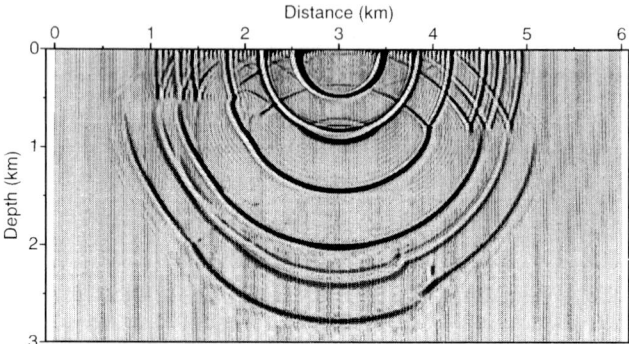

FIG. 2. Reverse-time migration impulses. The migration is based on the same input data and the same velocity model as used in Figure 1.

Table 1. Differences between Kirchhoff and reverse-time migrations for shot gathers.

	Kirchhoff	Reverse-Time
Recursive	No	Yes
Selective	Yes	No
Accuracy	Very good (except near surface)	Excellent
Favor frequency	High f	Low f
Inclusion of topography	Easy	Reasonably easy
Computation cost	$O(N_x' \cdot N_z \cdot N)$	$O(N_x' \cdot N_z \cdot N_t')$
Cost and frequency	$\propto f \sim f^2$	$\propto f^3$
Vectorization	Good	Excellent
Parallelization	Excellent	Excellent
Data preparation	Easy	With some effort

to image some prespecified targets because it is trace based (Gray and May, 1994; van der Schoot et al., 1989). This also makes the Kirchhoff method easy to use in areas with rough topography. Thus, near-surface topographic corrections can be included easily in the Kirchhoff shot migration (Gray and Marfurt, 1995; Lines et al., 1996). Furthermore, the preparation of the model and data in Kirchhoff migration is much simpler than in other methods. The selectivity of the data, high computation efficiency, as well as the easy preparation of data sets render the Kirchhoff migration the preferred method, especially in the process of recursive migration and velocity analysis (Jervis et al., 1996).

Reverse-time migration is recursive in time and represents a general wave-equation–based method. It can be a very accurate method because the only possible error other than that caused by the velocity model is the discretization error that occurs when differentials are approximated by finite differences. Its high accuracy is nevertheless offset by its very extensive computations. For a model of $N_x \times N_z$, reverse-time migrating a single shot of N traces with each consisting of N_t samples will take $O(N'_x \cdot N_z \cdot N'_t)$ operations. N'_x is generally chosen, as in the Kirchhoff case, depending on the nature of the earth model and the recorded wavefields. N'_t is the number of extrapolation time steps. It is often much larger than the value of N_t, because N'_t is determined by the stability condition of the finite-difference scheme. Compared to the $O(N'_x \cdot N_z \cdot N)$ operations involved in the Kirchhoff scheme, reverse-time migration generally will require many more computations, because N'_t would be much larger than N in most cases. It is apparent from this estimation that the computations involved in reverse-time migration are independent of the number of traces in each shot, which is definitely in contrast to the Kirchhoff method. Therefore, reverse-time migration for a gather of a single trace is computationally just the same as migration of a gather with many traces. For example, the migration of Figure 2 on a gather of only one trace takes 41 minutes on the same SPARC station 10/30. This estimation also implies that when the grid size is halved for a given model, the computation time will increase to eight times the original for the reverse-time migration, and four times the original for the Kirchhoff method. Moreover, the preprocessing for reverse-time migration previously was also considered a bit more complicated. Our recent study, however, indicates that interpolation of missing traces, which is very difficult in complicated areas, can be bypassed in many cases, because wavefields are capable of healing themselves during the reverse-time extrapolation procedure (Zhu and Lines, 1997).

Despite higher computational demands, reverse-time migrations tend to have a wider range of applications. This is because of the recent progress in the computer sciences and the preferred high accuracy of the method. Compared to the Kirchhoff method, its independence of accuracy and computations on the complexity of the geological model also is an advantage. These characteristics, in addition to its implicit ability of static corrections, filtering, and self-healing of wavefields, make reverse-time migration a very powerful method for imaging geologically complex areas (McMechan and Chen, 1990; McMechan and Sun, 1991; Reshef, 1991; Lines et al., 1996).

In addition to the differences we discuss above, Table 1 shows a more complete summary of performance comparisons between Kirchhoff and reverse-time migrations.

DATA EXAMPLES

In this section, we will show the performance of both Kirchhoff and reverse-time migrations on two data sets. The first example is the well-known Marmousi model data. This model data set is intended particularly to show the effectiveness of each migration on data from complicated structures where a correct velocity model is available. The second is the Husky Foothills data set, provided by Christof Stork of Advance Geophysical and Larry Mewhort of Husky Oil. We use this real data set to evaluate the accuracy and efficiency of the two methods in the real world where the true velocity model is never known exactly.

Migration of the Marmousi data

The Marmousi model data presents a challenge to exploration geophysicists in imaging complicated geological provinces (Versteeg, 1993). The model contains very complicated geological features, especially the shallow steep faults and the underlying high-velocity salt creeps. It has served as a standard test data set for migration, inversion algorithms, and velocity analysis methods (Versteeg, 1993; Gray and May, 1994; Nichols, 1996). Both the prestack Kirchhoff and the reverse-time migration algorithms described in the last section have been tested extensively on this model data. Figure 3 shows three selected migration shots by prestack Kirchhoff integral schemes. In contrast, Figure 4 shows the corresponding shot migrations produced by reverse-time migration techniques. Both sets of migrations illustrate basically the same geological zone. However, there are still noticeable differences between them. The most obvious difference probably lies at the near-source area where the Kirchhoff result lacks detail in the migration shots compared to the reverse-time migration result. In addition, in the reverse-time migration shots, it seems that the direct waves have masked the images somewhat. This reflects the fact that the reverse-time migration uses the complete recorded wavefields, whereas the Kirchhoff method deals essentially with reflections and diffractions. Though there exist other striking differences between these two sets, it is still not obvious which method provides better image, except in the near-source zone. Figure 5 shows the migration image with the Kirchhoff integral method when a 12.5-m grid size is used. It takes about 2.44 hours of CPU time on Memorial University's campus computer DEC AlphaServer 1000 with a clock frequency of 200 MHz. Figure 6, on the other hand, shows the migrated section with the reverse-time migration algorithm using the same gridded velocity model as in the Kirchhoff migration. However, this migration takes 21.43 hours of CPU time on the same machine. These two plots are displayed with the same plotting parameters without cosmetic processing applied, so a direct comparison should be applicable. It is apparent that both methods have restored the geological features of the model fairly well despite the striking differences in the migration shots. Nevertheless, as we notice, there are several places where the two images are different. The left and the middle faults in the Kirchhoff result are not as sharply defined as in the reverse-time migration image. The reverse-time migration section presents a sharp image of the right fault, but the integral result smears the image of the fault around a depth of

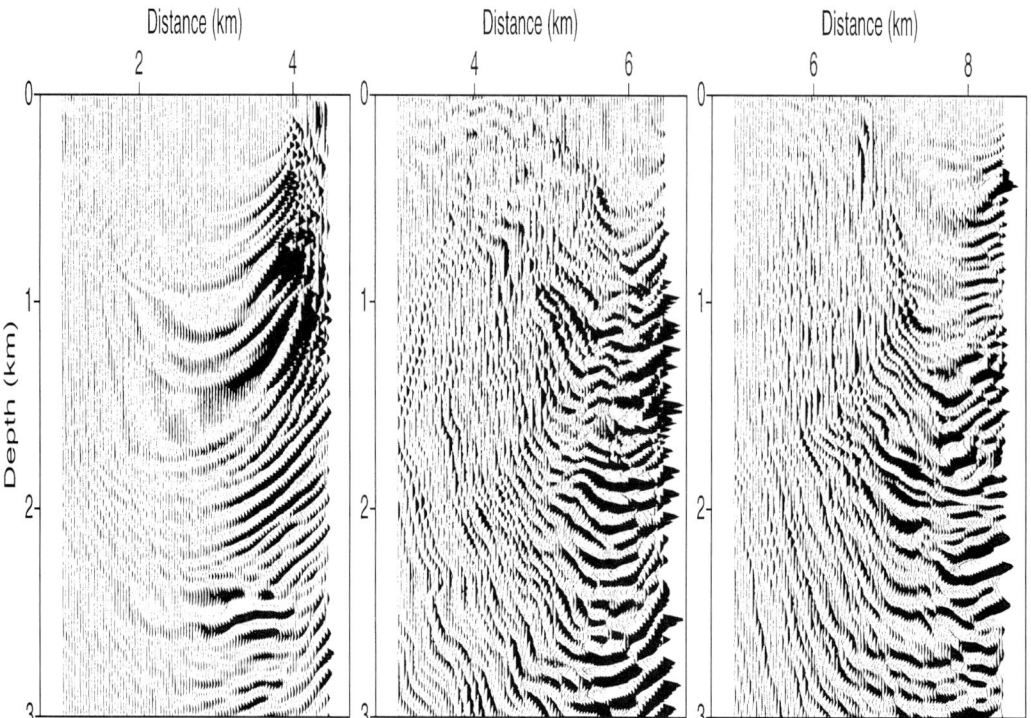

FIG. 3. Selected migration of shot records from the Marmousi model data set produced by the Kirchhoff integral method. A velocity model of 12.5 × 12.5 m is used in the migration.

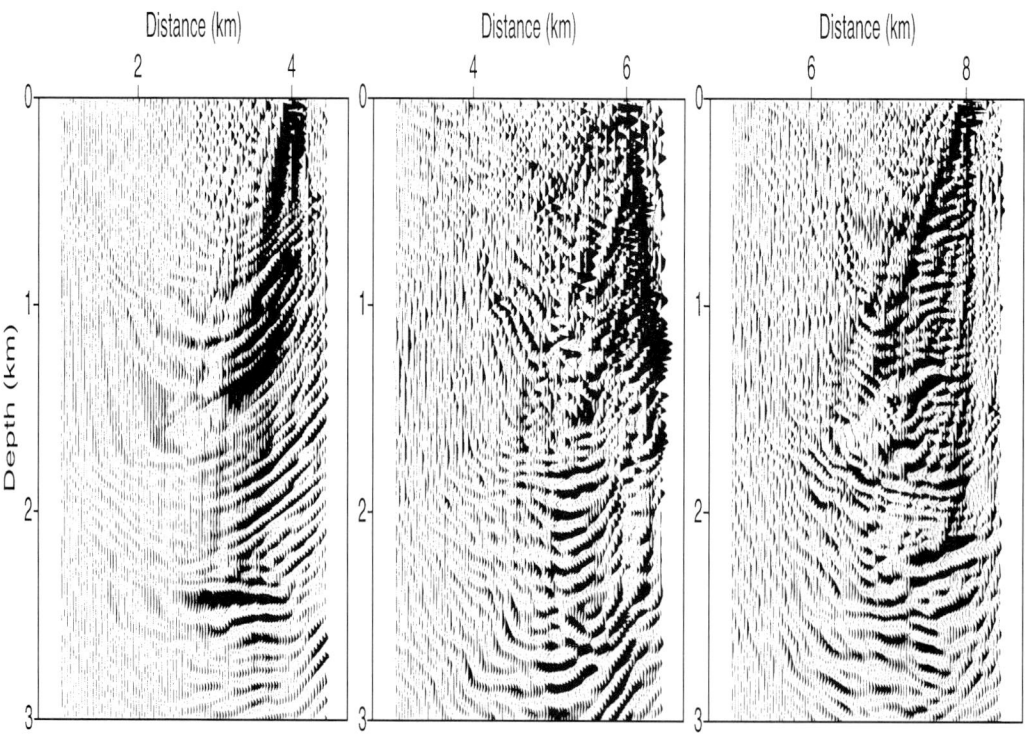

FIG. 4. Selected migration of shot records from the Marmousi model data set produced by the reverse-time migration method. A velocity model of 12.5 × 12.5 m is used in the migration.

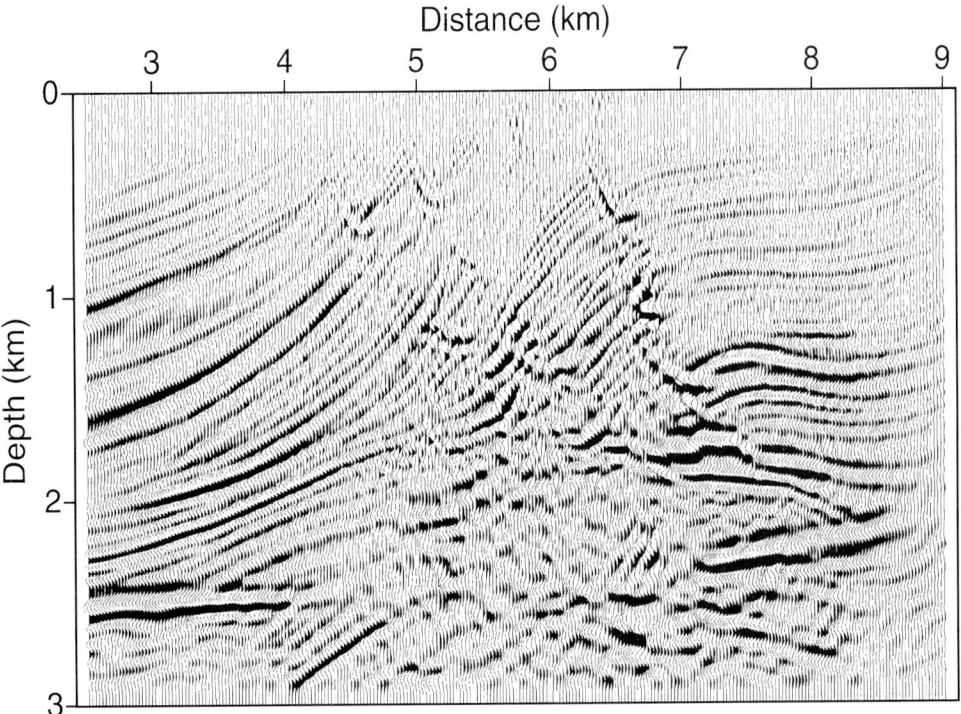

FIG. 5. The final migration section of the Marmousi model data set produced by the Kirchhoff integral method.

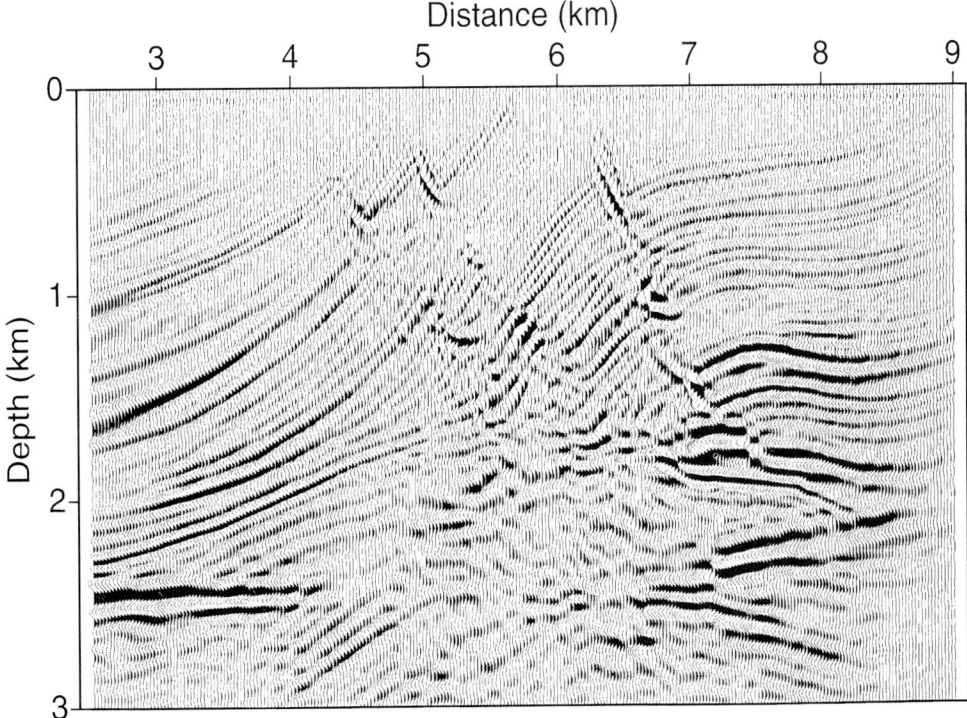

FIG. 6. The final migration section of the Marmousi model data set produced by the reverse-time migration.

1.5 km. In addition, reverse-time migration provides a slightly more continuous definition of the subsalt anticline structure than does the Kirchhoff method. These differences are mainly caused by the algorithmic details involved in the two methods, especially the negligence of the obliquity and the use of first-arrival times in our Kirchhoff method (Nichols, 1996).

Thus, in the example of the Marmousi data, the reverse-time migration gives a more accurate migration image than the Kirchhoff method. Its higher accuracy is nevertheless based on its use of the true velocity model and is achieved by requiring more computation time.

Migration of the Husky Foothills data

The Husky Foothills data have been assembled and distributed to many research groups. It is anticipated that these data will serve as an excellent test case for imaging complicated structures (Stork et al., 1995). The Canadian Foothills are characterized by overthrust structures of great variety (Skuce, 1995). Generally in these rough terrains, it is difficult in the field to acquire high-quality records. Problems exist such as poor geophone coupling and little signal penetration because of the exposition of older high-velocity rocks over the young low-velocity rocks. In addition, lines from such mountainous areas tend to suffer from serious near-surface problems. Nevertheless, the distributed foothills line is of excellent signal quality (Stork et al., 1995).

The line consists of 143 shots with most of them having 300 groups of records. Figure 7 shows a comparison of migration shot 142 produced by Kirchhoff integral and reverse-time migrations when a 10×10 m gridded velocity model is used. The velocity model was originally created based on structural geological information and the stacked section. The well-log information nearby provides good constraints to the velocity model. The model is then updated by an interpretive imaging procedure consisting of iterative prestack depth migration, migration velocity analysis, and geological interpretation. Both results image the shallow dipping layers very well. They are very similar in many respects, especially considering the fact that they are only based on a single shot gather. Nevertheless, there are differences identifiable between the two shot migrations. As in the Marmousi example, the greatest differences occur in the near-field areas. Figure 8 shows the final Kirchhoff migrated section. In this migration image, the shallow dipping formations at the upper left side of the section clearly are seen to be detached from the underlying gentle formations at the depth of about 2600 m. Two main thrust faults are well defined around CDP numbers 580 and 810, respectively. Overall, this migration result offers a very encouraging result that is relatively easy to interpret. The migration of this foothills line, however, only takes about 22.91 hours of CPU time (see Table 2). In fact, in our early stages of studying this line, the migration was done on a much coarser grid (20×20 m). In that case, the Kirchhoff migration took only about 5.51 hours of CPU time with quite similar results. Our impression of the difference is that the coarser grid lacks a bit of continuity at shallow parts of the earth model. From the CPU times, it is seen that use of a twice fine grid will increase the CPU time to about 4.2 times, which is pretty close to our theoretical estimate of 4 times, considering the overhead of computations involved in the migration.

In contrast, Figure 9 shows the reverse-time migration section. It is based on the same velocity model as used in Figure 8. It essentially reveals the same salient features of the geological province as Figure 8 with a little improvement in the triangular zone around CDP 800–1200. In the enlarged view of the

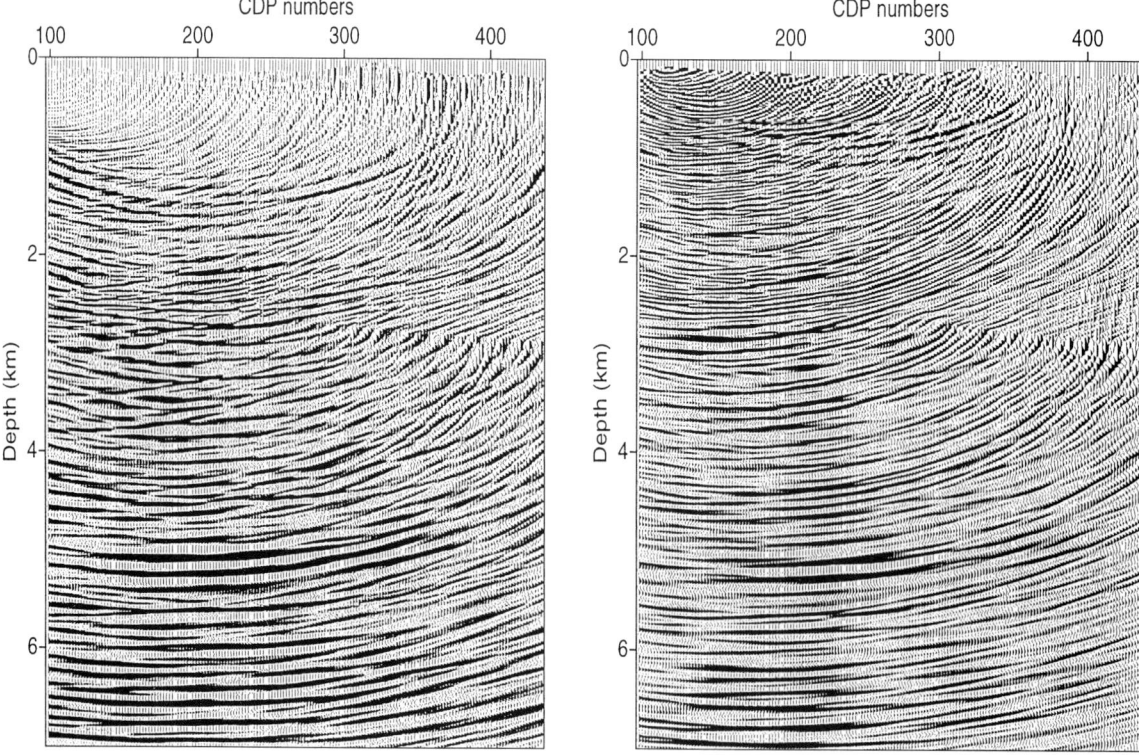

FIG. 7. A comparison of a representative migration shot from the Husky foothills line. The left corresponds to the Kirchhoff result; the right is the reverse-time migration result.

Table 2. Computation cost examples of Kirchhoff and reverse-time depth migrations.

	Marmousi data		Husky foothills line	
	$h = 25.0$ m	$h = 12.5$ m	$h = 20$ m	$h = 10$ m
Kirchhoff	37.28 minutes	2.44 hours	5.51 hours	22.91 hours
RT	2.41 hours	21.31 hours	13.59 hours	135.55 hours

Note: Computations were done on an AlphaServer 1000 with a clock frequency of 200 MHz.

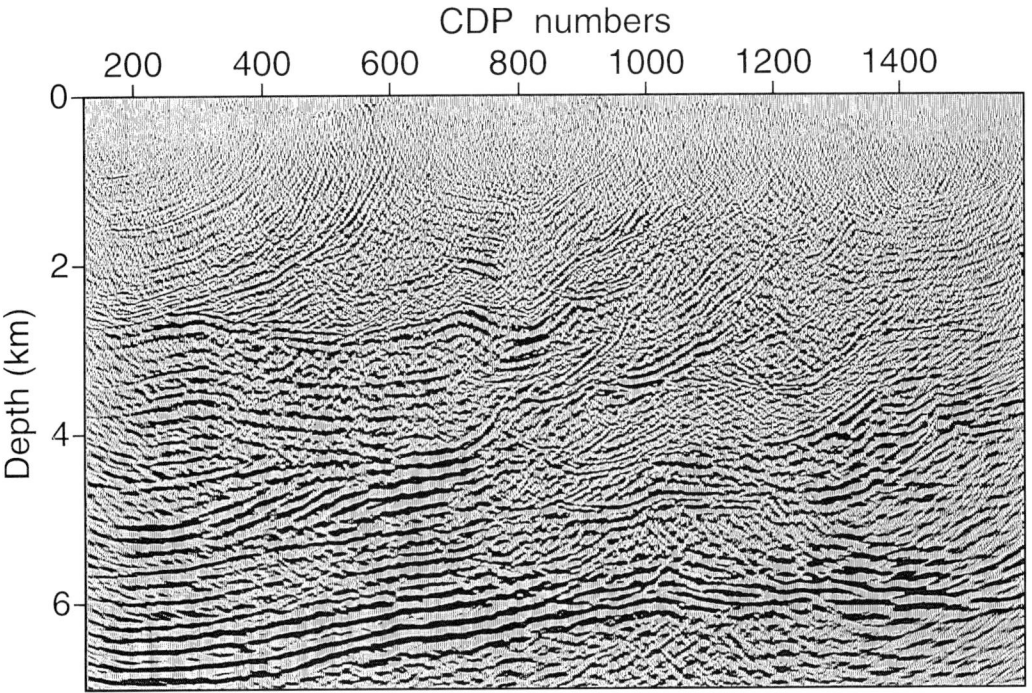

FIG. 8. The final migration section of the Husky foothills line produced by the Kirchhoff integral technique.

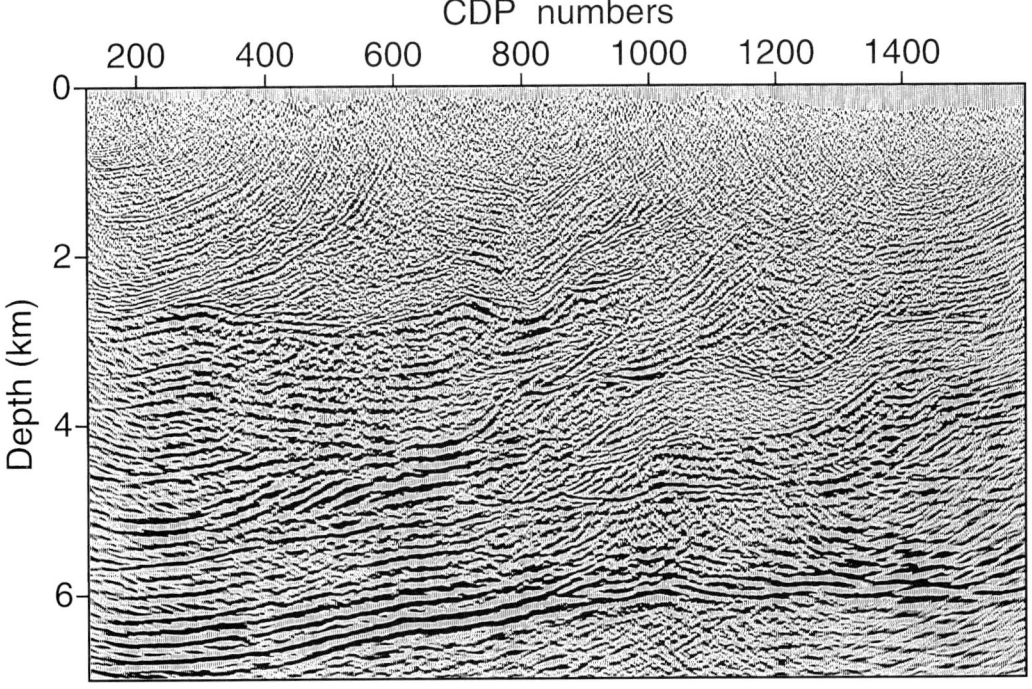

FIG. 9. The final migration section of the Husky foothills line produced by the reverse-time migration technique.

migration image that corresponds to the upper left portion of the whole (shown in Figures 10 and 11), it is still difficult to tell one migration image from the other. However, the production of this image with the reverse-time migration method requires about 135.55 hours of CPU time! This is definitely a large amount of computer time compared to the 22.91 hours taken by the Kirchhoff method. For comparison purposes, we have also tested reverse-time migration technique on this line using a coarser gridded velocity model of 20×20 m. It required 13.59 hours of CPU time. Thus, for the reverse-time migration in the case of this foothills line, when the model grid is twice as fine, the computation time nearly increases by a factor of 10, which is a bit higher than our theoretical estimate of a factor of 8. This possibly results from the fact that a larger proportion

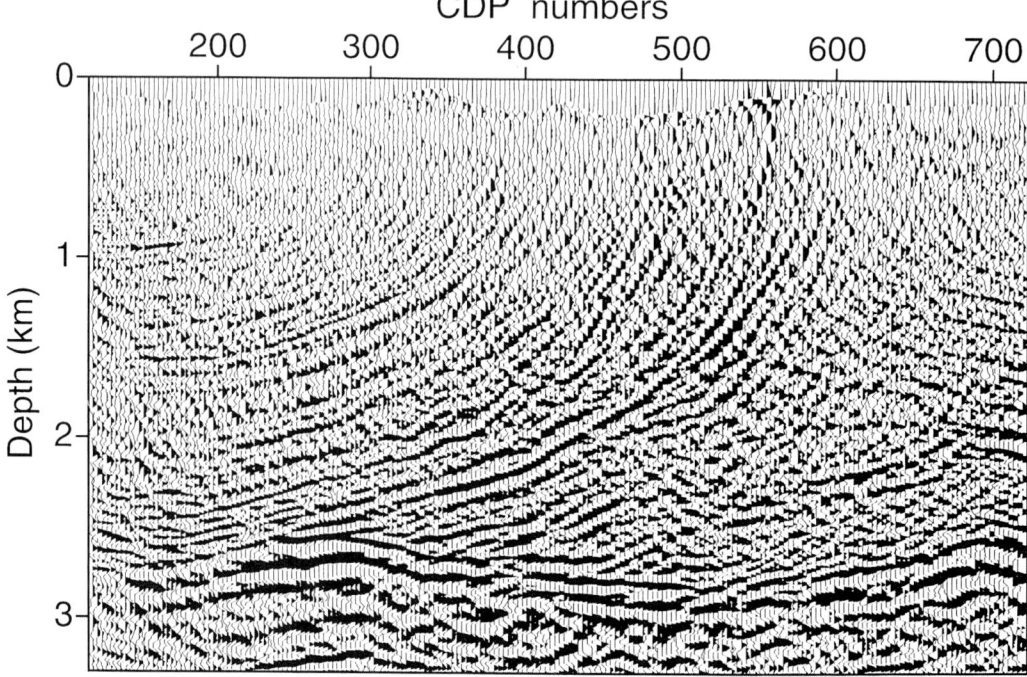

FIG. 10. The enlarged view of the final migration section of the Husky foothills line produced by the Kirchhoff integral technique.

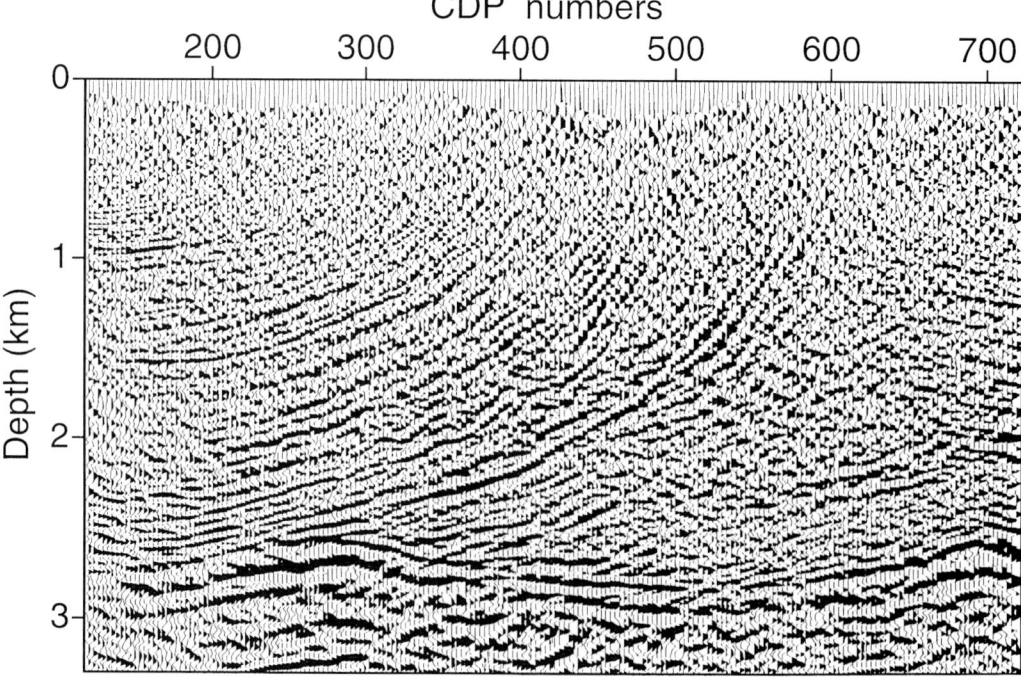

FIG. 11. The enlarged view of the final migration section of the Husky foothills line produced by the reverse-time migration technique.

of CPU time has been involved in swapping data as the digital model gets larger. Nevertheless, the similarity of the migration results between the Kirchhoff and reverse-time migration methods does not indicate that our Kirchhoff method is as accurate as the reverse-time method. It only implies that there are still errors in the estimated velocity model. Because of these errors, it is not evident which method works better in achieving migration accuracy. This possibly shows that in the real world of exploration seismology, where only an approximation of the true geological and velocity model is available, even the approximate version of the Kirchhoff method may work as well as the accurate reverse-time migration.

CONCLUSIONS

Prestack depth migration is currently a viable means for imaging complex geological scenarios. Kirchhoff and reverse-time migrations are two of the most widely used methods for prestack seismic depth imaging of complex structures. Both methods are based on the wave equation, and theoretically have no limit in migrating steep dips. In the sense of prestack migration, either one can be directly applied to areas with rough topography. They both have the ability to image real complicated data sets.

However, their performance is different. These differences spring mainly from the fact that one is based on the integral solution, whereas the other is based on the finite-difference solution of the wave equation. First of all, the Kirchhoff scheme used here is nonrecursive, whereas the reverse-time scheme is recursive in time. Second, our Kirchhoff method is not exact in the sense that the obliquity factor is neglected, in addition to the high-frequency approximation. Third, the Kirchhoff migration of a gather of N traces using an $N_x \times N_z$ gridded mesh takes only $O(N'_x \cdot N_z \cdot N)$ operations, while the reverse-time migration requires $O(N'_x \cdot N_z \cdot N'_t)$ operations. Last but not least, the Kirchhoff method is capable of migrating selective shots and traces to focus illumination of a prespecified zone because the method is trace based. In contrast, reverse-time migration treats a gather of only one trace just as a gather of numerous traces. Its computation is relatively independent of the number of traces and the complexity of the earth model. Thus, it is hoped that reverse-time migration can be used for migration of seismic data acquired with a very long cable, possibly with comparable computational costs to the Kirchhoff method. Use of long cables is currently the industrial tendency.

Both synthetic and real data examples demonstrate that more significant differences exist in individual shot migrations for the two methods than for the final migrated sections. The application of the two methods on the Marmousi model data strongly supports the idea that reverse-time migration does a better job of imaging the steeply dipping faults. This possibly results from neglecting the obliquity factor and from the use of first-break times in the Kirchhoff integral. This higher accuracy of the reverse-time migration is based nevertheless on its use of the exact velocity model, and is traded off by its large amount of computations. In contrast, the migration images of the Husky Foothills data, obtained separately by the two methods, are very similar. For this real data set, even though we have a reasonably good estimate of the model, it is still quite difficult to differentiate one result from the other. The similarity of the results suggests that the methods are similar in performance in the case of an estimated velocity model.

ACKNOWLEDGMENTS

We appreciate Andrew Ehinger of Institut Français du Pétrole for providing us the Marmousi model data package. We also thank Larry Mewhort of Husky Oil and Christof Stork of Advance Geophysical for providing the Husky Foothills line to us. Andrew Burton and Jamie Jamison are acknowledged for their superb data preprocessing and the geological interpretations. Thanks are also extended to Paul Fardy for his help during our use of the computer facilities. We thank NSERC, Petro-Canada, and the sponsors of the MUSIC project for their generous financial support of this research.

REFERENCES

Aki, K., and Richards, P. G., 1980, Quantitative seismology: Theory and methods, vol. 1: W.H. Freeman and Co.
Baysal, E., Kosloff, D. D., and Sherwood, J. W. C., 1983, Reverse-time migration: Geophysics, **48**, 1514–1524.
——— 1984, A two-way nonreflecting wave equation: Geophysics, **49**, 132–141.
Berkhout, A. J., 1982, Seismic migration imaging of acoustic energy by wave field extrapolation: Elsevier Science Publ. Co., Inc.
Bleistein, N., Cohen, J. K., and Hagin, F. G., 1987, Two and one-half dimensional Born inversion with an arbitrary reference: Geophysics, **52**, 26–36.
Chang, W. F., and McMechan, G. A., 1986, Reverse-time migration of offset vertical seismic profiling data using the excitation-time imaging condition: Geophysics, **51**, 67–84.
Dablain, M. A., 1986, The application of high-order differencing to the scalar wave equation: Geophysics, **51**, 54–66.
Docherty, P. C., 1991, A brief comparison of some Kirchhoff integral formulas for migration and inversion: Geophysics, **56**, 1164–1169.
French, W. S., 1975, Computer migration of oblique seismic reflection profiles: Geophysics, **40**, 961–980.
Gray, S. H., 1986, Efficient traveltime calculation for Kirchhoff migration: Geophysics, **51**, 1685–1688.
Gray, S. H., and May, W. P., 1994, Kirchhoff migration using eikonal equation traveltimes: Geophysics, **59**, 810–817.
Gray, S. H., and Marfurt, K. J., 1995, Migration from topography: improving the near-surface image: Can. J. Expl. Geophys., **31**, 18–24.
Hemon, C., 1978, Equations d'onde et modeles: Geophys. Prosp. **26**, 790–821.
Jervis, M., Sen, M., and Stoffa, P. L., 1996, Prestack migration velocity estimation using nonlinear methods: Geophysics, **61**, 138–150.
Keho, T. H., and Beydoun, W. B., 1988, Paraxial ray Kirchhoff migration: Geophysics, **53**, 1540–1546.
Larner, K., and Hatton, L., 1990, Wave equation migration: two approaches: First Break, **8**, 433–448.
Levin, S. A., 1984, Principle of reverse-time migration: Geophysics, **49**, 581–583.
Lines, L. R., Wu, W., Lu, H., Burton, A., and Zhu, J., 1996, Migration from topography: Experience with an Alberta Foothills data set: Can. J. Expl. Geophys., **32**, 24–30.
Loewenthal, D., and Mufti, I. R., 1983, Reversed time migration in spatial frequency domain: Geophysics, **48**, 627–635.
Loewenthal, D., Stoffa, P. L., and Faria, E. L., 1987, Suppressing the unwanted reflections of the full wave equation: Geophysics, **52**, 1007–1012.
McMechan, G. A., 1983, Migration by extrapolation of time-dependent boundary values: Geophys. Prosp., **31**, 413–420.
McMechan, G. A., and Chen, H. W., 1990, Implicit static corrections in prestack migration of common-source data: Geophysics, **55**, 757–760.
McMechan, G. A., and Sun, R., 1991, Depth filtering of first breaks and ground roll: Geophysics, **56**, 390–396.
Mufti, I. R., Pita, J. A., and Huntley, R. W., 1996, Finite-difference depth migration of exploration-scale 3-D seismic data: Geophysics, **61**, 776–794.
Nichols, D. E., 1996, Maximum energy traveltimes calculated in the seismic frequency band: Geophysics, **61**, 253–263.
Rajasekaran, S., and McMechan, G. A., 1995, Prestack processing of land data with complex topography: Geophysics, **60**, 1875–1886.
Ratcliff, D. W., Gray, S. H., and Whitmore, N. D., 1992, Seismic imaging of salt structures in the Gulf of Mexico: The Leading Edge, **11**, No. 4, 15–31.
Reshef, M., 1991, Depth migration from irregular surfacs with depth extrapolation methods: Geophysics, **56**, 119–122.
Schneider, W. A., 1971, Developments in seismic data processing and analysis (1968–1970): Geophysics: **36**, 1043–1073.

——— 1978, Integral formulation for migration in two and three dimensions: Geophysics: **43**, 49–76.

Skuce, A., 1995, Seismic imaging in the Canadian Rocky Mountain Foothills: Paper presented at the SEG workshop #6.

Stork, C., Welsh, C., and Skuce, A., 1995, Demonstration of processing and model building methods on a real complex structure data set: Presented at the SEG workshop #6.

Van der Schoot, A., Romijn, R., Larson, D. E., and Berkhout, A. J., 1989, Prestack migration by shot record inversion and common depth point stacking: A case study: First Break, **7**, 293–304.

Versteeg, R., 1993, Sensitivity of prestack depth migration to the velocity model: Geophysics, **58**, 873–882.

Vidale, J. E., 1988, Finite-difference calculation of traveltimes: Bull. Seis. Soc. Am., **78**, 2062–2076.

Whitmore, N. D., 1983, Iterative depth migration by backward time propagation: 53rd Ann. Internat. Mtg.: Soc. Expl. Geophys, Expanded Abstracts, 827–830.

Whitmore, N. D., and Lines, L. R., 1986, Vertical seismic profiling depth migration of a salt dome flank: Geophysics, **51**, 1087–1109.

Whitmore, N. D., Gray, S. H., and Gersztenkorn, A., 1988, Two-dimensional post-stack depth migration: A survey of methods: First Break, **6**, 189–197.

Zhu, J., and Lines, L. R., 1994, Imaging of complex subsurface structures by VSP migration: Can. J. Expl. Geophys. **30**, 73–83.

——— 1997, Implicit interpolation in reverse-time migration: Geophysics, **62**, 906–917.

Chapter 3 – Migration in Structurally Complex Areas

Many conventional assumptions of prestack migration are rendered invalid by the geology of the Canadian Foothills. Conventional processing often uses static corrections and common midpoint stacking. These assumptions will generally cause smearing of seismic images as shown by *Gray and Marfurt (1995)*. An alternative approach, which does not violate these assumptions, is the "migration from topography" approach proposed by *Gray and Marfurt (1995)* and applied to real data by *Lines et al (1996)*. This chapter discusses these and other methods, such as discussed in *Margrave and Ferguson (1999)*, for dealing with complex geology and topographic relief.

References:

Gray, S.H., 1998, Interpretative seismic imaging in structurally complex areas: Geo-Triad '98, June 15-18, Calgary, Alberta, Expanded Abstracts, 282-283.

Gray, S.H., and Marfurt, K.J., 1995, Migration from topography: improving the near-surface image: *Canadian Journal of Exploration Geophysicists*, 31, Nos. 1 & 2, 18-24.

Lines, L., Wu, W., Lu, H., Burton, A., and Zhu, J., 1996, Migration from topography: experience with an Alberta foothills data set: *Canadian Journal of Exploration Geophysicists*, 32, No. 2, 24-30.

Margrave, G.F., and Ferguson, R.F., 1999, Wavefield extrapolation by nonstationary phase shift: *Geophysics*, 64, 1067-1078.

Interpretive seismic imaging in structurally complex areas
Samuel H. Gray, Amoco Canada Petroleum Company

Introduction
In geologically simple areas, seismic imaging has the relatively simple objectives of improving the resolution of lateral truncations and sequence variations. Little or no interpretation is needed to produce the image; the interpretation begins after the image is produced. Structurally complicated areas, on the other hand, impose more stringent demands on seismic imaging: velocity becomes important for both vertical and lateral positioning of reflectors, and geologic consistency places constraints on the reflector geometry. As a result, the interpreter needs to be involved in processing seismic data from structural areas. Indeed, in such areas, producing the seismic image is at least partly an interpretive process. In the most ambitious cases where the imaging objectives are to obtain both a geologically consistent image and interval velocity field, seismic imaging is *essentially* interpretive.

In this paper, I describe some of the interpretive tools available for seismic imaging in structurally complex areas. I argue that the "standard flow" consisting of NMO/DMO/stack/migration is obsolete and should be replaced by prestack migration. I describe advantages and limitations of prestack time migration and prestack depth migration, and I discuss 2-D vs. 3-D imaging, as well as emerging issues in interpretive processing.

Interpretive imaging tools
Why is the standard flow obsolete in structurally complex areas? For two reasons: its speed advantage over prestack migration is now small, and it is not accurate enough. For speed, modern prestack migration methods, which allow velocity analysis to be performed after migration, can be made nearly as efficient as NMO/DMO/stack/migration, which typically requires a number of data sorts and velocity analysis steps. For accuracy, consider Figures 1 and 2. Figure 1 shows a poststack depth migration (using correct interval velocities) of a synthetic data set that was acquired over an area with severe topographic variations and strong lateral velocity variations, and was processed using NMO/DMO/stack. This migrated image is poor compared with the image shown in Figure 2, which shows a prestack depth migration, again using the correct interval velocities, of the same data set. Both migrations were performed using the same algorithm, so the poor image quality in Figure 1 is due not to the migration, but rather to the inability of the stacking process to accommodate the variations of topography and velocity.

Given the need to migrate data before stack, what are the interpretive trade-offs between time migration and depth migration? In a nutshell, prestack time migration is not iterative and is less interpretive, and prestack depth migration is iterative and is more interpretive. Prestack depth migration is also potentially more rewarding, in allowing an interpreter to investigate geologic hypotheses (and not just image quality) and obtain estimates for depth conversion that can be revised as the interpreter gains more geologic knowledge. To obtain the extra benefits of prestack migration, the interpreter must do more work, feeding more geologic knowledge into the process and evaluating the effects of velocity updates. However, modern velocity analysis tools based on removing residual moveout from migrated CDP gathers have removed much of the uncertainty and drudgery from this interpretive chore, at least in 2-D.

When can we get away with 2-D imaging in structural areas, and when do we need to do 3-D imaging? Most present-day marine (streamer) data is 3-D, and economies of scale have evolved in processing capability to the point where marine 3-D prestack depth migration is being applied routinely, although the interpretive tools have not been developed at the same rate as the imaging tools. For land data, the situation is different for both acquisition and processing. Record for record, land acquisition is far more expensive than marine acquisition, and play economics often dictate an unhappy choice between oversampled 2-D acquisition and undersampled 3-D acquisition in areas with significant structural variation in one direction and gentle structural variation in the orthogonal direction. Generally, the deeper the target, the more likely it is to be influenced by non-dip structural variation in the overburden, and the greater the need for 3-D imaging. Luckily, these deep targets suffer less than shallow targets from the ill effects of sparse surface sampling, such as statics problems. Also luckily, out-of-plane events that interfere with shallow targets on 2-D images can often be detected during the velocity analysis procedure.

On the horizon
Two issues that will have an affect on interpretive imaging are emerging: anisotropy and shear-wave imaging. Anisotropy, the dependence of seismic velocity on direction of propagation, affects our ability to image accurately, for example, beneath steeply-dipping shales. Estimating anisotropy is entwined with velocity estimation; this will force us to revise our imaging velocity analysis tools, in both time and depth. Shear waves, typically in the form of mode converted data, are becoming recognized as signal that should be processed along with compressional waves, and not thrown away. In some geometries, for example base-of-salt imaging, shear arrivals appear to be the only events that contain information from the reflectors.

Conclusions
Seismic imaging remains the basic tool for building geologic structure maps. As imaging tools are developed, allowing us to see increasingly subtle hydrocarbon traps beneath complex geologic structures, interpreters and imaging developers need to work closely together. This will insure that the tools are technically up to the task and fit the interpreters' needs.

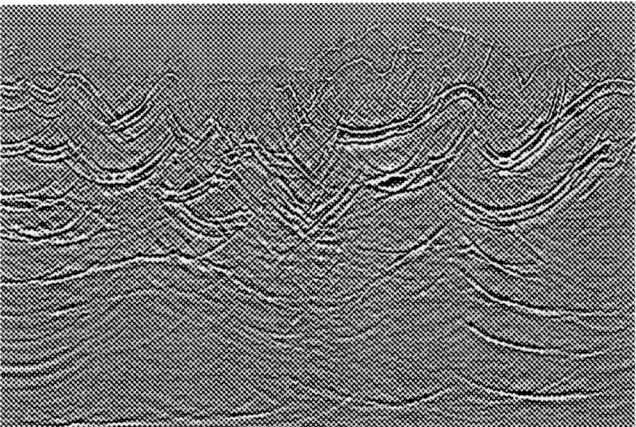

Figure 1. Poststack depth migration of a synthetic data set from a structurally complex area. The failure of the standard flow NMO/DMO/stack has led to a poor migrated image.

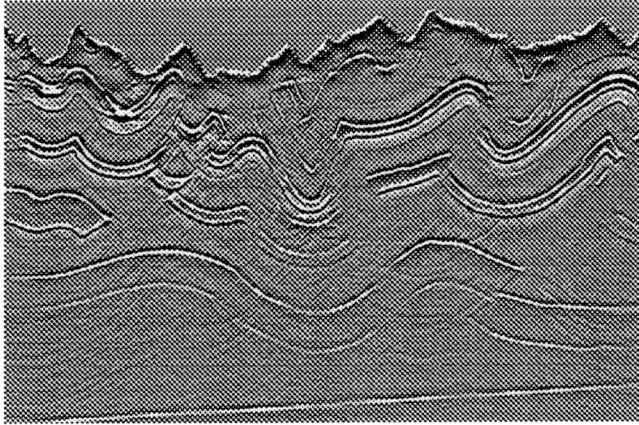

Figure 2. Prestack depth migration of the synthetic data set. The migration is the same as in Figure 1, but removing the approximations inherent in the standard flow has resulted in a much clearer and more accurate image.

MIGRATION FROM TOPOGRAPHY: IMPROVING THE NEAR-SURFACE IMAGE

Samuel H. Gray[1] and Kurt J. Marfurt[2]

Abstract

In mountainous areas, such as the foothills of the Canadian Rockies, careful processing of near-surface seismic data is essential, for without an accurate interpretation of the near surface, a reliable map of the deeper reflectors is unlikely. Although statics corrections are the primary focus of attention for near-surface imaging in mountainous areas, other factors can also affect the quality of the image near the earth's surface. In this paper, we investigate the effects of migration amplitudes on imaging quality. We use a synthetic seismic data set that models the geology of northeastern British Columbia to test the application of Kirchhoff depth migration directly from an irregular surface. We show that the near-surface image obtained using migration weights referenced to the recording surface is significantly better than one obtained using migration weights referenced to a flat datum above the surface. In addition, prestack migration velocity analysis, used to estimate velocities for accurate imaging, is more interpretable when the migration weights are referenced to the recording surface.

Introduction

Poststack seismic migration has two major uses: to image structure and to delineate stratigraphy. Prestack migration commonly has a third use: to estimate migration velocities. Less frequently, prestack migration is used to estimate reflection amplitude versus offset in the presence of complex overburden. For all these applications, poststack and prestack migration must preserve the integrity of the bandpass-filtered seismic wavelet. Some migration methods, such as finite-difference and Stolt migration, automatically preserve the character of the input wavelet. In order to accomplish the same goal, Kirchhoff migration requires the explicit application of a phase rotation to the data, as well as amplitude factors ("weights") to the summation along diffraction curves. The phase rotation occurs in differentiating the traces before or after migration, and is very straightforward to apply. The weights correct for geometrical spreading loss and obliquity factors, and are not so easy to apply. Schneider (1978) presented the simplest expression for these weights in a constant-velocity medium; this expression is often satisfactory in poststack migration even when the migration velocity varies significantly. For prestack migration, the weights account for nonzero source-receiver separation and domain (common shot, common midpoint, or common offset) and are consequently more complicated than the poststack migration weights. Several authors (e.g., Hanitzsch et al., 1994) have presented expressions for "true-amplitude" prestack migration weights. Whether one chooses a simplified, constant-velocity weight function appropriate for poststack migration or the more complicated weights appropriate for prestack migration, it is imperative to use a weight that tapers to zero at the farthest offsets of the migration impulse response operator (Figure 1). For poststack migration, this weight will approximate Schneider's obliquity factor $\cos \theta$, where θ is the angle between the normal to the earth's surface and the ray from the image point to the location of the input trace at the earth's surface. For prestack migration, the weight involves both $\cos \theta_S$ and $\cos \theta_R$, where θ_S (θ_R) is the angle between the normal to the earth's surface and the ray from the image point to the source (receiver) location (Hanitzsch et al., 1994). When migration weights are carefully applied in combination with the necessary phase rotation, the migration will preserve the character of the input wavelet, even if the seismic energy has propagated through one or more caustics. On the other hand, careless application of the migration weights can cause the migration operator to alias, or image high-frequency data improperly at steep dips. Migration operator aliasing, at its most benign, causes the output wavelet to be distorted and, when it is severe, can cause irreparable damage to an otherwise correctly migrated section.

Although the importance of migration weights is well appreciated for marine data and for land data acquired over

---

Manuscript received by the Editor June 22, 1995; revised manuscript received June 22, 1995.
[1] Amoco Exploration and Technology Group, P.O. Box 3385, Tulsa, OK 74102; also, Amoco Canada Petroleum Ltd., P.O. Box 200, Station M, Calgary, Alberta T2P 2H8
[2] Amoco Exploration and Technology Group, P.O. Box 3385, Tulsa, OK 74102
We thank Gary MacLean for generating the synthetic data set used in this study.

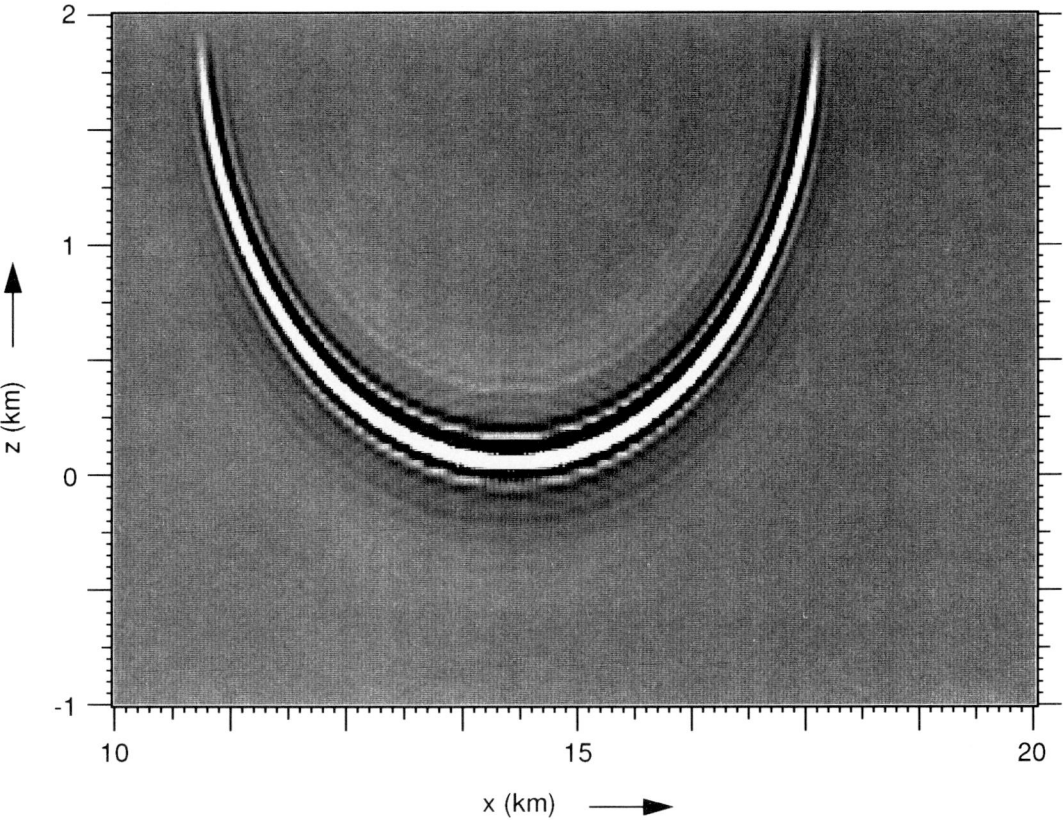

Fig. 1. Migration impulse response for a source/receiver pair located at *x* = 14 km along a flat, constant-velocity recording surface whose elevation is 2 km.

fairly flat terrain, it is perhaps not so well appreciated for data acquired over a surface with significant topographic variations. In this note, we examine the effects of two choices of weights, one simple but inappropriate and the other more appropriate, for Kirchhoff migration of seismic data acquired along an irregular surface and migrated directly from that surface (Wiggins, 1984). When the acquisition surface is irregular, there are considerable advantages to processing seismic data directly from topography, without shifting them to a flat or floating datum. For example, irregular topography is usually associated with highly contorted near-surface formations. Imaging these formations is crucial to tying geologic (e.g., outcrop) control to the seismic data. On the other hand, imaging these beds requires a level of precision in processing the near-surface data that datuming usually does not allow; thus, our preference for processing, including migration, directly from topography whenever possible. However, we shall show that, even when the data are migrated directly from topography, the migration weights must be computed from topography and not from an arbitrarily chosen flat datum to avoid operator aliasing and its consequences for prestack migration. Fortunately, computing the weights from topography can be accomplished as efficiently as computing the weights from a flat datum and, therefore, need not add to the cost of the migration.

THE EFFECT OF MIGRATION WEIGHTS ON THE MIGRATED IMAGES

Figure 2 shows a geologic cross-section, consisting of a number of faulted and folded layers typical of mountainous thrust regions. The top layer is air, with the surface of the earth indicated by the irregular white curve near the top of the model. The model is 25 000 m long: the top of the model in 2000 m above sea level and the bottom of the model is 8000 m below sea level. The total relief of the earth's surface along this cross-section is approximately 1600 m. This figure also indicates velocities, which range from 3500 m/sec (the darkest areas near the top) to 5900 m/sec (the lightest areas near the bottom). The horizontal bands indicate a vertical velocity gradient. This velocity model was used to generate 278 two-dimensional (2-D) acoustic synthetic shot records from the earth's surface. The data were recorded by a split spread of 480 receivers with offsets ranging from 15 m to 3600 m on both sides of the shotpoints. The shot spacing was 90 m. The 2-D wave equation used to generate the data set caused cylindrical spreading loss (roughly proportional to $1/\sqrt{t}$), while the Kirchhoff migration used to migrate the data assumed that energy from the sources underwent spherical spreading loss (roughly proportional to $1/t$) while propagating in the earth. To accommodate this difference in spreading loss, the traces were multiplied by $1/\sqrt{t}$ before migration.

Figure 3 shows the migration of a single, near-offset trace.

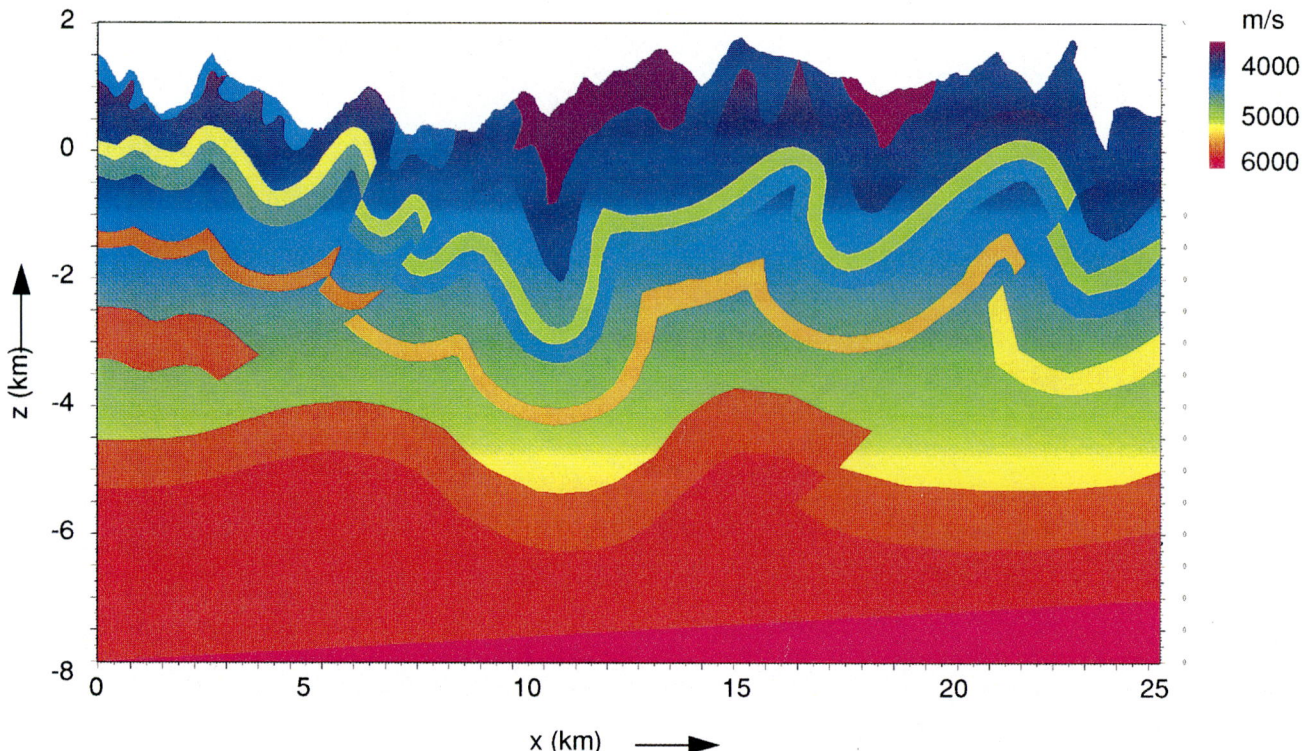

Fig. 2. A velocity/depth model representative of northeastern British Columbia. There is roughly 1600 m of elevation relief along the seismic cross-section.

The correct source and receiver elevations were used in generating traveltime curves to migrate this trace. In Figure 3a, the migration weights, modified from the constant-velocity common offset migration expression found in Bleistein et al. (1987), were computed from the flat datum above the earth's surface. That is, the depth z of an image point, which appears as a factor in the migration weight, was measured from the flat datum at the top of the model (2000 m elevation), well above the source and receiver locations. In Figure 3b, the migration weights were calculated using the same formula, except that the depth of an image point was measured from the lesser of the source and receiver elevations. Thus, the quantity $z-z_{topo}$, where z is again the depth of an image point below the flat datum and z_{topo} is the greater of the source and receiver depths below the flat datum, replaced z when the weight was applied. For the shallow part of the section, $z-z_{topo}$ is nearly zero, resulting in much weaker amplitudes near the earth's surface for the shallow elliptical event in Figure 3b. Because the migrated amplitudes shown in Figure 3a do not decay to zero along the acquisition surface, we can expect some problems when these migration weights are used in migrating the entire data set. Of course, migrating a single trace gives no visual indication of the problems that will arise when inappropriate migration weights are used; it is only through the interference of many migrated traces, as shown in the next figure, that the effects of migration operator aliasing become apparent.

Figure 4 shows the result of a prestack depth migration; all the traces were migrated from topography and accumulated into a final stacked section. In Figure 4a, as in Figure 3a, the migration weights were calculated with depth measured from the flat datum. This section, accurate though it is at depths well below the topography, is excessively blurred out near the earth's surface (which is precisely where we would want to tie to any available outcrop control!). This is the manifestation of migration operator aliasing; the blurring occurred because the migration weights were incorrect. In turn, the migration weights were incorrect because the depth was measured from the wrong elevation. By contrast, Figure 4b shows a very clear image near the earth's surface. [A very small amount of operator aliasing is evident in Figure 4b. This is due to a cause entirely different from the use of incorrect migration weights, namely, the application of common offset migration to seismic data whose source spacing is very different from the receiver spacing. The common offset migration formula, due to Bleistein et al. (1987), combines the common shot and common receiver migration expressions and, thus, depends on the source spacing as well as the receiver spacing. These spacings, and the maximum frequency present in the data, determine the maximum well-sampled apparent dip (Vermeer, 1990), which is transformed by migration into the maximum well-imaged near-surface dip. The effects of this type of operator aliasing can be eliminated, but only with the loss of steep dips near the earth's surface.] As noted in the previous paragraph, the depths used in the migration weights were only approximately correct, being most incorrect for large offset traces with substantially different source and receiver elevations. However, the effects of this approximation on the image are minimal.

Figure 5 shows a detail of Figure 4, chosen from the left side

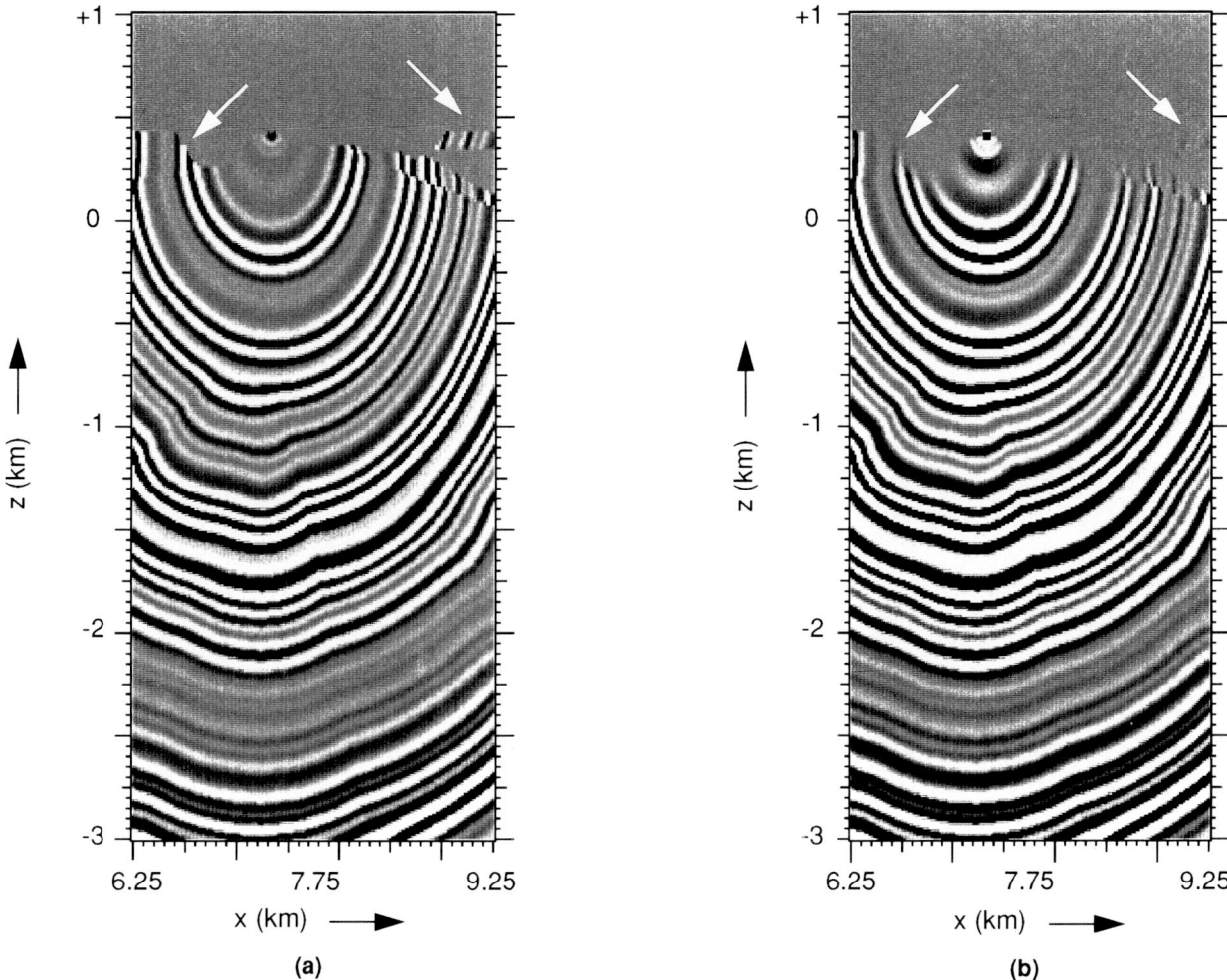

Fig. 3. Prestack Kirchhoff depth migration of a single trace corresponding to a source at 7755 m and receiver at 7770 m. Migration weights are calculated from **(a)** a flat datum at $z = 2000$ m and **(b)** topography, or $z = 375$ m. Note the variation in shallow amplitudes indicated by the arrows.

of the section, where the elevation changes are the most rapid. Although the anticlinal reflector clearly evident in Figure 5b is also evident in Figure 5a, the high-amplitude migration smile and the coherent noise created by the aliased migration operator have degraded the character of the wavelet in Figure 5a as well as our confidence in the event's interpretability.

It might be tempting to simplify our strategy of using the local source and receiver elevations in computing the migration weights; we might, for example, choose to compute the weights from a flat datum z lying between the minimum and maximum elevations. As long as elevation changes along the line are not too severe such an approach will be accurate, but drastic elevation changes will give rise to large differences $z - z_{topo}$, leading to poor images such as those in Figures 4a and 5a. Also, it is as efficient to compute weights from $z - z_{topo}$ as from any datum z.

In addition to compromising the imaging capabilities of a migration, using incorrect migration weights can have an unexpected negative effect on prestack migration velocity analysis. One of the objectives of prestack depth migration is to allow the estimation of seismic velocities by inspection of common-reflection point (CRP) gathers. A CRP gather displays traces at a particular earth location taken from each migrated common-offset section. By comparing the depths of a migrated event on all the traces in a gather, one can determine whether the velocities used in the migration are correct. If a particular event appears to be flat on a gather, then all the common-offset migrations have imaged that reflector at the same depth, indicating that the velocity used is correct, at least in an average sense. If an event shows residual curvature up or down, then the event has been imaged at different depths from different offsets, indicating that the migration velocity is, on the average, too low or too high. Figure 6 shows the effects of the two choices of migration weight calculations on the CRP gathers. The same (correct) velocity function was used for both migrations. In Figure 6a (using migration weights with depths computed from the flat datum), the near-surface amplitudes are skewed significantly towards the mid and far offsets. This has resulted in a high noise level on the shallow events relative to Figure 6b, where the weights were computed from topography. The extra noise in Figure 6a makes the difficult task

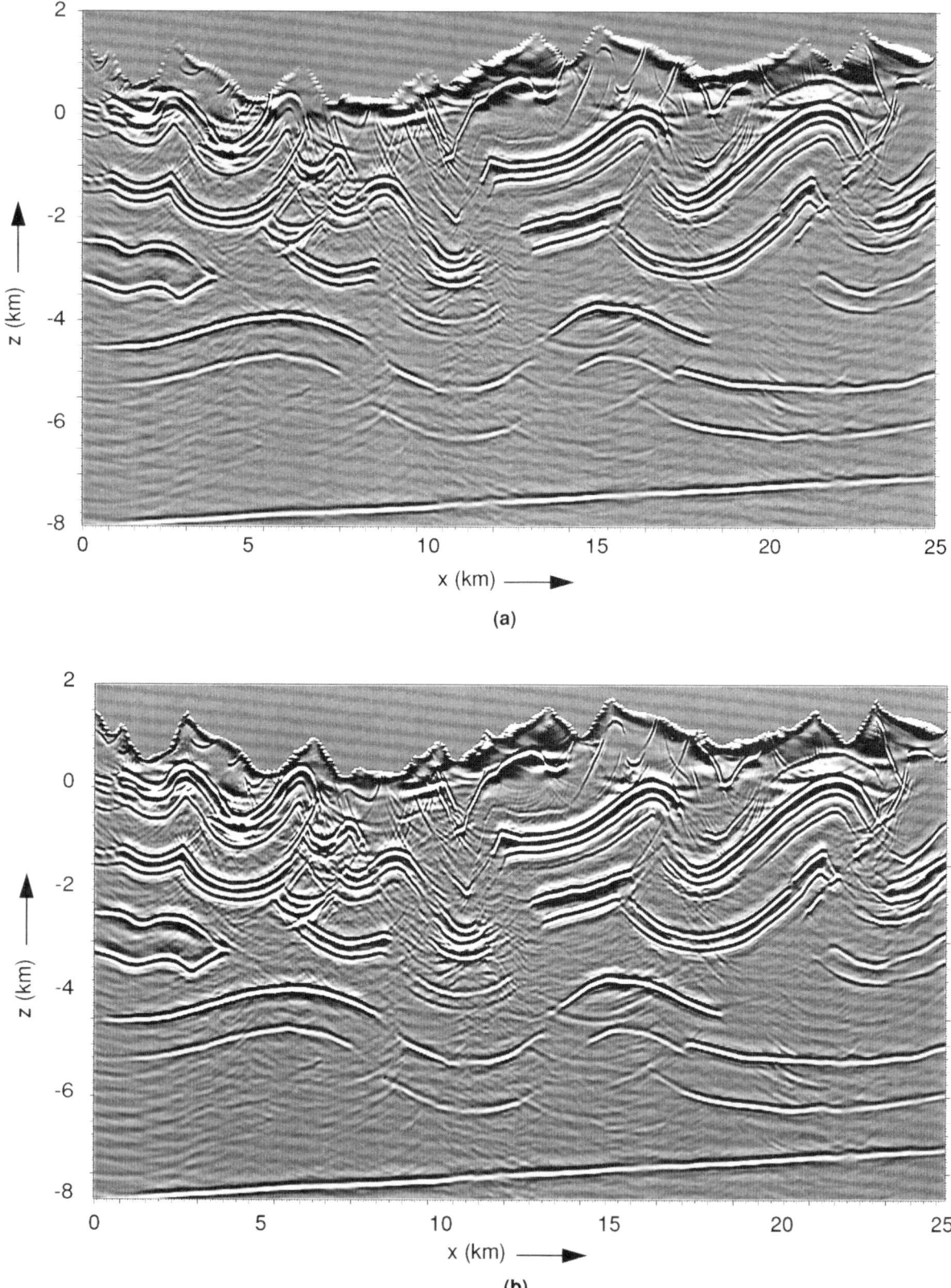

Fig. 4. Prestack Kirchhoff depth migration of 278 common-shot gathers, each with 480 traces. Migration weights are calculated from **(a)** a flat datum at $z = 2000$ m and **(b)** topography.

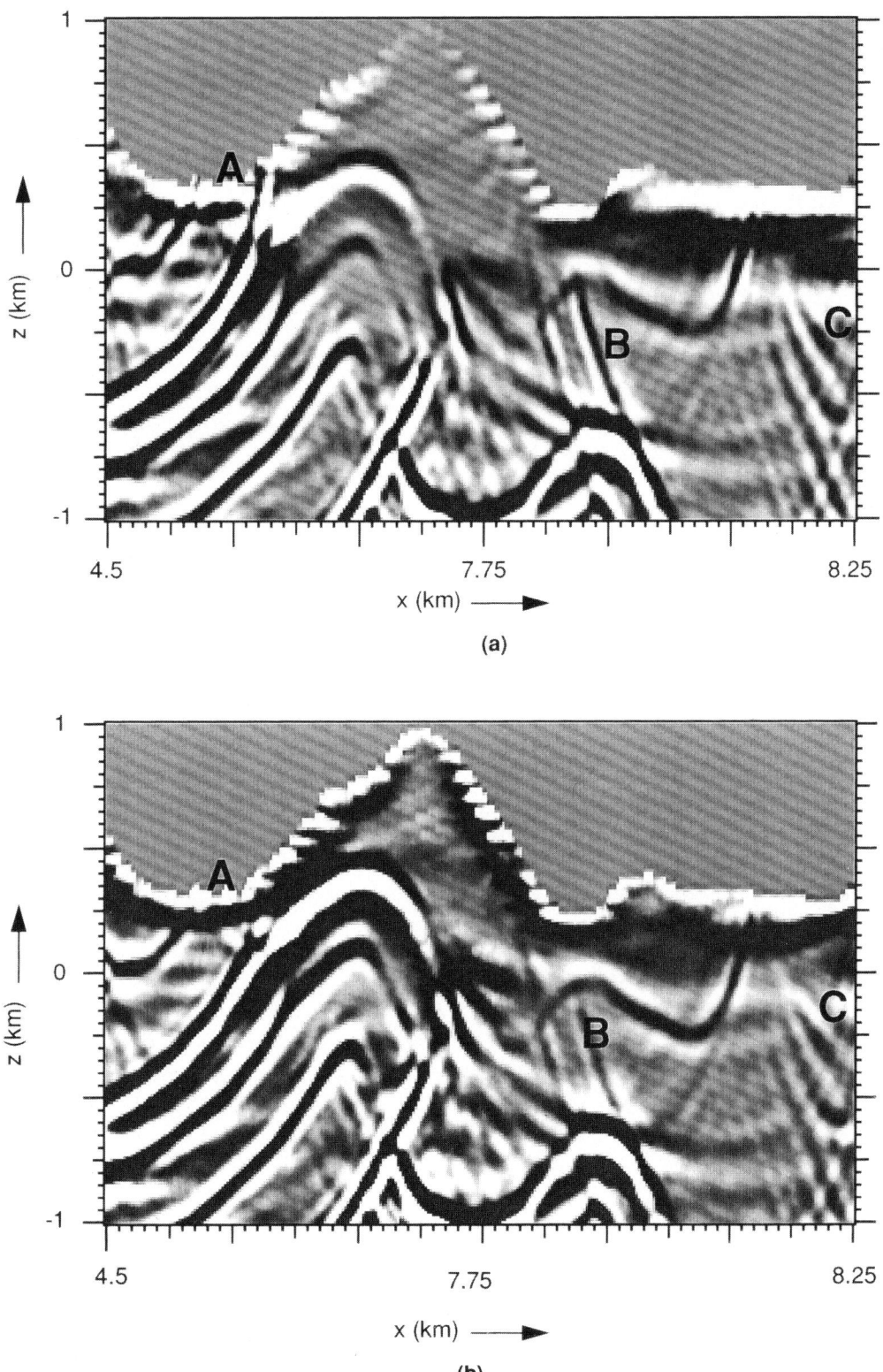

Fig. 5. Enlarged details of images shown in Figure 4. Surface "ties" corresponding to formation A are less ambiguous in Figure 5b. Smiles (event B) due to migration aperture limitations are diminished. Migration operator aliasing of steep events (C) due to the coarse shot spacing of 90 m remains but is attenuated.

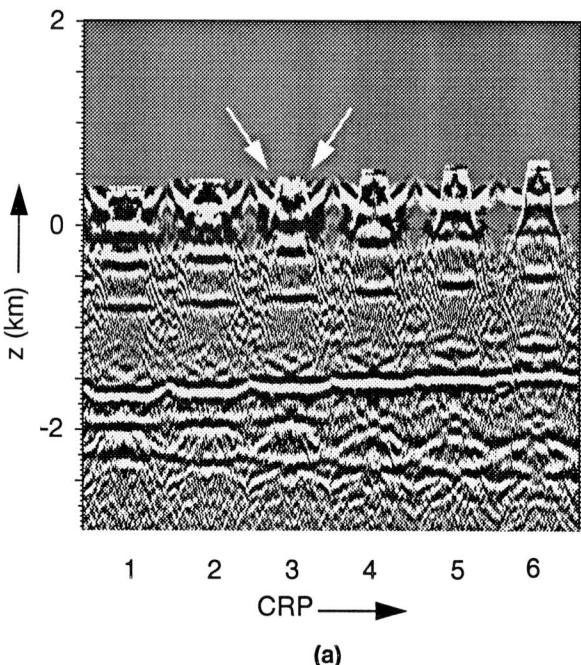
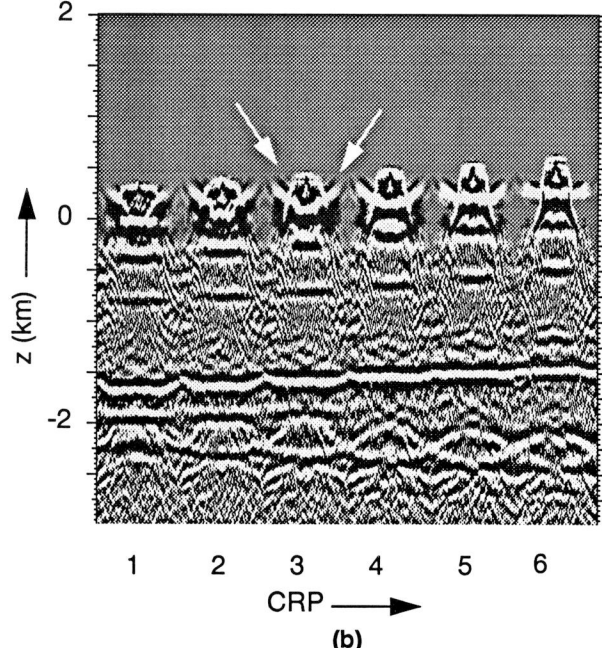

Fig. 6. Common-reflection point (CRP) gathers corresponding to the window in Figure 5, containing offsets ranging from −3600 m to 3600 m. The ambiguity of the event indicated by the arrows is due to significant diving wave or refraction strength at the far offsets.

of estimating near-surface migration velocities nearly impossible. To make matters worse, errors or uncertainties in estimating the near-surface interval velocity affect the reliability of deeper velocity estimation, so that using incorrect migration weights can easily lead to unreliable migration velocities for all depths.

We note the presence of nonflat events on the CRP gathers in Figure 6, even though we used the correct migration velocity. These result from the migration of events that have undergone some refractions before or after reflection. Such events are presently a source of annoyance to the interpreter.

Conclusions

When seismic data have been acquired along an irregular topographic surface, it is desirable to honour the topographic variations while processing the data. We have shown the effects on an important processing step (migration) of only partially honouring the topographic variations. Even if the data are migrated directly from topography, without using any processing steps taken to "move" the data from topography to a flat datum (either by wave equation datuming or by the simpler but less correct procedure of static-shifting traces upward to the datum), the quality of the imaging can be seriously compromised if the weights used in the migration have not also been computed from topography. If, in addition, the migration is performed before stack, with the two-fold objective of imaging and velocity estimation, the use of migration weights where depth is computed from the surface topography will result in a more complete accomplishment of both objectives than when depth is measured from a flat datum above the topography.

References

Bleistein, N., Cohen, J.K. and Hagin, F.G., 1987, Two and one-half dimensional Born inversion with an arbitrary reference: Geophysics **52**, 26-36.

Hanitzsch, C., Schleicher, J. and Hubral, P., 1994, True-amplitude migration of 2D synthetic data: Geophys. Prosp. **42**, 445-462.

Schneider, W.A., 1978, Integral formulation for migration in two and three dimensions: Geophysics **43**, 49-76.

Vermeer, G., 1990, Seismic wavefield sampling: Society of Exploration Geophysicists, Tulsa.

Wiggins, J.W., 1984, Kirchhoff integral extrapolation and migration of nonplanar data: Geophysics **49**, 1239-1258.

MIGRATION FROM TOPOGRAPHY: EXPERIENCE WITH AN ALBERTA FOOTHILLS DATA SET

Larry Lines[1], Wen-Jing Wu[2], Han-Xing Lu[3], Andrew Burton[1] and Jinming Zhu[1]

Abstract

In imaging steeply dipping layers of a Foothills data set, it is apparent that thrust belt geology can violate the conventional assumptions of elevation datum corrections and CMP stacking. In order to circumvent these problems, we use "migration from topography" as introduced by Wiggins (1984) and advocated by Gray and Marfurt (1995), in which we perform prestack migration on the data using correct source and receiver elevations. Migration from topography produces enhanced images of steep shallow reflectors when compared to conventional processing.

Introduction

Recent Canadian Foothills synthetic model studies by Gray and Marfurt (1995) have shown the effectiveness of migration from topography. We show results from a real data set that agree with these model results.

The data set is the Husky structural data set which was provided for the 1995 SEG Convention Workshop on Structural Imaging by Larry Mewhort of Husky Oil and by Christof Stork of Advance Geophysical. It is a data set intended for analyzing the current state of technology for imaging complex 2-D structures. The line is from the Benjamin Creek area of southern Alberta, and many of the preliminary results of the studies were compiled by Stork, Welsh and Skuce (1995). It is anticipated that this real data set will serve as a standard for many future processing tests, in the same way that the Marmousi data (Versteeg, 1994) has provided a means for testing seismic imaging algorithms on model data.

It is our intention to demostrate use of the "migration from topography" approach of Gray and Marfurt (1995) in processing the Husky data set. Thus far, we have processed this line using various depth migration algorithms including poststack reverse-time migration, Kirchhoff depth migration and prestack reverse-time migration. For the purposes of this discussion, we shall focus on the marked contrast between the quality of near-surface images for the poststack and prestack reverse-time migration algorithms. (Our prestack Kirchhoff depth migration results were similar to those obtained by prestack reverse-time migration.) As evidenced by the subsalt imaging results of Ratcliff et al. (1994), Kirchhoff prestack depth migrations also give improvements over poststack Kirchhoff migrations.

Methodologies

Poststack reverse-time migration has proven to be a useful and general migration method for imaging both 2-D and 3-D data as evidenced by the many papers including those of McMechan (1983), Whitmore (1983), Baysal et al. (1983) and Mufti et al. (1994). Reverse-time methods have also been successfully used for prestack depth migration as exhibited by the research of Chang and McMechan (1986, 1990), Whitmore and Lines (1986) and Wu et al. (1995). Reverse-time migration is a general method which is able to successfully define complex structures such as overhanging salt domes and overthrust folded layers. We shall show a comparison between the poststack and prestack forms of reverse-time migration for steeply dipping beds and variable topography.

Major problems with the conventional seismic processing of Canadian Foothills data occur due to the wide variation in surface elevations and the complex geology of reverse faulting in steeply dipping sedimentary layers. Thrust belt geology can cause problems with conventional methods that attempt to apply statics corrections to a datum plane. Figure 1a illustrates the conventional elevation statics "correction" from the actual recording surface to the processing datum as described by Dobrin (1976). Although the raypaths as shown in Figure 1a are steep but not vertical, this "correction"

Manuscript received by the Editor December 6, 1995; revised manuscript received January 10, 1996.
[1]Department of Earth Sciences, Memorial University of Newfoundland, St. John's, Newfoundland A1B 3X5
[2]Geo-X Systems Ltd., 900, 425 1st Street S.W., Calgary, Alberta T2P 3L8
[3]Lithoprobe Seismic Processing Facility, The University of Calgary, Calgary, Alberta T2N 1N4
We wish to thank the Natural Sciences and Engineering Research Council (NSERC), Petro-Canada and sponsors of the MUSIC project for their financial support of this research. We also acknowledge the support of Larry Mewhort of Husky Oil Ltd. and Christof Stork of Advance Geophysical for making this Alberta Foothills data set available for our research.

assumes vertically travelling energy (raypaths) between the datum plane and the surface. (The correction is only slightly more complicated for a source at depth.) If the surface layer velocity is considerably less than the velocity of deeper layers and reflections from deeper layers are travelling in a near-vertical direction, then the vertical ray statics assumption is reasonably valid. This assumption is a good one for much of the Western Canadian Basin which is covered by a low-velocity unconsolidated glacial drift layer. However, this assumption is often violated in thrust belt environments where steeply dipping layers of high seismic velocity outcrop at the surface. In foothills geology, seismic reflection energy travels at oblique angles through steeply dipping surface layers.

Secondly, thrust belt geology often causes a violation of the common-reflection point (CRP) assumptions. As shown by the sketch in Figure 1b, source-receiver combinations having a common midpoint (CMP) will generally not share a common-reflection point for a steeply dipping event. Therefore, the sorting and stacking of common-midpoint traces will cause a smearing of reflection energy. This smearing effect will be particularly deleterious for shallow steeply dipping events which often occur in foothills geology. Due to the breakdown of these assumptions, it is not surprising that shallow steeply dipping events are not clearly imaged by conventional processing.

Both the statics correction problem and the CMP problem can be obviated by the use of prestack migration with sources and receivers at their correct elevation – provided we have an accurate velocity model. This concept is illustrated for synthetic examples by McMechan and Chen (1990). Figure 1c shows the basic model for prestack migration where the source and receiver positions are correctly located. This model follows the discussion of Liner and Lines (1994) who describe prestack migration in terms of aplanatic surfaces. For a particular traveltime, the aplanatic surface defines the locus of possible reflection points for particular locations of source and receiver. For a constant velocity medium, the aplanatic surfaces would be ellipses. For variable velocity media, the aplanatic surface is a distortion of an ellipse which can be computed by solutions to the eikonal equation or by solutions to the wave equation. The use of eikonal solvers to define aplanatic surfaces is used in Kirchhoff migration (Gray and May, 1994; Zhu and Lines, 1995), whereas reverse-time prestack migration methods can use eikonal solvers and finite-difference wave equation solutions to effectively define aplanatic surfaces (Chang and McMechan, 1986, 1990; Botelho and Stoffa, 1988). Seismic trace energy is spread over these aplanatic surfaces for a given trace. We then repeat this procedure for a multitude of traces and then sum the trace migrations to obtain the migrated image in depth.

Prestack migration from topography does not suffer from either the shortcomings of datum statics corrections or the assumptions that CMPs are the same as CRPs. As we shall demonstrate, these shortcomings are particularly problematic for Foothills data. The key ingredient for the success of prestack migration is the availability of an accurate velocity model. As pointed out by Lines et al. (1993), there are various methods for velocity analysis. Some of the most general (and useful) techniques for complicated geological structures involve the iterative use of prestack depth migration.

One prestack migration technique uses the criterion that the depth image at a CRP should be independent of offset. That is, the correct depth estimate of the reflector should not depend on the geometry of the seismic experiment. In order to test the validity of the migration result and the velocity model, we apply prestack depth migration to the data and for a CRP we examine the offset variation of the depth migrations. A variation in the depth estimates versus offset leads to velocity adjustment of the model in order to eliminate "smiles" or "frowns" as a function of offset for migrated CRP depth gathers. This approach is very similar to normal-moveout analysis of CMP data. As explained by Whitmore and Garing (1993), the same kind of analysis can be performed using plane-wave angles instead of source-receiver offset.

A second velocity analysis technique attempts to improve the focusing of the prestack migration by velocity adjustment. This interpretive method is similar to the focusing of a camera in that the correct velocity model is assumed to be the one which provides the most focused image. This form of migration velocity analysis is analogous to velocity analysis by CMP stacking.

Traveltime tomography represents yet another general method for velocity model estimation. This method can utilize prestack traveltimes from direct arrivals, refracted arrivals or reflected arrivals. Tomography sometimes experiences difficulties with complex geology since one has to be careful that the picked traveltimes are the traveltimes which one is modelling by ray tracing.

For the Husky data set, there is also additional information in the form of well logs and VSP data which can provide constraints on the velocity model. One of these constraint methods developed by Lines (1993) attempts to match depth migrations to formation tops. Although this was helpful for this study, it had the shortcoming that many layers in the model do not intersect the logged portions of the wells. Therefore, this migration/inversion method will not provide the entire velocity model picture in this case. Iterative prestack depth migration was the main tool in our development of a velocity model. Geological interpretation of the complex structural models by Jamie Jamison also provided a key part of the velocity model definition.

RESULTS

The differences between conventional processing from a flat datum (including poststack depth migration) as opposed to prestack depth migration from topography are most pronounced on the shallow steeply dipping reflections. This can be seen in a comparison between the poststack migration with datum corrections in Figure 2 and a prestack migration

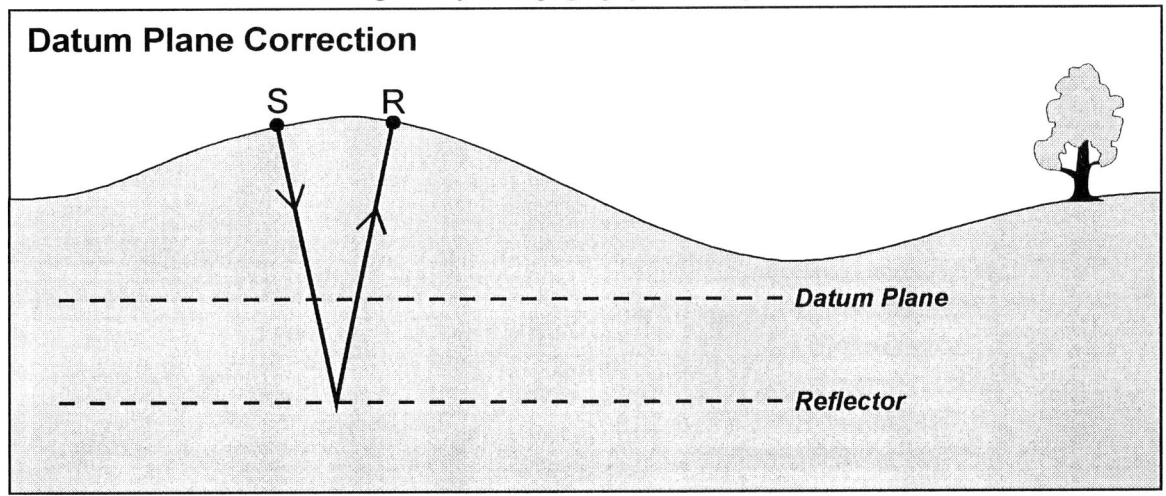
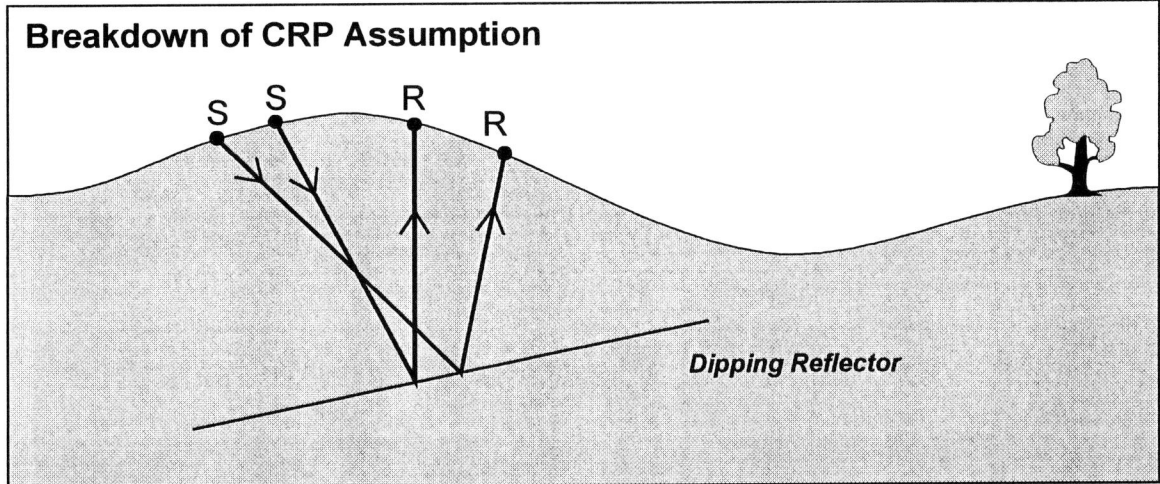
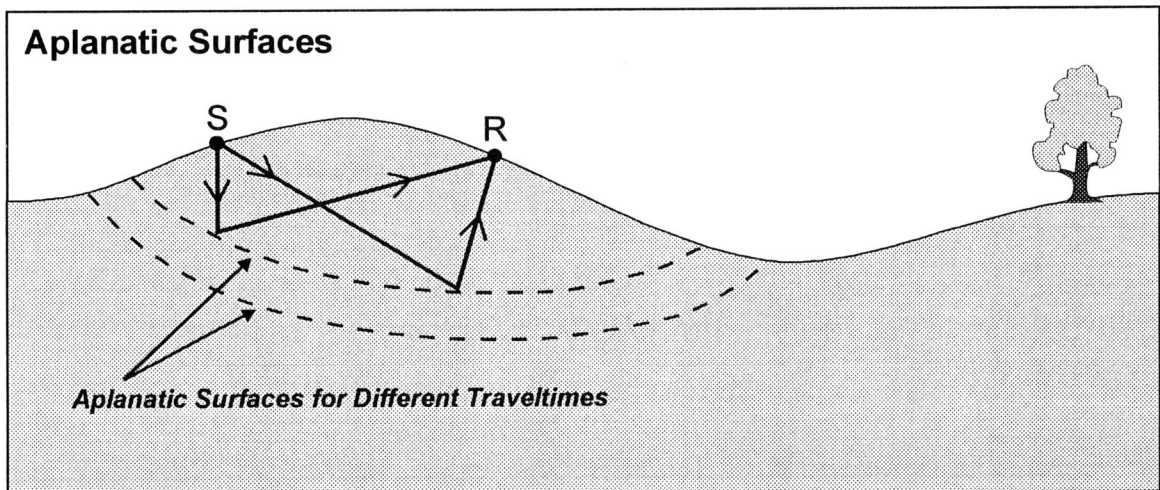

Fig. 1. (a) The elevation statics "correction" is an approximation which assumes vertical raypaths between the datum plane and the surface, although, as shown here, the raypaths are often not quite vertical (Dobrin, 1976). **(b)** Raypaths for common midpoint (CMP) do not coincide with a common-reflection point (CRP) for dipping reflectors. **(c)** Aplanatic surfaces define isochrons of possible reflection points for arbitrary source-receiver elevations.

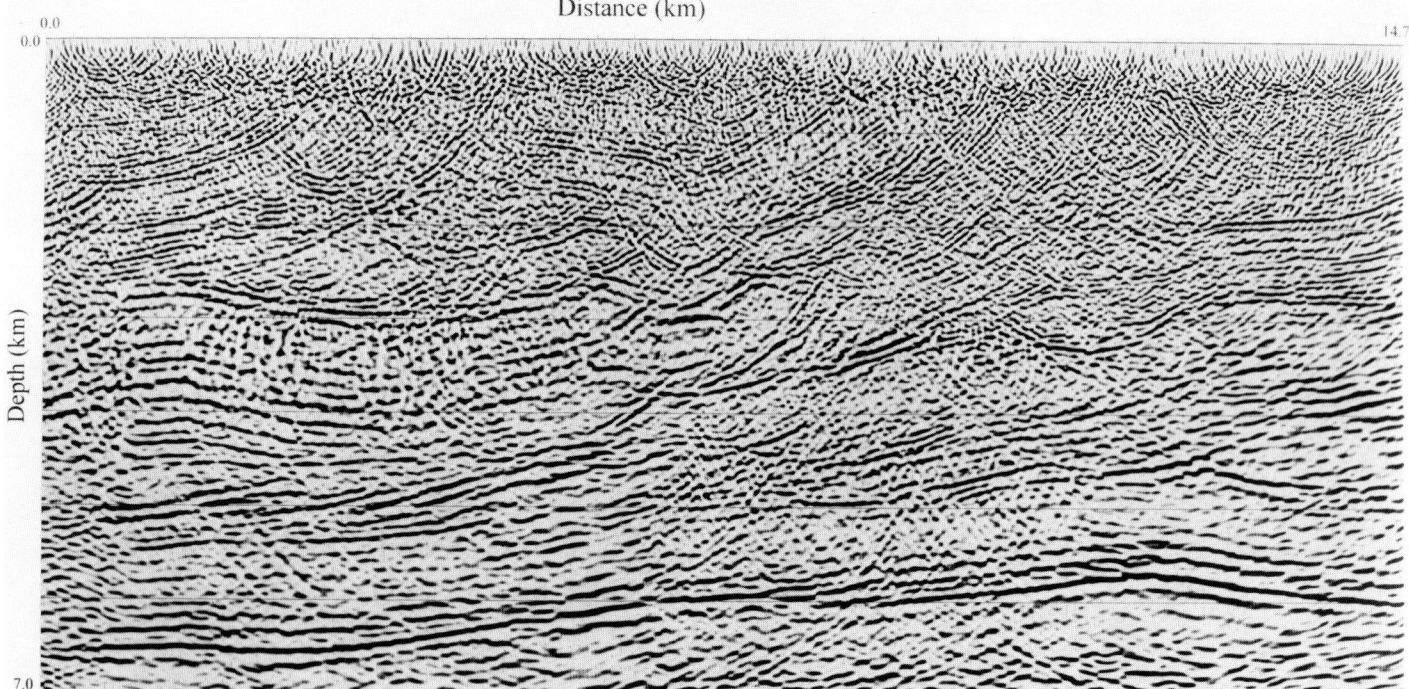

Fig. 2. Poststack migration of data with datum corrections. Section width = 14.7 km. Section depth = 7 km.

of the data in Figure 3 which uses the same velocity model. We see this difference very clearly in the upper left portion of the sections (upper 3 km and westmost 4.5 km of the section) where there is a syncline containing dipping beds of the Brazeau, Wapiabi, Cardium, Blackstone and Blairmore formations. An enlargement of this zone in Figures 4a and 4b shows a distinct difference between the poststack and prestack migrations. The poststack migration with elevation statics correction to a datum lacks near-surface coherent energy while the prestack migration from topography shows a clearer definition of near-surface reflections. In the middle of the section in Figure 4a, the poststack migration also shows some reflection energy with conflicting dips; these conflicting dips do not appear in Figure 4b. Trace density in Figure 4b is less than in Figure 4a due to trace interpolation used in the reverse-time algorithm.

In comparing the poststack reverse-time migration of Figure 2 to the prestack reverse-time migration of Figure 3,

Fig. 3. Prestack migration from topography. Section width = 14.7 km. Section depth = 7 km.

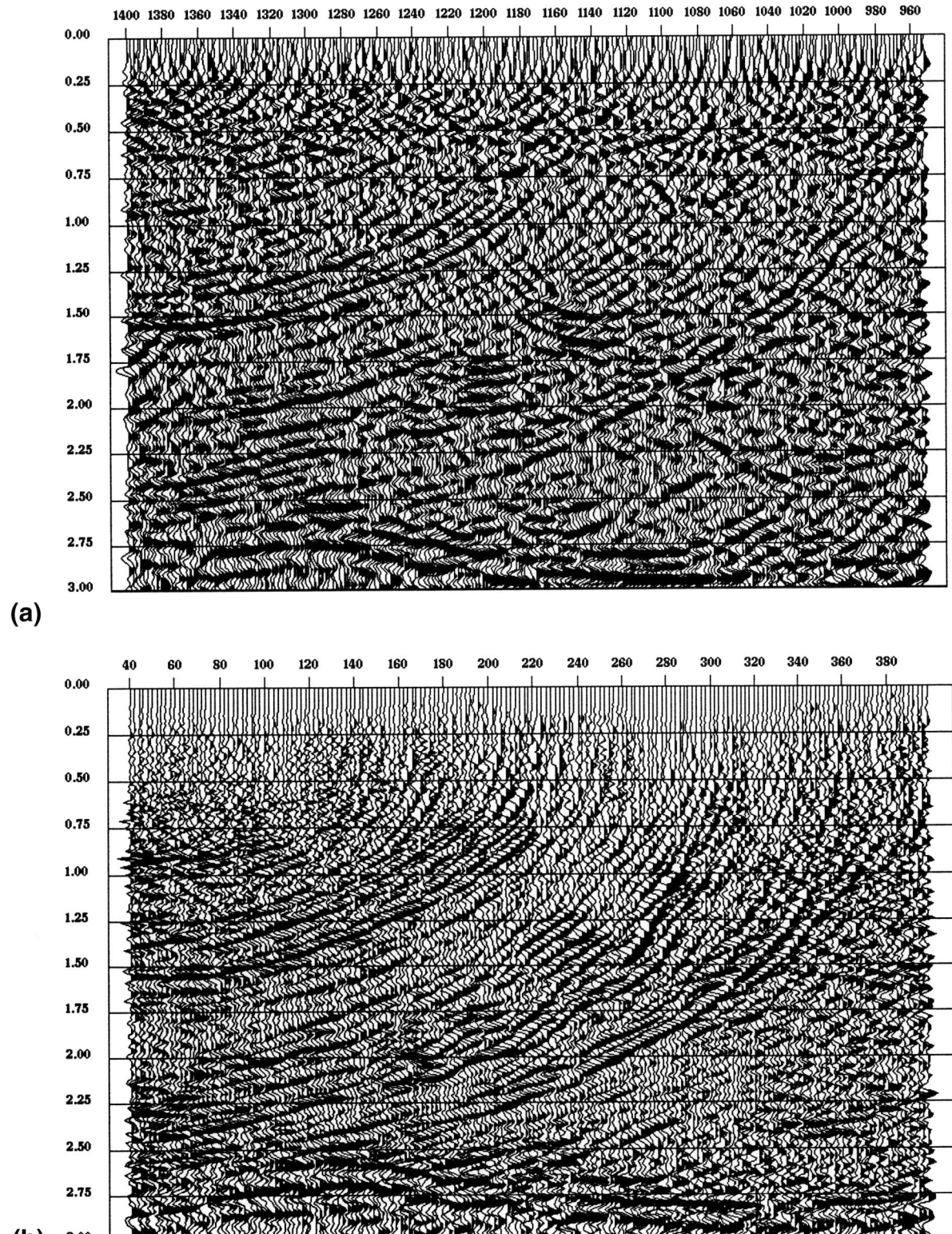

Fig. 4. (a) Enlarged portion of poststack reverse-time migration. **(b)** Enlarged portion of prestack reverse-time migration. Section width = 4.5 km. Section depth = 3 km.

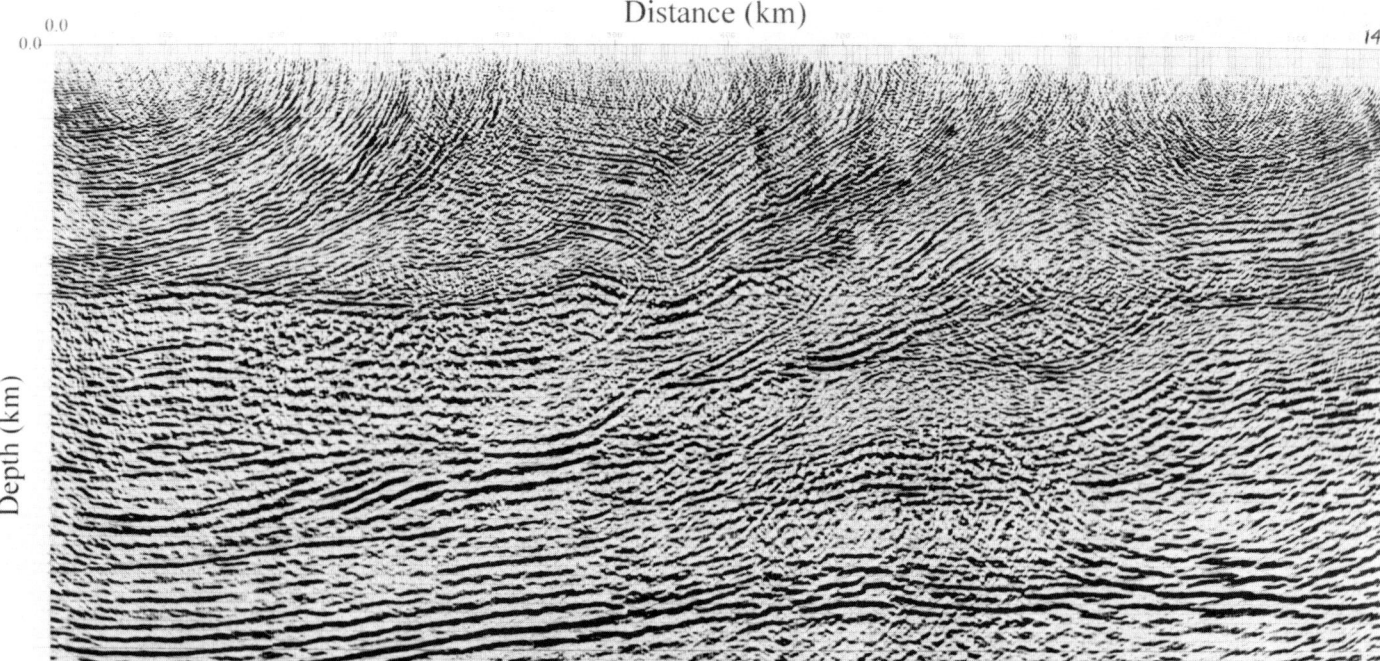

Fig. 5. Prestack migration from topography using improved velocity model. Section width = 14.7 km. Section depth = 7 km.

we note that the depth migration differences between the two sections becomes less obvious for deeper reflections. This is an expected result since as offset/reflection depths decrease, raypath differences between prestack and poststack models become less pronounced. In almost all cases, the prestack migration result is the same or better than the poststack migration. There are still improvements which can be made by adjustment of the velocity to produce the prestack depth migration result of Figure 5. (Further improvements could possibly be made by further adjustment of velocities on the eastern segment of the line.) According to Skuce (1995) velocity models should also take anisotropy into account, especially in shale layers. This appears to be a very difficult inversion problem for complicated structures unless rock property anisotropy measurements are available as constraints on the problem.

CONCLUSIONS

Useful depth migrations of Foothills data depend on accurate velocity models, general migration and velocity analysis algorithms, and detailed interpretation. The results of the Husky data set experiments for the 1995 SEG workshop will undoubtedly see future publications. Thus far, one of the largest improvements in processing that we have seen is due to the use of prestack migration from topography. In this sense, our results for the Husky data set basically agree with the synthetic model results of Gray and Marfurt (1995). Our conclusions are that for shallow, steeply dipping Foothills reflections, prestack migration from topography will provide a substantial improvement over the conventional processing flow of elevation datuming, CMP stacking and poststack migration.

REFERENCES

Baysal, E., Kosloff, D.D. and Sherwood, J.W.C., 1983, Reverse time migration: Geophysics **48**, 1514-1524.

Botelho, M.A.B. and Stoffa, P.L., 1988, Velocity analysis using reverse-time migration: EOS **49**, 1326.

Chang, W.F. and McMechan, G.A., 1986, Reverse-time migration of offset vertical seismic profiling data using the excitation-time imaging condition: Geophysics **51**, 67- 84.

_____ and _____, 1990, 3D acoustic prestack reverse-time migration: Geophys. Prosp. **38**, 737-755.

Dobrin, M., 1976, Introduction to geophysical prospecting: McGraw-Hill Book Co.

Gray, S.H. and Marfurt, K.J., 1995, Migration from topography: improving the near-surface image: Can. J. Expl. Geophys. **31**, 18-24.

_____ and May, W.P., 1994, Kirchhoff migration using eikonal equation traveltimes: Geophysics **59**, 810-187.

Liner, C.L. and Lines, L.R., 1994, Simple prestack migration of crosswell seismic data: J. Seis. Expl. **3**, 101-112.

Lines, L.R., 1993, Optimization of seismic migration through use of well information: Can. J. Expl. Geophys. **29**, 419-428.

_____, Rahimian, F. and Kelly, K.R., 1993, A model-based comparison of modern velocity analysis methods: The Leading Edge **12**, 750-754.

McMechan, G.A., 1983, Migration by extrapolation of time-dependent boundary values: Geophys. Prosp. **31**, 413-430.

_____ and Chen, H.W., 1990, Implicit static corrections in prestack migration of common-source data: Geophysics **55**, 757-760.

Mufti, I.R., Pita, J.A. and Huntley, R.W., 1994, Finite-difference depth migration of exploration-scale 3-D seismic data: 64th Ann. Internat. Mtg., Soc. Expl. Geophys.

Ratcliff, D.W., Jacewitz, C.A. and Gray, S.H., 1994, Subsalt imaging via target-oriented 3-D prestack depth migration: The Leading Edge **13**, 163-172.

Skuce, A., 1995, Seismic imaging in the Canadian Rocky Mountain Foothills: Ann. Conv., Workshop #6, Soc. Expl. Geophys., Paper.

Stork, C., Welsh, C. and Skuce, A., 1995, Demonstration of processing and model building methods on a real complex structure data set: Ann. Conv., Workshop #6, Soc. Expl. Geophys., Proceedings.

Versteeg, R., 1994, The Marmousi experience: velocity model determination on a synthetic complex data set: The Leading Edge **13**, 927-936.

Wiggins, W., 1984, Extrapolation and migration of nonplanar data: Geophysics **49**, 1239-1248.

Whitmore, N.D., 1983, Iterative depth migration by backward time propagation: 1983 Mtg., Soc. Expl. Geophys., Exp. Abstr., 827-830.

_____ and Garing, J.D., 1993, Interval velocity estimation using iterative prestack depth migration in the constant angle domain: The Leading Edge **12**, 757-762.

_____ and Lines, L.R., 1986, Vertical seismic profiling depth migration of a salt dome flank: Geophysics **51**, 1087-1109.

Wu, W., Lines, L.R. and Lu, H., 1995, Reverse-time migration with finite-difference solutions of the acoustic wave equation: Ann. Mtg., Soc. Expl. Geophys., Exp. Abstr., 1149-1152, and accepted for publication in Geophysics.

Zhu, J. and Lines, L.R., 1995, Practical subsurface imaging by prestack depth migration and velocity anaylysis: Ann. Mtg., Can. Soc. Expl. Geophys., Exp. Abstr., 5-6.

Wavefield extrapolation by nonstationary phase shift

Gary F. Margrave* and Robert J. Ferguson*

ABSTRACT

The phase-shift method of wavefield extrapolation applies a phase shift in the Fourier domain to deduce a scalar wavefield at one depth level given its value at another. The phase-shift operator varies with frequency and wavenumber, and assumes constant velocity across the extrapolation step. We use nonstationary filter theory to generalize this method to nonstationary phase shift (NSPS), which allows the phase shift to vary laterally depending upon the local propagation velocity. For comparison, we derive an analytic form for the popular phase shift plus interpolation (PSPI) method in the limit of an exhaustive set of reference velocities. NSPS and this limiting form of PSPI can be written as generalized Fourier integrals which reduce to ordinary phase shift in the constant velocity limit. In the (x, ω) domain, these processes are the transpose of each other; however, only NSPS has the physical interpretation of forming the scaled, linear superposition of laterally-variable impulse responses (i.e., Huygen's wavelets).

The difference between NSPS and PSPI is clear when they are compared in the case of a piecewise constant velocity variation. Define a set of windows such that the jth window is unity when the propagation velocity is the jth distinct velocity and is zero otherwise. NSPS can be computed by applying the window set to the input data to create a set of windowed wavefields, which are individually phase-shift extrapolated with the corresponding constant velocity, and the extrapolated set is superimposed. PSPI proceeds by phase-shift extrapolating the input data for each distinct velocity, applying the jth window to the jth extrapolation, and superimposing. Though neither process is fully correct, PSPI has the unphysical limit that discontinuities in the lateral velocity variation cause discontinuities in the wavefield, whereas NSPS shows the expected wavefront "healing."

We then formulate a finite aperture compensation for NSPS which has the practical result of absorbing lateral boundaries for all incidence angles. Wavefield extrapolation can be regarded as the crosscorrelation of the wavefield with the expected response of a point diffractor at the new depth level. Aperture compensation simply applies a laterally varying window to the infinite, theoretical diffraction response. The crosscorrelation becomes spatially variant, even for constant velocity, and hence is a nonstationary filter. The nonstationary effects of aperture compensation can be simultaneously applied with the NSPS extrapolation through a laterally variable velocity field.

INTRODUCTION

In a general context, wavefield extrapolation refers to the mathematical technique of advancing a wavefield through space or time. Such techniques can be used in both seismic migration and seismic modeling. In this paper, we will restrict the scope of wavefield extrapolation to the problem of deducing a scalar wavefield at one depth level in the earth given knowledge of its properties at another level. We also assume that the wave propagation velocity, v, depends only on the lateral spatial coordinates, (x, y), and not on the depth, z. Consequently, our technique is intended for use in a recursive scheme in which vertical velocity variations are handled in the usual manner through an appropriate choice of depth levels, and only lateral velocity variations are directly addressed by our theory.

Wavefield extrapolation by phase shift (Gazdag, 1978) has many desirable properties and one overriding difficulty. On the positive side, the phase-shift operator is theoretically exact for constant velocity, unconditionally stable, shows no grid dispersion, and is accurate for all scattering angles. (We prefer the term "scattering angle" to the more commonly used "dip" because the latter is often confused with the geologic dip of reflectors.) The major difficulty is that it is not immediately apparent how lateral velocity variations can be incorporated into

---

Manuscript received by the Editor October 16, 1997; revised manuscript received December 18, 1998.
*University of Calgary, Consortium for Research in Elastic Wave Exploration Seismology (CREWES) Project, 2500 University Drive N.W., Calgary, Alberta T2N 1N4, Canada. E-mail: gary@geo.ucalgary.ca; rferguso@geo.ucalgary.ca.
© 1999 Society of Exploration Geophysicists. All rights reserved.

a phase-shift method because the space coordinate has been Fourier transformed. As a result, extrapolation techniques for $v(x)$ [we use "$v(x)$" as synonymous with the phrase "a laterally variable velocity field"] are usually formulated in the space-frequency domain (e.g., Gazdag, 1980; Berkhout, 1984; Holberg, 1988; Hale, 1991) as a (scattering) angle-limited approximation to the inverse Fourier transform of the phase-shift operator. The velocity dependence of such a local space-domain extrapolator is then varied with the local velocity of the computation grid. However, since the multidimensional Fourier transform is a complete description of a wavefield, it follows that it must be possible to extrapolate a wavefield through lateral velocity variations with a Fourier domain technique. We present such a technique here and illustrate its relation to established methods. Black et al. (1984) and Wapenaar (1992) have presented similar Fourier methods (see also Wapenaar and Dessing, 1995, and Grimbergen et al., 1995).

The split-step Fourier method (Stoffa et al., 1990) and the similar phase-screen methods (Wu, 1994; Huang and Wu, 1996; Huang and Fehler, 1997) are related but distinct from the methods described here. These phase-screen methods split the extrapolation into two parts: an angle-independent thin-lens delay (accomplished in the space-frequency domain) and an angle-dependent scattering (accomplished in the full-Fourier domain). The thin-lens delay is computed with the actual lateral velocity variations, whereas the Fourier domain scattering is done using a constant background velocity. The method presented here does not separate scattering from thin-lens delay nor does it require a constant velocity background. Instead, wavefield extrapolation by nonstationary phase shift (NSPS) can be done entirely in the Fourier domain with the actual lateral velocity variation.

We present our work in the context of nonstationary filter theory (Margrave, 1998), and show its analytic link to the popular phase shift plus interpolation (PSPI) method of Gazdag and Squazzero (1984). NSPS is presented as an explicit closed-form expression for one-way wavefield extrapolation through $v(x)$ and has the physical interpretation of a laterally varying, or nonstationary, phase shift. Next, we give a detailed comparison between NSPS and PSPI for the case of a step velocity model. As a further demonstration of the utility of our approach, we conclude with a modification of NSPS which has perfectly absorbing (that is, reflections are suppressed at all dips) lateral boundaries. This is achieved through the compensation of the NSPS operator for finite recording aperture. Finally, we discuss practical extensions of this technique to 3-D wavefield extrapolation.

THEORETICAL DEVELOPMENT

We begin with a summary of PSPI and show how to formulate the most accurate, limiting form of PSPI as a generalized Fourier integral. Then, using results from the theory of nonstationary linear filters, we show that the PSPI limiting form is a type of nonstationary filter called a combination filter. Such filters are linear and have definable properties; however, they do not form the linear superposition of impulse responses which Huygen's principle suggests is desirable in wave propagation. This motivates the use of a nonstationary convolution filter that does form the desired linear superposition and is the basis for our NSPS algorithm. We give expressions for NSPS and PSPI in the dual (space-wavenumber) domain and in the full-Fourier domain.

The PSPI method

PSPI (Gazdag and Squazzero, 1984) is a rational attempt to build an approximate extrapolation through $v(x)$ from a set of constant velocity phase-shift extrapolations using a suitable set of reference velocities, $\{v_j\}$. For simplicity, we present the theory in two dimensions as the extrapolation of a wavefield from $z = 0$ to $z = \Delta z$. (A summary of our mathematical notation appears in Appendix A.) After an initial Fourier transform over time, we denote the wavefield at $z = 0$ as $\Psi(x, 0, \omega)$, where ω is temporal frequency, and the desired extrapolated wavefield at $z = \Delta z$ as $\Psi_{v(x)}(x, \Delta z, \omega)$, where the subscript provides information about the velocity field. Phase-shift extrapolation with each v_j produces a reference wavefield, $\Psi_{vj}(x, \Delta z, \omega)$, given by

$$\Psi_{vj}(x, \Delta z, \omega) = \int_{-\infty}^{\infty} \varphi(k_x, 0, \omega) \alpha_{vj}(k_x, \omega) e^{ik_x x} \, dk_x, \quad (1)$$

where

$$\varphi(k_x, 0, \omega) = \frac{1}{2\pi} \int_{-\infty}^{\infty} \Psi(x, 0, \omega) e^{-ik_x x} \, dx \quad (2)$$

is the forward spatial Fourier transform of the input data, the phase-shift operator, α_{vj}, is given by

$$\alpha_{vj}(k_x, \omega) = \begin{cases} e^{i\Delta z k_{zj}}, & |k_x| \leq \dfrac{\omega}{v_j} \\ e^{-|\Delta z k_{zj}|}, & |k_x| > \dfrac{\omega}{v_j} \end{cases}, \quad k_{zj} = \sqrt{\dfrac{\omega^2}{v_j^2} - k_x^2}, \quad (3)$$

and k_x and k_z are horizontal and vertical wavenumbers, respectively. This definition of α_{vj} ensures that evanescent energy suffers exponential decay. Note that each reference wavefield, Ψ_{vj} is a complete phase-shift extrapolation defined at all x and ω although we do not expect it to contribute to $\Psi_{v(x)}$ where $v(x)$ differs significantly from v_j. It is a fundamental assumption of PSPI that the desired extrapolation is equivalent to a reference wavefield wherever the actual velocity equals the reference velocity. That is,

$$\Psi_{v(x)}(x_j, \Delta z, \omega) = \Psi_{vj}(x_j, \Delta z, \omega), \quad \text{if } v(x_j) = v_j. \quad (4)$$

PSPI proceeds by choosing a small set of reference velocities that bracket the extremes of $v(x)$ and sample its fluctuations. Once the set $\{\Psi_{vj}\}$ is determined, an approximation to $\Psi_{v(x)}$ is formed by some sort of linear (in velocity) interpolation (LI),

$$\Psi_{v(x)}(x, \Delta z, \omega) \approx \text{LI}(\Psi_{vj}(x, \Delta z, \omega), \Psi_{vj+1}(x, \Delta z, \omega)),$$

$$v_j \leq v(x) \leq v_{j+1}. \quad (5)$$

The choice of the reference velocities and the details of the interpolation process symbolized by equation (5) are major technical design questions because they control the accuracy of the final result. However, we are not concerned with them here because we wish to proceed to the most accurate limiting case of PSPI, when a reference wavefield is computed for every distinct velocity. In this case, the PSPI algorithm converges to

$$\Psi_{v(x)}(x, \Delta z, \omega) \approx \Psi_{\text{PSPI}}(x, \Delta z, \omega)$$
$$= \int_{-\infty}^{\infty} \varphi(k_x, 0, \omega) \alpha_{v(x)}(k_x, x, \omega) e^{ik_x x} \, dk_x, \quad (6)$$

where

$$\alpha_{v(x)}(k_x, x, \omega) =$$
$$\begin{cases} e^{i\Delta z k_z(x)}, & |k_x| \leq \dfrac{\omega}{v(x)} \\ e^{-|\Delta z k_z(x)|}, & |k_x| > \dfrac{\omega}{v(x)} \end{cases}, \quad k_z(x) = \sqrt{\dfrac{\omega^2}{v(x)^2} - k_x^2}.$$
(7)

Note that we reserve the symbol Ψ_{PSPI} to refer specifically to the most accurate limiting form of PSPI as expressed by equation (6). For the remainder of this paper, when we use the term PSPI, we will be referring to a computation done with equation (6) (or a mathematical equivalent). Equation (6) is essentially similar to equation (1) except that the constant velocity, v, in the latter has become $v(x)$ in the former. This means that equation (6) is no longer an inverse Fourier transform but is a more general Fourier integral. It can be interpreted as a prescription which applies the nonstationary filter of equation (7) simultaneously with the transformation from k_x to x. In order to appreciate the validity of this result, it is useful to explicitly verify that equation (4) is satisfied:

$$\Psi_{\text{PSPI}}(x_j, \Delta z, \omega) = \Psi_{v_j}(x_j, \Delta z, \omega), x_j \Rightarrow v(x_j) = v_j.$$
(8)

Thus the limiting PSPI wavefield, as given by equation (6), is equivalent to producing a complete set of reference velocities and extrapolated wavefields, then slicing through the wavefields such that each is used only where its velocity equals $v(x)$. This is illustrated in Figure 1. An alternative to this slicing process, is the direct numerical integration of equation (6). In this limiting case, the problems of reference velocity selection and choice of interpolation algorithm vanish.

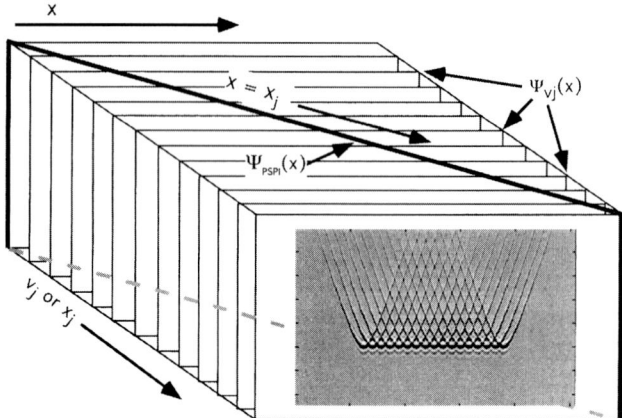

FIG. 1. The limiting form of PSPI produces a continuous set of extrapolated wavefields $\Psi_{v_j}(x)$, one for each v_j. The final extrapolated wavefield $\Psi_{\text{PSPI}}(x)$ is the set of traces found along a slice at $x = x_j$ through the data volume.

The NSPS method

The theory of nonstationary linear filters, as presented in Margrave (1998), shows that at least two distinct forms of nonstationary filters are possible. Termed combination and convolution filters, both filter forms are equivalent in the stationary limit—in this context, stationary means constant velocity—but otherwise they can differ dramatically. The theory gives explicit prescriptions for filter application in the space, Fourier, or dual domains as well as formulas to move the filter prescription between domains. (A dual-domain filter expression is one which changes the data domain from space to Fourier, or the reverse, in the process of applying the filter.)

As described above, Ψ_{PSPI} is computed by an ordinary forward Fourier transform, equation (2), and then the generalized inverse Fourier integral, equation (6), and is an example of a nonstationary, dual-domain, combination filter. The nonstationarity of the filter is evidenced by the fact that the filter description, $\alpha_{v(x)}(k_x, x, \omega)$, is dependent upon both wavenumber and spatial location. (The spatial dependence vanishes for a stationary filter.)

The distinction between combination and convolution filter forms is most apparent in the dual-domain form. Given equation (6) and following the nonstationary filter theory, it is now a simple matter to write the equations describing the related nonstationary convolution filter. The first step applies the nonstationary wavefield extrapolator, $\alpha_{v(x)}$, given by equation (7), simultaneously with the forward Fourier transform

$$\varphi_{\text{NSPS}}(k_x, \Delta z, \omega)$$
$$= \frac{1}{2\pi} \int_{-\infty}^{\infty} \Psi(x, 0, \omega) \alpha_{v(x)}(k_x, x, \omega) e^{-ik_x x} \, dx. \quad (9)$$

The final step is an ordinary inverse Fourier transform:

$$\Psi_{\text{NSPS}}(x, \Delta z, \omega) = \int_{-\infty}^{\infty} \varphi_{\text{NSPS}}(k_x, \Delta z, \omega) e^{ik_x x} \, dk_x. \quad (10)$$

Equations (9) and (10) form the basis of our method of wavefield extrapolation by nonstationary phase shift. In comparison with the limiting form of PSPI, both methods apply the same nonstationary filter, $\alpha_{v(x)}$ as given by equation (7), but NSPS applies it simultaneously with the forward Fourier transform from x to k_x, whereas in PSPI it is applied simultaneously with the inverse Fourier transform from k_x to x. In the stationary limit, when $\alpha_{v(x)}$ becomes independent of x, it is a simple matter to verify that both expressions reduce to the constant velocity phase-shift extrapolation.

Although we have based our presentation on the concepts of nonstationary filter theory, we note that these are closely linked to the theory of pseudodifferential operators (Saint Raymond, 1991) and the related Fourier integral operators (Duistermaat, 1996). In fact, equation (6) is a Fourier integral operator in standard form, and equation (9) can be shown to be the adjoint form.

Fourier domain formulation

At this point, both the PSPI limiting process and NSPS have been presented as dual-domain algorithms which have the characteristic that the nonstationary extrapolation filter is applied simultaneously with a data transformation from wavenumber to space or the reverse. Nonstationary filter theory provides the mathematical formulas to move either process fully into the Fourier domain [where the input and output wavefields are in (k_x, ω)] or into the space domain [where the wavefields are in (x, ω)]; however, we present only the Fourier domain expressions here.

The space-domain (x, ω) expressions for NSPS and PSPI can be derived through mathematics similar to that presented here. For discrete data, it can be shown that NSPS and PSPI are accomplished with extrapolation matrices that are the transpose of one another. Curiously, this transpose symmetry does not hold in the full-Fourier domain. This is because $\alpha_{v(x)}(k_x, x, \omega)$, as given by equation (7), is symmetric in k_x [i.e., $\alpha(k_x) = \alpha(-k_x)$] but not in x.

PSPI can be moved into the Fourier domain by performing the forward Fourier transform of equation (6) (Appendix B). This results in

$$\varphi_{\text{PSPI}}(k_x, \Delta z, \omega) = \int_{-\infty}^{\infty} \varphi(k'_x, 0, \omega) A(k'_x, k_x - k'_x, \omega) \, dk'_x, \quad (11)$$

where

$$A(p, q, \omega) = \frac{1}{2\pi} \int_{-\infty}^{\infty} \alpha_{v(x)}(p, u, \omega) e^{-iqu} \, du. \quad (12)$$

In equation (12), p and q are wavenumber variables and u is a space coordinate. The wavenumber connection function, A, is seen to be the ordinary forward Fourier transform over the spatial coordinate of $\alpha_{v(x)}$.

The Fourier expression for NSPS (Appendix C) is derived from equation (9) by substituting for Ψ its expression as an inverse Fourier transform of its spectrum, φ. The result is

$$\varphi_{\text{NSPS}}(k_x, \Delta z, \omega) = \int_{-\infty}^{\infty} \varphi(k'_x, 0, \omega) A(k_x, k_x - k'_x, \omega) \, dk'_x, \quad (13)$$

where A is given by equation (12).

Equations (11) and (13) are very similar, differing only in how the p dependence of $A(p, q)$ is mapped into (k_x, k'_x) space. For discretely sampled data, both of these extrapolation equations can be represented as matrix operations in which an extrapolation matrix populated from $A(p, q)$ is multiplied into a column vector containing samples of φ. In the stationary limit [i.e., $v(x) =$ constant], both of the extrapolation matrices become diagonal with the phase-shift extrapolator appearing on the diagonal. As $v(x)$ is allowed to vary, off-diagonal terms appear in the matrices and, when multiplied into the data vector, cause a "mixing" of the wavenumbers of φ to produce each wavenumber of $\varphi_{v(x)}$. An alternative perspective is that $\alpha_{v(x)}$ represents a phase-shift model based on the velocity model $v(x)$. These formulas [equations (11) and (13)] prescribe how the wavenumbers of the phase-shift model, and hence indirectly the velocity model, mix with the wavenumbers of the data during wavefield extrapolation.

The strong similarity of equations (11) and (13) suggests that the computational effort for PSPI and NSPS is nearly identical [recall that we speak of the generalized PSPI here as defined by equation (6)]. Both methods require a forward Fourier transform of the input wavefield and an inverse Fourier transform of the extrapolated field. The wavenumber connection function, A, is constructed identically for both as the forward Fourier transform of α [equation (12)]. The only difference is in the application of A. In a digital application using matrices, the efficiency of the calculation of equation (12) is increased if α is stored with u as the row coordinate so that each Fourier transform operates on a vector in contiguous memory. The resulting matrix is optimal for computation of equation (11) by matrix-vector multiplication but must be transposed to compute equation (12). The conclusion that there may be a slight efficiency advantage for PSPI depends on the assumption that the Fourier transform of contiguous data is significantly more efficient than for separated data elements. This may not always be the case, especially for highly vectorized hardware. Our experience (with Matlab on Sun workstations) shows the two methods have identical floating point operations counts and only slightly different CPU times.

We emphasize that these Fourier-domain expressions will give theoretically identical results to the dual-domain formulas or to space-domain results. However, the formulas are distinct from a numerical perspective because each domain has its potential strengths and weaknesses in a particular computational setting. A potential advantage of this Fourier approach is the possibility of gaining efficiency for smooth velocity models by computing and applying only a limited number of off-diagonal terms.

COMPARISON OF NSPS AND PSPI

The formal demonstration that nonstationary convolution forms the linear superposition of the nonstationary filter impulse response, whereas nonstationary combination does not, is given in Margrave (1998). Here, we will take a more conceptual approach. Consider the computation of both Ψ_{NSPS} and Ψ_{PSPI} in the case when the nonstationary phase-shift operator is given by

$$\alpha_{v(x)}(k_x, x, \omega) = \begin{cases} \alpha_{v1}(k_x, \omega), & x < 0 \\ \alpha_{v2}(k_x, \omega), & x \geq 0 \end{cases}, \quad (14)$$

where α_{v1} and α_{v2} are two different constant velocity phase-shift operators corresponding to velocities v_1 and v_2 as given by equation (3). We will give an analytic analysis and show numerical examples. (All of our numerical examples were computed using the full-Fourier method just discussed.)

Figure 2 shows the numerical test case that we will use to illustrate the conceptual results. The seismic section shown contains a horizontal line of impulses with a zero pad attached to both sides to avoid operator wraparound, as is customary for Fourier methods. The velocity model is 5000 m/s on the left, and changes discontinuously in the middle of the section to 2000 m/s. The wavefield extrapolations to be shown will all use a 50-m downward extrapolation step. For comparison with NSPS and PSPI, Figure 3a shows an ordinary phase-shift extrapolation using the intermediate velocity of 3500 m/s. Figure 3b shows the amplitude spectrum of the Fourier extrapolation matrix for a particular frequency, ω. As discussed previously, it is a purely diagonal matrix whose nonzero elements contain the phase-shift extrapolator, α_{vj} [equation (3)]. Multiplication of the input wavefield, represented as a column vector of wavenumber components for a single frequency, results in a column vector of the output wavefield with no wavenumber mixing.

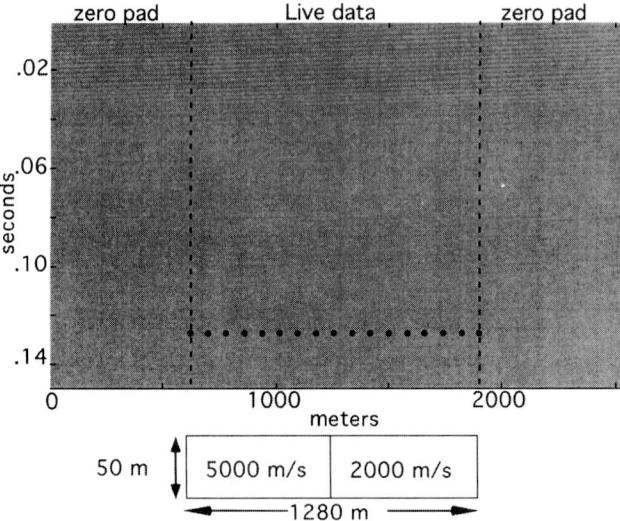

FIG. 2. Numerical test case showing impulses to be extrapolated through a discontinuous velocity model. "Live data" refers to the input wavefield; "zero pad" refers to the zero pad in x required by the Fourier domain extrapolation.

Next, we compute Ψ_{PSPI} by substituting equation (14) into equation (6). After some elementary manipulations, we obtain

$$\Psi_{\text{PSPI}}(x, \Delta z, \omega) = \begin{cases} \Psi_{v1}(x, \Delta z, \omega), & x < 0 \\ \Psi_{v2}(x, \Delta z, \omega), & x \geq 0 \end{cases}, \quad (15)$$

where Ψ_{v1} and Ψ_{v2} are reference wavefields for α_{v1} and α_{v2} computed from equation (1). Equation (15) shows that Ψ_{PSPI} is the discontinuous juxtaposition of two reference wavefields. Margrave (1998) shows that nonstationary combination filters generally have the property that lateral discontinuities in filter specifications will cause similar discontinuities in the filtered result. This is a nonphysical behavior since a superposition of Huygen's wavelets should always smooth over discontinuities.

Figure 4a shows Ψ_{PSPI} for our numerical test case. The central discontinuity is clearly obvious, as is the dramatic difference in traveltime delay between the left and right sides. If this result were input into a subsequent extrapolation step, the discontinuity would cause objectionable wavefronting. Note also that the hyperbolic impulse responses show two different curvatures. Figure 4b is the amplitude spectrum of the Fourier extrapolation matrix for the same frequency as in Figure 3b. The nonzero off-diagonal terms are clearly evident, although it is interesting to note that, even for this discontinuous velocity model, they quickly decrease away from the diagonal.

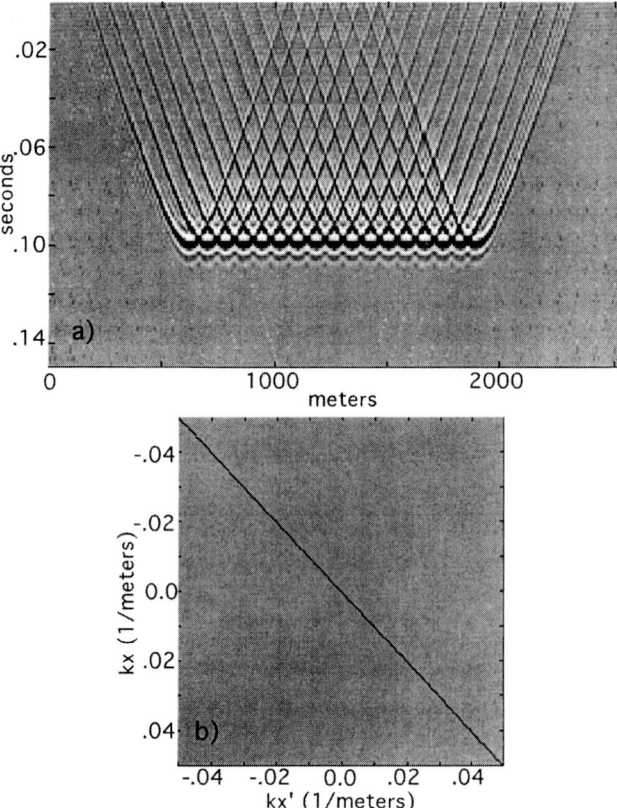

FIG. 3. (a) Phase-shift extrapolation for the numerical test case of Figure 2 using constant velocity ($v = 3500$ m/s). (b) Amplitude spectrum of Fourier extrapolation matrix for a particular ω in the constant velocity case.

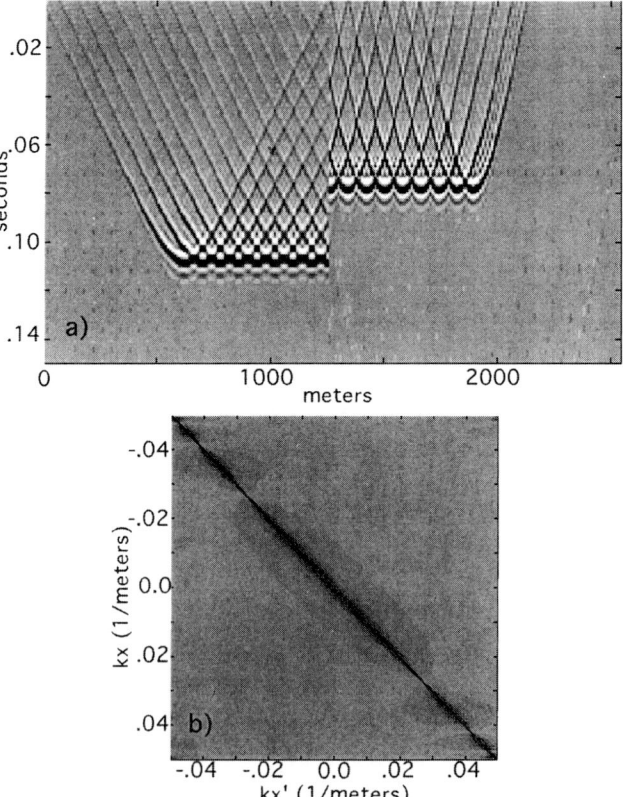

FIG. 4. (a) Ψ_{PSPI} for the numerical test case of Figure 2. The discontinuity in the output wavefield in the figure corresponds to the discontinuity in the velocity field. (b) Amplitude spectrum of the PSPI extrapolation matrix for a particular ω. Laterally varying velocities generate off-diagonal terms.

Next, consider Ψ_{NSPS} by substituting equation (14) into equation (9) and breaking the integral into two parts to get

$$\varphi_{\text{NSPS}}(k_x, \Delta z, \omega) = \frac{1}{2\pi}\Bigg[\alpha_{v1}(k_x, \omega)$$
$$\times \int_{-\infty}^{0-} \Psi(x, 0, \omega)e^{ik_xx}\,dx + \alpha_{v2}(k_x, \omega)$$
$$\times \int_{0}^{\infty} \Psi(x, 0, \omega)e^{-ik_xx}\,dx\Bigg]. \quad (16)$$

Now define two differently windowed versions of the input wavefield:

$$\Psi|_{v1}(x, 0, \omega) = \begin{cases} \Psi(x, 0, \omega), & x < 0 \\ 0, & x \geq 0 \end{cases}$$

and $\quad (17)$

$$\Psi|_{v2}(x, 0, \omega) = \begin{cases} 0, & x < 0 \\ \Psi(x, 0, \omega), & x \geq 0 \end{cases}.$$

Then, equation (16) can be written

$$\varphi_{\text{NSPS}}(k_x, \Delta z, \omega) = \alpha_{v1}(k_x, \omega)\varphi|_{v1}(k_x, 0, \omega)$$
$$+ \alpha_{v1}(k_x, \omega)\varphi|_{v2}(k_x, 0, \omega), \quad (18)$$

where $\varphi|_{v1}$ and $\varphi|_{v2}$ are the ordinary Fourier transforms of $\Psi|_{v1}$ and $\Psi|_{v2}$, respectively. Ψ_{NSPS} is simply the inverse Fourier transform of φ_{NSPS} as in equation (10). Since the inverse Fourier transform is linear, it can be distributed over the sum in equation (18). This analysis shows that Ψ_{NSPS} may be computed by "windowing" the input wavefield as in equation (17) to isolate those portions spatially coincident with each distinct velocity, extrapolating the windowed wavefields with phase shifts, and superimposing the results.

Figure 5a shows Ψ_{NSPS} for the numerical test case. Unlike Figure 4a, there is no central discontinuity, and each input impulse has been replaced by the time-reversed diffraction response characteristic of the local velocity. This is a more physically plausible result than that of Figure 4a and can be seen to be in qualitative agreement with Huygen's principle. However, it is still not fully correct as the diffraction responses do not show refractions at the velocity boundary. In comparison with Figure 4a, it is apparent that PSPI does refract the diffraction limbs but at the cost of introducing discontinuities in the wavefield. Figure 5b is the Fourier matrix that achieves NSPS extrapolation. As mentioned previously, the NSPS and PSPI matrices are not quite transposes of one another in the full-Fourier domain. The NSPS matrix can be formed by transposing the PSPI matrix and then flipping each row about the diagonal [compare equations (11) and (13)].

To accentuate the comparison, Ψ_{PSPI} [equation 15] can be rewritten to incorporate an explicit windowing step as well by defining

$$\Psi_{v1}|_{v1}(x, \Delta z, \omega) = \begin{cases} \Psi_{v1}(x, \Delta z, \omega), & x < 0 \\ 0, & x \geq 0 \end{cases}$$

and $\quad (19)$

$$\Psi_{v2}|_{v2}(x, \Delta z, \omega) = \begin{cases} 0, & x < 0 \\ \Psi_{v2}(x, \Delta z, \omega), & x \geq 0 \end{cases},$$

then

$$\Psi_{\text{PSPI}}(x, \Delta z, \omega) = \Psi_{v1}|_{v1}(x, \Delta z, \omega) + \Psi_{v2}|_{v2}(x, \Delta z, \omega). \quad (20)$$

So, PSPI and NSPS can be contrasted by where, in the process, the windowing step occurs. In NSPS, the input dataset is windowed to create the set $\{\Psi|_{v_j}\}$, each member of the set is phase-shift extrapolated with the corresponding member of $\{v_j\}$, and the results are superimposed. In PSPI, the set $\{\Psi_{v_j}\}$ is created by phase-shift extrapolating Ψ with each member of $\{v_j\}$, each member of $\{\Psi_{v_j}\}$ is then windowed giving the set $\{\Psi_{v_j}|_{v_j}\}$, and the results superimposed. The windowing functions are the same in both algorithms. This computation

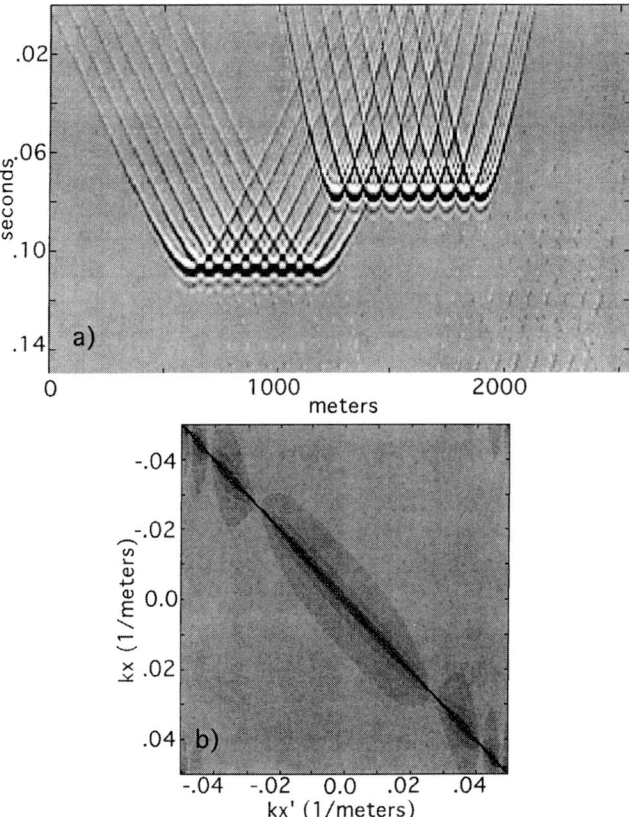

FIG. 5. (a) Ψ_{NSPS} for the numerical test case of Figure 2. The discontinuity in the velocity field is not imposed on the output wavefield. Instead, the response is a smooth superposition of wavefields. (b) Amplitude spectrum of the NSPS extrapolation matrix for a particular ω. Laterally varying velocities generate off-diagonal terms. Note the similarity of the spectrum to that of the PSPI extrapolator; in fact, they differ only by a matrix transpose and reversal of each resulting row about the center diagonal.

procedure is exact for both Ψ_{PSPI} and Ψ_{NSPS} whenever the velocity variation is piecewise constant, and illustrates again that the computational effort required for NSPS is very similar to that required for PSPI.

This analysis can be generalized to nearly arbitrarily complicated velocity variations (as long as the number of distinct velocities is countable) by defining the windowing function:

$$\Omega_j(x) = \begin{cases} 1, & v(x) = v_j \\ 0, & \text{otherwise} \end{cases} \quad (21)$$

Then, Ψ_{NSPS} can be written

$$\Psi_{NSPS}(x, \Delta z, \omega) = \underset{k_x \Rightarrow x}{\text{IFT}} \left[\sum_j \alpha_{vj}(k_x, \omega) \underset{x \Rightarrow k_x}{\text{FT}} (\Omega_j(x)\Psi(x, 0, \omega)) \right], \quad (22)$$

while Ψ_{PSPI} is

$$\Psi_{PSPI}(x, \Delta z, \omega) = \sum_j \Omega_j(x) \underset{k_x \Rightarrow x}{\text{IFT}} [\alpha_{vj}(k_x, \omega) \underset{x \Rightarrow k_x}{\text{FT}} (\Psi(x, 0, \omega))]. \quad (23)$$

In these expressions, FT and IFT are forward and inverse Fourier transforms, and the sum is over the complete set of distinct velocities. Equations (22) and (23) define the windowing analog for the computation of NSPS and PSPI, respectively. In the constant velocity case, the equivalence of both methods with ordinary phase shift can be easily appreciated since Ω_j becomes unity and the sums collapse to a single term.

Figure 6 shows Ψ_{NSPS} and Ψ_{PSPI} for the complicated velocity function shown in Figure 6c. The NSPS result is clearly more coherent than that from PSPI. [In fairness, we note that a practical implementation PSPI would never be run with such rapid lateral velocity variations. Instead, a few reference wavefields would be computed, and a smoothed interpolated result would be obtained from equation (5). Thus the result would be less chaotic than that shown in Figure 6b but also less accurate than that shown in Figure 6a.]

The reason for the relatively chaotic nature of the PSPI extrapolations compared with NSPS can be appreciated from the windowing analog. In the former case, phase-shift wavefield extrapolation (with evanescent filtering) precedes windowing and the windowing creates evanescent energy. In the NSPS case, the final step is phase-shift extrapolation so the result is properly filtered for evanescent energy. In a multistep scheme, the PSPI result of Figure 6b would be immediately subjected to an evanescent filter in a second step that will reduce its chaotic appearance. After multiple steps through complex velocity models, the differences between NSPS and PSPI become more subtle.

APERTURE COMPENSATION

Intuitively, the reason that nonstationary theory is required for vertical wavefield extrapolation through $v(x)$ is that the wavefield extrapolation operator changes spatially as $v(x)$ varies. It follows that any other space and wavenumber variant processes may be incorporated into the extrapolation operator in similar fashion. One such process is the implementation of absorbing lateral boundaries. Absorbing boundaries have been developed quite successfully for finite difference and other space-domain methods (Clayton and Engquist, 1977; Keys, 1985), and we extend them to Fourier methods here. The usual concept is to alter the dispersion relation of waves near the boundary such that only outward traveling wavefronts are allowed; however, this is usually not possible for all propagation angles (Claerbout, 1985). We achieve absorbing boundaries for all propagation angles from the viewpoint of developing an extrapolation operator that is compensated for finite recording aperture.

Aperture compensation follows from an understanding of the downward extrapolation of upward traveling waves as a

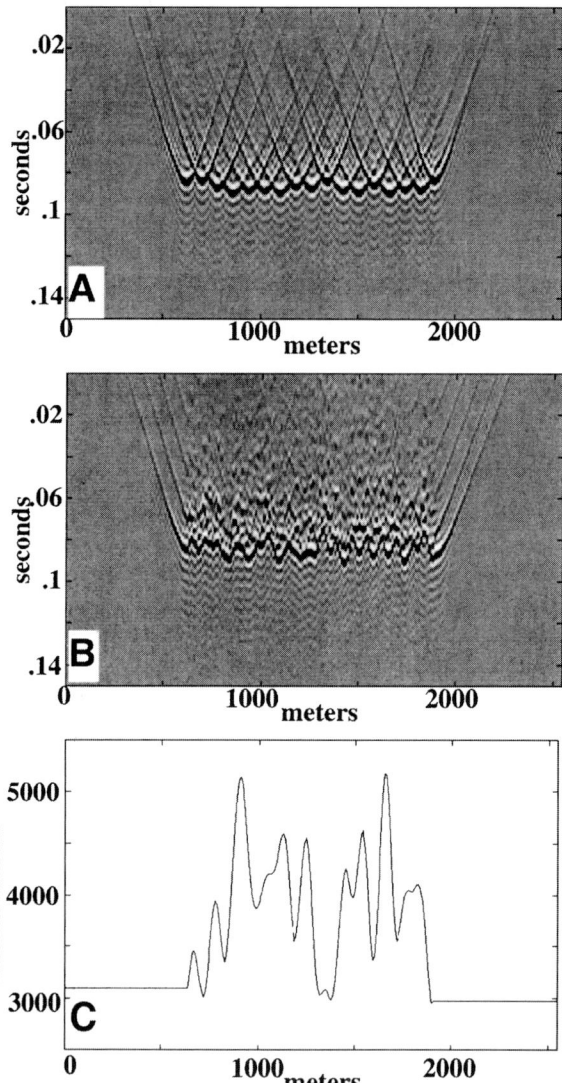

FIG. 6. (a) Ψ_{NSPS} for complicated velocity variation. The wavefield of Figure 2 was used as input to NSPS extrapolation through the complicated (though arbitrary) velocity of (c). The resulting NSPS wavefield is a continuous superposition of diffraction responses. (b) Ψ_{PSPI} for complicated velocity variation. The wavefield of Figure 2 was used as input to PSPI extrapolation through the complicated velocity of (c). The chaotic response of the extrapolation is conceptually the result of windowing a set of constant velocity extrapolations and combining them into an output section. (c) Complicated velocity function used to compare NSPS and PSPI.

process of crosscorrelation with an appropriate diffraction response. The inverse Fourier transform of the phase-shift operator is essentially the diffraction response of the scalar wave equation (Robinson and Silvia, 1981, p. 370). From here, it is not difficult to show that the space-time equivalent of phase-shift downward continuation is a convolution with a time-reversed diffraction response (hyperbola), as shown in Figure 7a. Equivalently, this can be regarded as a crosscorrelation with the time-normal diffraction response. Thus a very appealing picture emerges: the downward continuation of upward traveling waves from depth z_1 to depth z_2 can be done by crosscorrelation of the wavefield recorded at z_1 with the expected response of a point scatterer at z_2.

We can regard any seismic line as a spatial window that allows only a portion of the response of a point scatterer at z_2 to be recorded at z_1. We deduce that a better crosscorrelation operator than the normal infinite, symmetric operator would be that operator with an appropriate spatial window applied. It follows immediately that aperture-compensated downward continuation must be a nonstationary process even in the constant velocity case because the expected windowed diffraction response must vary laterally.

Consider a seismic line, recorded at z_1, where the only reflecting element is a point scatterer at z_2 near the left edge of the line (Figure 7b). The expected zero-offset response is the right-hand limb of a diffraction hyperbola. Downward continuation by crosscorrelation with a symmetric hyperbola (simulating, perhaps, a limited scattering angle operator) is shown in Figure 7c (the temporal delay of the operator is not shown so that the focusing effects can be more clearly appreciated). The use of an aperture-compensated operator is shown in Figure 7d, where the crosscorrelation is done with the expected windowed diffraction response. The crosscorrelation is shown as a convolution-by-replacement with the time-and-space-reversed diffraction response.

An extrapolation operator that has been aperture compensated varies from completely left-sided on the left end of a seismic line, to symmetric in the middle, and then to completely right-sided on the right end. This means that the operator has a scalar wave-dispersion relation that varies smoothly from a left or right quarter circle on either end to symmetric in the middle. This is exactly the "Engquist boundary condition" for absorbing lateral boundaries discussed by Claerbout (1985). Thus absorbing boundaries arise as a natural consequence of aperture compensation and, additionally, a smooth lateral variation of the dispersion relation is obtained.

A one-sided diffraction response has a one-sided ω-k_x spectrum. Viewing crosscorrelation as a multiplication of ω-k_x spectra, it is easy to appreciate that the crosscorrelation of a one-sided diffraction with a two-sided diffraction will produce the same result as the crosscorrelation of the one-sided diffraction with itself. The problem with the symmetric operator stems from the fact that seismic data generally contains energy at all k_x values, even near the aperture boundaries, due primarily to noise. Thus, the symmetric operator can produce "false correlations" near the boundaries which appear as wavefronts "reflecting" from the boundary. Such events are unphysical as they represent reflector dips that could not possibly have been recorded by the finite-aperture seismic line. The one-sided operator cannot produce such events.

We formulate an aperture-compensated extrapolation operator by directly limiting its spectral content as a function of position. As shown in Figure 8, the finite aperture can be regarded as a space-variant scattering-angle filter where the

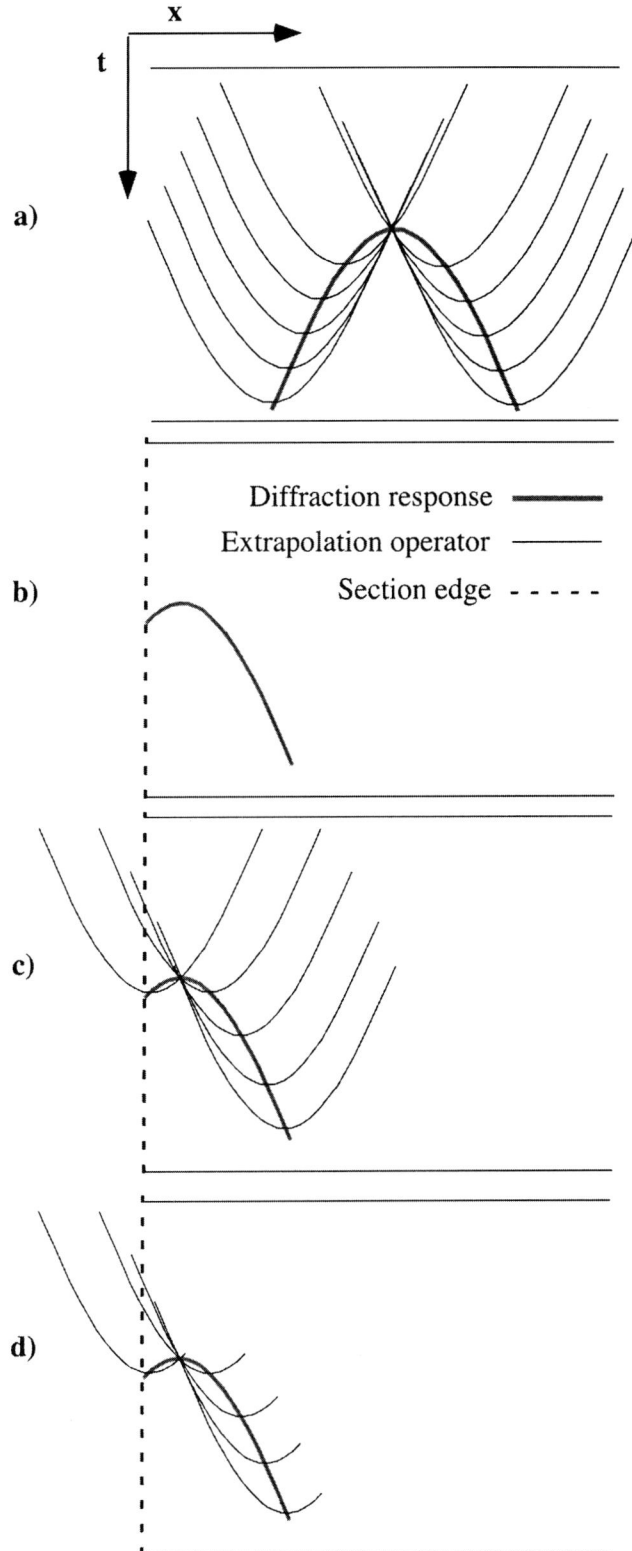

FIG. 7. (a) Application of the downward continuation operator considered as a convolution of the recorded wavefield at one depth with the time and space reversed response of a point scatterer at another depth. (The extrapolation time shift has been ignored to simplify comparison.) (b) Diffraction response near the edge of the recording aperture. (c) Downward continuation of a diffraction response at the edge of the recording aperture using a symmetric operator. Residual wavefronts will be generated by this process and will appear as boundary reflections. (d) Downward continuation using an aperture-compensated operator, resulting in a perfectly absorbing boundary.

left and right scattering-angle limits correspond to raypaths from a scatterpoint to either end of the line. Approximating these raypaths as straight rays, this filter can be expressed as

$$\beta(k_x, x, \omega) = \begin{cases} 1, & -\omega \sin(\theta_L) \leq v(x)k_x \leq \omega \sin(\theta_R) \\ 0, & \text{otherwise} \end{cases} \quad (24)$$

where θ_L and θ_R are left and right scattering angles as defined in Figure 8. Then, the aperture-compensated operator can be written:

$$\alpha_{v(x)}^{\text{aper}}(k_x, x, \omega) = \beta(k_x, x, \omega)\alpha_{v(x)}(k_x, x, \omega). \quad (25)$$

Using $\alpha_{v(x)}^{\text{aper}}$ in place of $\alpha_{v(x)}$ in equation (9) or equation (12) implements aperture compensation in either the dual or Fourier domains.

Figure 9a shows Ψ_{NSPS} computed with aperture compensation where the aperture is defined as the live data zone of Figure 2. Careful inspection shows that the impulse responses on both edges are completely one-sided, having only an outgoing wavefront. The second impulse response in from each edge is also slightly modified. Figure 9b shows amplitude spectrum of the Fourier extrapolation matrix, and it is obvious that aperture compensation has been purchased at the expense of a considerable increase in off-diagonal power.

Finally, we note that a Fourier method actually has two mechanisms that can lead to similar wavefronting near the boundary. In addition to the effect discussed above, there is the possibility of "operator wraparound" resulting from an insufficient lateral zero pad. As formulated here, our method of aperture compensation still requires an adequate zero pad, although its suppression with a further nonstationary operator appears possible.

EXTENSION TO THREE DIMENSIONS

To this point, we have considered only the extrapolation of 2-D wavefields and have shown results computed with the full-Fourier domain equations (11) and (13). As discussed, the digital implementation of these equations can be formulated as a matrix multiplication, which means that a 2-D (k_x, k_x') operator is required to handle the lateral variations along a single spatial axis. Therefore, it appears that a 3-D implementation of NSPS, addressing lateral variations in both x and y, will require a 4-D (k_x, k_x', k_y, k_y') operator. Though we have not yet produced a 3-D implementation, there are a number of possible strategies which circumvent this memory-intensive algorithm. We mention several of them here.

Perhaps the simplest approach is to directly implement the approximate windowing expression, equation (22), which is based on the assumption of a piecewise constant-velocity variation. This allows the extrapolation to be done with conventional 3-D phase-shift code and simple spatial windowing operations. Although the desired velocity variation may not be piecewise constant, it is a straightforward matter to approximate it with a piecewise constant function to any desired accuracy. A reasonable approach would be to find the piecewise constant approximation to $v(x, y)$ with the least number of segments which reduces the maximum phase error below a predetermined target. [Here, phase error refers to the phase difference between the exact phase as computed with a 3-D version of equation (7) using the exact velocity variation and the phase computed from the piecewise constant-velocity approximation.] The cost of an extrapolation step would be approximately the number of segments in the piecewise constant approximation times the cost of an ordinary 3-D phase shift. If the number of segments can be kept much smaller than the number of (x, y) locations, then this should be preferable to a direct numerical integration of a 3-D version of equation (9).

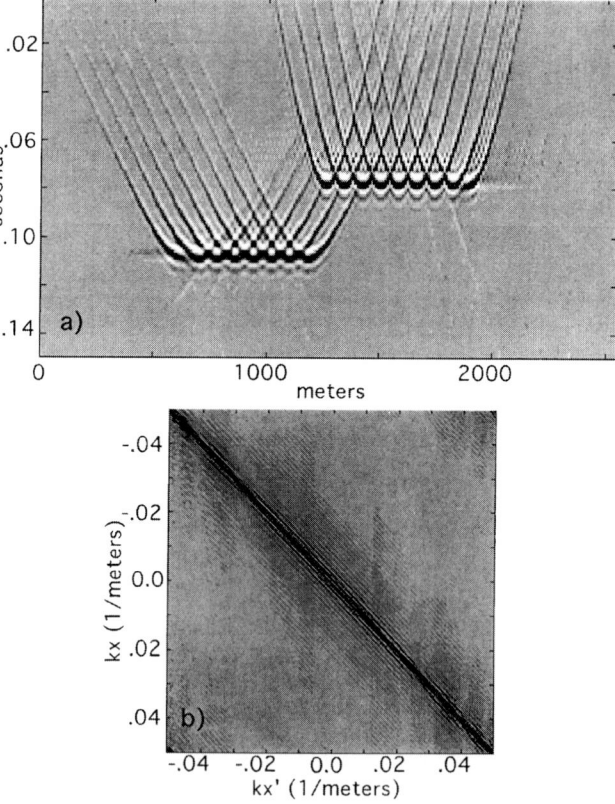

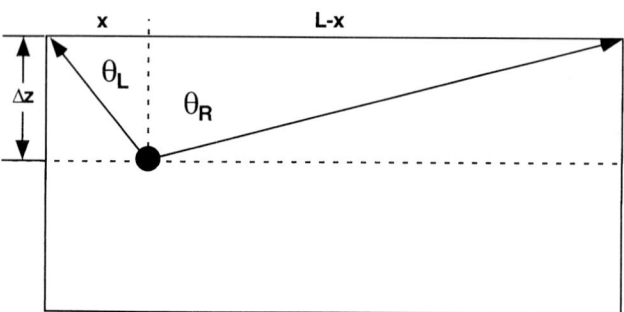

FIG. 8. An aperture-compensating (absorbing-boundary) filter is a laterally varying (nonstationary) ω–k_x fan filter defined by the maximum scattering angles allowed by the given aperture. Straight raypaths are assumed.

FIG. 9. (a) Ψ_{NSPS} computed including compensation for finite aperture. Impulse responses at the edges are one-sided grading to symmetric at the center of the section. (b) Amplitude spectrum of NSPS extrapolation matrix (Fourier domain). Comparison of this figure to Figure 5(b) shows the additional off-diagonal terms require for aperture compensation.

Another alternative involves developing a suitable analytic approximation to $A(p, q)$ as given by equation (12), for example perhaps by the method of stationary phase. Examination of equation (13) shows that the extrapolation operation can be considered as a weighted mix of input wavenumbers for each output wavenumber. The weights are prescribed by $A(p, q)$, which is obtained by a forward Fourier transform of $\alpha_{v(x)}(p, u)$, for which equation (7) is the analytic expression. In two dimensions, this is a 1-D mix whose weights change with output wavenumber (i.e., nonstationary). Thus, the entire A matrix need not be precomputed and held in memory; rather, it could be computed row-by-row, and each row discarded as it is used. In three dimensions, $A(p, q)$ becomes a 4-D function, say $A(p_x, q_x, p_y, q_y)$; however, its application could be viewed as a 2-D mixing operation where the 2-D array of mix weights changes with output wavenumber. Thus, the memory requirements could be reduced dramatically; however, this is computationally very demanding if A must be estimated numerically from the analytic form of α. If an approximate analytic form for A can be found, then this method becomes very attractive.

CONCLUSIONS

The vertical extrapolation of a scalar wavefield through a laterally variable velocity can be accomplished with high fidelity using a Fourier technique called nonstationary phase shift (NSPS). We assume that the wave propagation velocity, v, depends only on the lateral spatial coordinates and not on the depth. Vertical velocity variations can be addressed by using our method in a recursive progression through a series of depth levels.

The phase-shift method applies a frequency and wavenumber dependent phase shift in the Fourier domain to accomplish wavefield extrapolation though a constant velocity layer. Our NSPS method applies a similar phase shift but allows the shift to vary spatially depending upon the local propagation velocity. Both NSPS and the limiting form of phase shift plus interpolation (PSPI) can be written as generalized Fourier integrals which are examples of nonstationary linear filters and which reduce to ordinary phase shift in the constant velocity limit. However, only NSPS can be considered as a scaled, linear superposition of impulse responses (i.e., Huygen's wavelets). In the presence of strong velocity gradients, differences between the methods are dramatic though the computational effort required is similar.

When considered for the case of a piecewise constant velocity variation, NSPS can be formulated as a three-step process: (1) window the input data to isolate those portions coincident with each distinct velocity, (2) phase-shift extrapolate each windowed dataset, and (3) superimpose the results. PSPI follows a similar pattern except that the windowing is performed after each phase-shift extrapolation. Although neither extrapolation technique is fully correct, the PSPI result has discontinuities wherever the velocity is laterally discontinuous.

The nonstationary extrapolation formalism can be easily extended to include compensation for finite recording aperture. When wavefield extrapolation is viewed as the crosscorrelation of the input wavefield with the expected diffraction response at the new depth level, it becomes clear that the recording aperture applies a spatially variant window to the expected diffraction response. Aperture compensation can be implemented by applying a spatially variant scattering-angle filter (ω-k_x filter) to the infinite aperture operator. This can be done simultaneously with the NSPS extrapolation through a laterally variable velocity field. The result is an operator whose dispersion relation is completely one-sided on the boundaries (equivalent to completely absorbing lateral boundaries) and which grades smoothly to a symmetric response in the center of the acquisition aperture.

A straight forward 3-D implementation of NSPS can be done using a piecewise constant approximation of the actual lateral velocity variations. This requires only conventional 3-D phase-shift and spatial-windowing operations.

ACKNOWLEDGMENTS

We thank the sponsors of CREWES for their support. We also acknowledge the influence of E. V. Herbert of Chevron Corporation (now retired) for his pioneering work on an algorithm similar to PSPI.

REFERENCES

Berkhout, A. J., 1984, Seismic migration: Imaging of acoustic energy by wave field extrapolation: Elsevier.
Black, J. L., Su, C. B., and Wason, C. B., 1984, Steep-dip depth migration: 54th Ann. Internat. Mtg. Soc. Expl. Geophys., Expanded Abstracts, 456–457.
Claerbout, J. F., 1985, Imaging the earth's interior: Blackwell Scientific Publications.
Clayton, R., and Engquist, B., 1977, Absorbing boundary conditions for acoustic and elastic wave equations: Bull. Seis. Soc. Am., **67**, 1529–1540.
Duistermaat, J. J., 1996, Fourier integral operators: Birkhauser.
Gazdag, J., 1978, Wave equation migration with the phase-shift method: Geophysics, **43**, 1342–1352.
——— 1980, Wave equation migration with the accurate space derivative method: Geophys. Prosp., **28**, 60–70.
Gazdag, J., and Squazzero, P., 1984, Migration of seismic data by phase shift plus interpolation: Geophysics, **49**, 124–131.
Grimbergen, J. L. T., Wapenaar, C. P. A., and Dessing, F. J., 1995, One-way operators in laterally varying media: 57th Conf., Eur. Assn. Geosci. Eng., Extended Abstracts, C032.
Hale, D., 1991, Stable explicit depth extrapolation of seismic wavefields: Geophysics, **56**, 1770–1777.
Holberg, O., 1988, Towards optimum one-way wave propagation: Geophys. Prosp., **36**, 99–114.
Huang, L. J., and Wu, R. S., 1996, Prestack depth migration with acoustic screen propagators: 66th Ann. Internat. Mtg. Soc. Expl. Geophys., Expanded Abstracts, 415–418.
Huang, L. J., and Fehler, M. C., 1997, Extended pseudo-screen migration with multiple reference velocities: 66th Ann. Internat. Mtg. Soc. Expl. Geophys., Expanded Abstracts, 1742–1745.
Keys, R. G., 1985, Absorbing boundary conditions for acoustic media: Geophysics, **50**, 892–902.
Margrave, G. F., 1998, Theory of nonstationary linear filtering in the Fourier domain with application to time variant filtering: Geophysics, **63**, 244–259.
Robinson, E. A., and Silvia, M. T., 1981, Digital foundations of time series analysis: Wave equation space-time processing: Holden-Day.
Saint Raymond, X., 1991, Elementary introduction to the theory of pseudodifferential operators: CRC Press.
Stoffa, P. L., Fokkema, J. T., de Luna Freire, R. M., and Kessinger, W. P., 1990, Split-step Fourier migration: Geophysics, **55**, 410–421.
Wapenaar, C. P. A., 1992, Wave equation based seismic processing: In which domain?: 54th Conf., Eur. Assn. Geosci. Eng., Extended Abstracts, B019.
Wapenaar, C. P. A., and Dessing, F. J., 1995, Decomposition of one-way representations and one-way operators: 57th Conf., Eur. Assn. Geosci. Eng., Extended Abstracts, C031.
Wu, R. S., 1994, Wide-angle elastic wave one-way propagation in heterogeneous media and an elastic wave complex-screen method: J. Geophys. Res., **99**, 751–766.

APPENDIX A
NOTATION

Short	Full	Description
Ψ	$\Psi(x, 0, \omega)$	Space-domain wavefield at $z = 0$.
φ	$\varphi(k_x, 0, \omega)$	Wavenumber domain wavefield at $z = 0$.
Ψ_{vj}	$\Psi_{vj}(x, \Delta z, \omega)$	Space-domain wavefield at $z = \Delta z$, extrapolated with v_j.
φ_{vj}	$\varphi_{vj}(k_x, \Delta z, \omega)$	Wavenumber domain wavefield at $z = \Delta z$, extrapolated with v_j.
$\Psi_{v(x)}$	$\Psi_{v(x)}(x, \Delta z, \omega)$	Space-domain wavefield at $z = \Delta z$, extrapolated with $v(x)$ using an unspecified algorithm.
$\varphi_{v(x)}$	$\varphi_{v(x)}(k_x, \Delta z, \omega)$	Wavenumber domain wavefield at $z = \Delta z$, extrapolated with $v(x)$ using an unspecified algorithm.
Ψ_{PSPI}	$\Psi_{\text{PSPI}}(x, \Delta z, \omega)$	Space-domain wavefield at $z = \Delta z$, extrapolated with $v(x)$ using the PSPI algorithm.
φ_{PSPI}	$\varphi_{\text{PSPI}}(k_x, \Delta z, \omega)$	Wavenumber domain wavefield at $z = \Delta z$, extrapolated with $v(x)$ using the PSPI algorithm.
Ψ_{NSPS}	$\Psi_{\text{NSPS}}(x, \Delta z, \omega)$	Space-domain wavefield at $z = \Delta z$, extrapolated with v(x) using the NSPS algorithm.
φ_{NSPS}	$\varphi_{\text{NSPS}}(k_x, \Delta z, \omega)$	Wavenumber domain wavefield at $z = \Delta z$, extrapolated with $v(x)$ using the NSPS algorithm.
α_{vj}	$\alpha_{vj}(k_x, \omega)$	Phase-shift extrapolator for constant velocity v_j.
$\alpha_{v(x)}$	$\alpha_{v(x)}(k_x, x, \omega)$	Phase-shift extrapolator for variable velocity $v(x)$.
A	$A(k_x, k'_x, \omega)$	Full-Fourier domain phase-shift extrapolator for variable velocity $v(x)$.
β	$\beta(k_x, x, \omega)$	Aperture compensation filter.
$\alpha_{v(x)}^{\text{aper}}$	$\alpha_{v(x)}^{\text{aper}}(k_x, x, \omega)$	Aperture-compensated phase-shift extrapolator for variable velocity $v(x)$.
k_{zj}	$k_{zj}(k_x, \omega)$	Vertical wavenumber for constant velocity v_j.
$k_z(x)$	$k_z(k_x, x, \omega)$	Vertical wavenumber for variable velocity $v(x)$.
$\Psi\vert_{vj}$	$\Psi\vert_{vj}(x, 0, \omega)$	Space-domain wavefield at $z = 0$, windowed to be nonzero only where $v(x) = vj$.
$\varphi\vert_{vj}$	$\varphi\vert_{vj}(k_x, 0, \omega)$	Wavenumber domain wavefield at $z = 0$, windowed to be nonzero only where $v(x) = v_j$.
$\Psi_{vj\vert vj}$	$\Psi_{vj\vert vj}(x, \Delta z, \omega)$	Space-domain wavefield at $z = \Delta z$, extrapolated with v_j, windowed to be nonzero only where $v(x) = v_j$.
Ω_j	$\Omega_j(x)$	Windowing function which is unity where $v(x) = v_j$ and zero otherwise.

APPENDIX B
FOURIER DOMAIN FORMULATION FOR PSPI

Equation (6) is repeated here as

$$\Psi_{\text{PSPI}}(x, \Delta z, \omega) = \int_{-\infty}^{\infty} \varphi(k_x, 0, \omega)\alpha(k_x, x, \omega)e^{ik_x x} dk_x. \quad \text{(B-1)}$$

The Fourier transform of equation (B-1) along the x axis is

$$\varphi_{\text{PSPI}}(k'_x, \Delta z, \omega) = \frac{1}{2\pi} \int_{-\infty}^{\infty}$$
$$\times \left[\int_{-\infty}^{\infty} \varphi(k_x, 0, \omega)\alpha(k_x, x, \omega)e^{ik_x x} dk_x \right] e^{-ik'_x x} dx. \quad \text{(B-2)}$$

The variable k'_x is used to distinguish the wavenumbers of the Fourier transform step from those of the extrapolation process. Next, switch the order of integration (see Margrave, 1998, for a discussion):

$$\varphi_{\text{PSPI}}(k'_x, \Delta z, \omega) = \int_{-\infty}^{\infty} \varphi(k_x, 0, \omega)$$
$$\times \left[\frac{1}{2\pi} \int_{-\infty}^{\infty} \alpha(k_x, x, \omega)e^{ik_x x} e^{-ik'_x x} dx \right] dk_x. \quad \text{(B-3)}$$

Then define

$$A(k_x, k'_x - k_x, \omega) = \frac{1}{2\pi} \int_{-\infty}^{\infty} \alpha(k_x, x, \omega)e^{-ix(k'_x - k_x)} dx, \quad \text{(B-4)}$$

and substitute into equation (B-3)

$$\varphi_{\text{PSPI}}(k'_x, \Delta z, \omega) = \int_{-\infty}^{\infty} \varphi(k_x, 0, \omega) A(k_x, k'_x - k_x, \omega) dk_x. \quad \text{(B-5)}$$

Rename the wavenumber variables so that equation (B-5) is the same form as equation (11):

$$\varphi_{\text{PSPI}}(k_x, \Delta z, \omega) = \int_{-\infty}^{\infty} \varphi(k'_x, 0, \omega) A(k'_x, k_x - k'_x, \omega) dk'_x. \quad \text{(B-6)}$$

APPENDIX C
FOURIER DOMAIN FORMULATION FOR NSPS

Equation (9) is repeated here as

$$\varphi_{\text{NSPS}}(k_x, \Delta z, \omega) = \frac{1}{2\pi} \int_{-\infty}^{\infty} \Psi(x, 0, \omega) \alpha(k_x, x, \omega) e^{-ik_x x} dx. \quad \text{(C-1)}$$

Replace $\Psi(x, 0, \omega)$ with its Fourier transform along the x axis:

$$\varphi_{\text{NSPS}}(k_x, \Delta z, \omega) = \frac{1}{2\pi}$$
$$\times \int_{-\infty}^{\infty} \left[\int_{-\infty}^{\infty} \varphi(k'_x, 0, \omega) e^{ik'_x x} dk'_x \right] \alpha(k_x, x, \omega) e^{-ik_x x} dx. \quad \text{(C-2)}$$

The variable k'_x is used to distinguish the wavenumbers of the Fourier transform of the input wavefield from those of the extrapolation process. The next step is to reverse the order of integration (see Margrave, 1998, for a discussion):

$$\varphi_{\text{NSPS}}(k_x, \Delta z, \omega) = \frac{1}{2\pi} \int_{-\infty}^{\infty} \varphi(k'_x, 0, \omega)$$
$$\times \left[\int_{-\infty}^{\infty} \alpha(k_x, x, \omega) e^{ik'_x x} e^{-ik_x x} dx \right] dk'_x. \quad \text{(C-3)}$$

Then define

$$A(k_x, k_x - k'_x, \omega) = \frac{1}{2\pi} \int_{-\infty}^{\infty} \alpha(k_x, x, \omega) e^{-ix(k_x - k'_x)} dx, \quad \text{(C-4)}$$

and substitute into equation (C-3)

$$\varphi_{\text{NSPS}}(k_x, \Delta z, \omega) = \int_{-\infty}^{\infty} \varphi(k'_x, 0, \omega) A(k_x, k_x - k'_x, \omega) dk'_x. \quad \text{(C-5)}$$

Chapter 4 – Case Histories

The test for seismic imaging is really made by applications to real data. Several lessons are learned from case histories. For this reason, we present case histories that illustrate methods and approaches. *Slawinski and Parkin (1996)* show how vertical seismic profiling images could be used to detect steeply sloping reflectors. *Wu et al. (1998)* show imaging results from a Benjamin Creek, Alberta line made available for industry testing by Husky Oil. A data set made available to *Yan and Lines (1998)* by Mobil Oil showed that the use of a simple approximate velocity model may prove more useful than structurally complicated incorrect velocity models. Finally, the problems of crooked lines in foothills seismic surveys are addressed in the case study by *Maclean et al. (1998)*. Finally, we examine the performance of depth migration methods in papers by *Grech et al. (1998)* and *Lines (1999)*.

References:

Kirtland Grech, M.G., Lawton, D.C., and Spratt, D.A., 1998, Numerical seismic modeling and imaging of complex fault-fold structures in a mountainous setting: Foothills Research Project Research Report, Volume 4, 1-1 – 1-35. (Revised).

Lines, L. R., 1999, Depth Migration Experiments with the Spratt Foothills Model: unpublished manuscript.

Maclean, G., Gray, S.H., and Marfurt, K., 1998, Crooked line, rough topography: advancing toward the correct seismic image: unpublished report, 3 p.

Slawinski, M.A., and Parkin, J.M., 1996, Migration of a multi-offset VSP: A case study in NE British Columbia: *Canadian Journal of Exploration Geophysicists*, 32, No. 2, 104-112.

Wu, W., Lines, L., Burton, A., Zhu, J., Jamison, W., and Bording, R.P., 1998, Prestack depth migration of an Alberta foothills data set – The Husky experience: *Geophysics*, 63, 392-298.

Yan, L., and Lines, L.R., 1999, An imaging comparison of three depth migration algorithms on Alberta Foothills datasets: CSEG Annual Meeting, Calgary, May 4-6, 1999 Technical Abstract Book, p. 27-30.

MIGRATION OF A MULTIOFFSET VSP: A CASE STUDY IN NE BRITISH COLUMBIA[1]

Michael A. Slawinski* and John M. Parkin**

Abstract

A walkaway vertical seismic profile (VSP) was performed in the Talisman-Ocelot c-54-J/93-P-4 well to image the Triassic anticlinal structure. Twenty sources, approximately paralleling the dip direction, were deployed. Seismic images acquired in the complex geological regimes of the foothills of NE British Columbia are often very difficult to interpret. In this study the VSP provided important information at a critical stage of the drilling program. The initial trajectory of the well failed to encounter the structure at the anticipated depth. The evaluation of the migrated VSP image, along with other data, including dipmeter, well logs and surface seismic, helped to direct the wellbore trajectory and successfully penetrate the structure.

The acquisition of all 20 offsets (ranging on either side of the wellhead from the near offset to about 1200 m), processing and interpretation was performed in about 50 hours. Vibroseis trucks were used as the energy source. Twenty-seven receivers in the borehole were positioned 20 metres apart at depth between 1400 m and 1940 m. Following processing, the final image of the subsurface was derived using the Kirchhoff migration scheme which combined all offsets. A relatively large number of source-receiver pairs as well as the acquisition geometry which had sources, receivers and raypaths contained in a single plane, resulted in the necessary conditions for effective use of our Kirchhoff migration method.

The VSP offers several advantages as compared to surface seismic surveys. Notably, positioning of receivers in the wellbore can allow for the location of targets with respect to the wellbore.

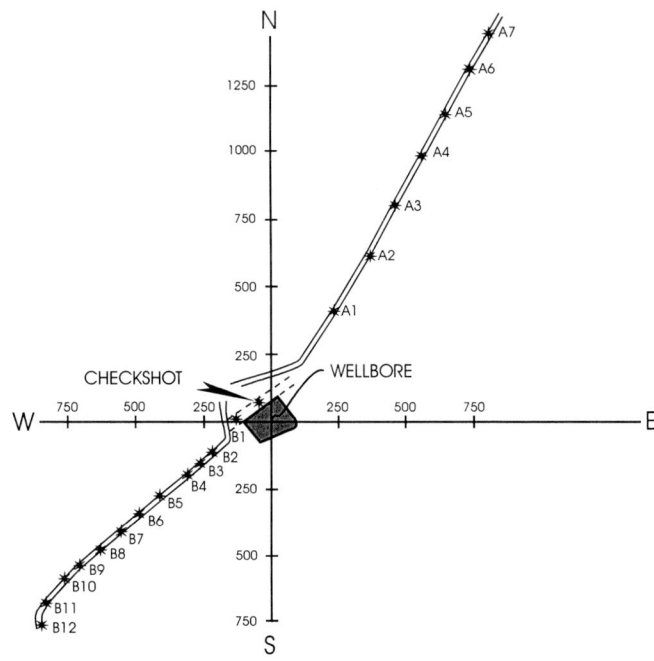

Fig. 1. Plan view of wellbore deviation and source locations A1-A7 and B1-B12. Axes are annotated in metres. The "lease road" on which sources are deployed approximately parallels the dip direction.

Introduction

In May 1994 a VSP was acquired in the c-54-J/93-P-4 Talisman/Ocelot well in northeastern British Columbia. Prior to this study there had been numerous other VSP surveys acquired in this area. The special characteristic of this study lies in the fact that 20 surface source locations were used in gathering the VSP data. The sources were all approximately collinear and coincided with the regional dip direction and borehole deviation as shown in Figure 1. This recording geometry allowed for a proper application of the VSP-Kirchhoff-Depth Migration algorithm, which, for optimal operation, requires that all sources, receivers and raypaths be coplanar.

Imaging of complex structures by VSP migration has been investigated by several researchers. For instance, an insightful synthetic study was presented by Payne et al. (1994). A study including field examples was described by Zhu and Lines (1994).

The purpose of this paper is to present a case study where a multioffset VSP was used successfully in an operational setting to help determine the location of the target and adjust the

---

[1]Presented at the CSEG National Convention, Calgary, Alberta, May 11, 1995.
*Talisman Energy Inc., 2400, 855 - 2nd Street S.W., Calgary, Alberta T2P 4J9; Department of Geology and Geophysics, University of Calgary, Calgary, Alberta T2N 1N4. Presently at: PanCanadian Petroleum Inc., 150 - 9th Avenue S.W., P.O. Box 2850, Calgary, Alberta T2P 2S5.
**Western Atlas Logging, 1200, 505 - 3rd Street S.W., Calgary, Alberta T2P 3E6.
The authors would like to thank Talisman Energy Inc. for supporting the study, Ocelot Energy Inc. for allowing release of the data, and Western Atlas for their support. In particular we would like to thank Messrs. M. Crous, P. Pelletier, C. Welsh, R. Green, B. Quartero of Talisman, Mr. M. Riemer of Ocelot, and Mr. D. Quinn of Western Atlas. Also, the authors would like to thank Dr. R.J. Brown for his critical review of the manuscript.

drilling program in progress. We believe this is an important example of the use of a VSP in an exploration setting.

EXPLORATION SETTING

Gas is trapped in Triassic (Late Carnian-Norion) Pardonnet and Baldonal carbonates in open to tight asymmetric anticlines, which may be detached from the underlying Charlie Lake anhydrites and Lower Triassic (Doig/Montney) siltstones and shales. Fractures, optimally developed in the fold hinge, allow the reservoir to produce at very prolific rates (20-80 mmcf/d). However, the asymmetry of the folds makes the optimal target narrow and easy to miss, resulting in either a low-productivity "backlimb" well or missing the crest and forelimb altogether.

Rugged surface topography, variation in thickness and velocity of the surface layer, and complex subsurface geology, make the interpretation of the seismic data difficult. To constrain the interpretation, we have used surface geology, well control, and seismic data. In addition, checkshot surveys and VSPs provide us with regional velocity for depth conversion.

PREDICAMENT OF C-54-J WELL

From initial drilling it appeared that, since the top of Triassic was not penetrated at the anticipated depth, the well trajectory had missed the apex and had paralleled the flank of the anticline. The drilling was stopped and the VSP data acquired, processed and interpreted. At the same time a suite of logs, including a dipmeter, was acquired. Based on all this information the most likely location of the apex of the anticline was inferred. When drilling resumed, the trajectory of the wellbore was deviated towards the new target.

RUDIMENTS OF VSP MIGRATION

The VSP provides information not contained in a surface seismic image. Firstly, the positioning of receivers (geophones) in the wellbore allows for deriving a reliable function relating the seismic velocity and depth from the traveltime of first arrivals. Secondly, the positioning of reflectors with respect to the wellbore can be inferred from the traveltime of reflected waves. Proper location of reflectors in a complex geological regime, e.g., non-planar reflectors, is greatly aided by multiple sources and receivers yielding a large number of raypaths. Since the VSP data set is in the time-depth domain, construction of an interpretable image in offset-depth domain must involve a process of mapping between the two domains.

Currently, there are two commonly used methods for transforming the VSP data into the offset-depth domain, namely VSP-to-CDP-Transform and Kirchhoff Migration. Our modelling studies, performed prior to the present study, show that the former is very dependent on accurate model of the subsurface, including proper location of interfaces, while the latter one requires only a reasonable velocity field for the algorithm. If the subsurface model is well known (as is often the case for the seismic data acquired on the plains) the VSP-to-CDP-Transform would often yield a sharper image without undesirable, yet intrinsic, artifacts of the Kirchhoff migration. If, on the other hand, the subsurface structure is not well known (as is often the case for the seismic data acquired in the foothills), the Kirchhoff migration can provide a reasonable image with a less accurate input model. Finally, it can be noted that both techniques may be used to arrive at the interpretation of the subsurface, with the VSP-to-CDP-Transform serving as a convenient verification for the model derived from the interpretation of the migration results. In the present study, the migration algorithm was used on its own.

The Kirchhoff migration of VSP data stems from the concept that if a point source, a point receiver and a point scatterer are contained in an infinite homogeneous, isotropic medium the location of the scatterer can be inferred from the traveltime of the signal taken to travel the path "source-scatterer-receiver". The distance travelled by the signal is the product of the traveltime and the speed of the signal in the medium. Thus it follows that the scatterer must be located on the ellipse whose string-length is equal to the distance travelled by the signal. Each source-receiver pair yields an ellipse corresponding to a single scattering point as shown in Figure 2. In a medium consisting of different velocity layers the term "ellipse" is used rather loosely, since a perfect ellipse corresponds only to a constant velocity medium.

A large number of source-receiver pairs allows the scatterer to be more accurately located. For example, Figures 3a and b show reflections from a planar dipping interface generated from a single source and captured by multiple receivers. In the migrated image, high amplitudes correspond to segment of the interface illuminated by the source. The high amplitudes are created by the superposition of many ellipses. The migration "smiles" are an inevitable by-product of the process with limited number of sources and receivers. Only in the case of very many sources and receivers will the migrated VSP image be comparable to a high-fold surface seismic image as illustrated by the synthetic study of Payne et al. (1994).

EXPERIMENTAL CONSIDERATIONS

In a geophysical application involving a complex subsurface, a limited frequency bandwidth signal, and other experimental limitations and errors, the generation of VSP migration requires careful acquisition, processing and interpretation. A crucial element in useful acquisition involves some knowledge of the study area to allow optimal deployment of sources and receivers.

The selection of suitable source locations faces the challenge posed by the rugged topography, and environmental considerations. The fortuitous location of the "lease road" in

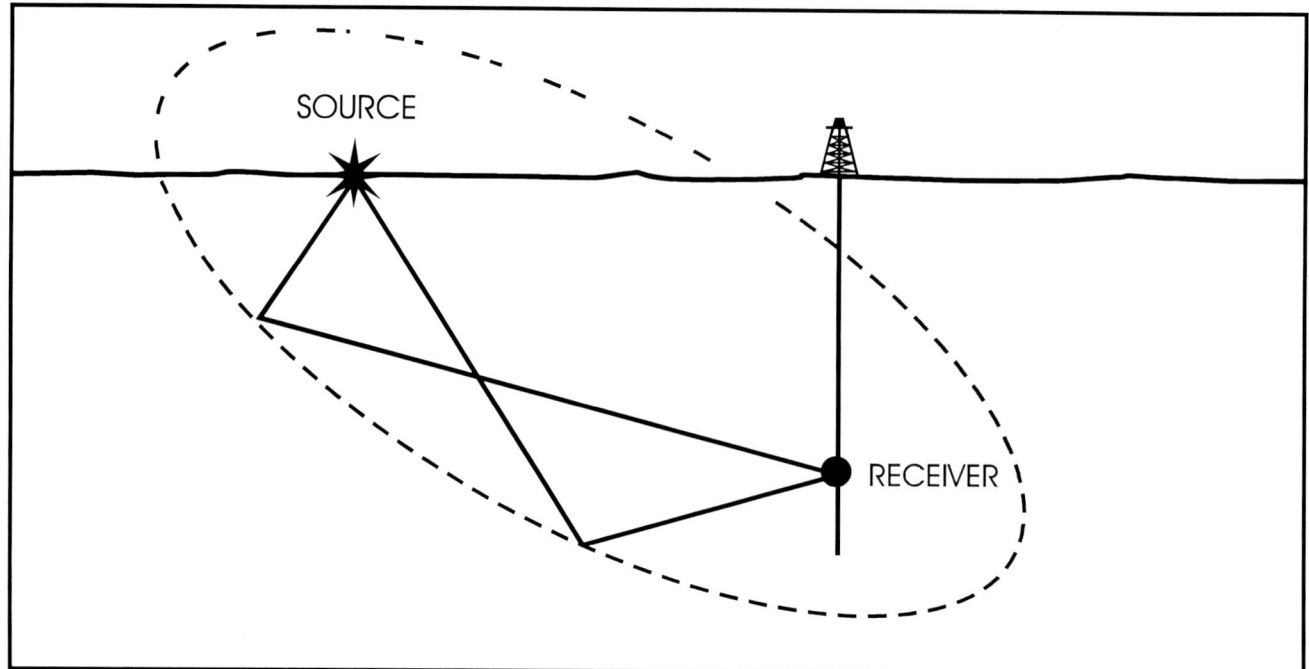

Fig. 2. Locus of all possible reflection points in a uniform velocity field for a given source-receiver configuration given the distance travelled by the signal, i.e., the product of traveltime and speed. The source and receiver constitute the focal points of the ellipse.

the vicinity of c-54-J which paralleled the dip direction, played an important role in the feasibility of this study. Although it would be unreasonable to expect such a convenient situation in all cases, one should consider the construction of a "lease road" with a possible "walk-away" VSP in mind.

Proper consideration of limitations imposed by both physical and economic constraints must be taken into account for optimal acquisition. It is important to realize that there are numerous cases where either of the above-mentioned factors rules out the usefulness of the VSP survey. Obviously, it is preferable to assess the usefulness of a tool prior to investing a significant effort.

Pre-survey Planning

Several considerations had to be taken into account in order to limit the time of acquisition while still acquiring a useful image. Firstly, one must consider the expected distribution of reflectors in the subsurface, including depth to target, reflectors dips and approximate seismic velocities in order to optimize the acquisition by properly locating sources and receivers. To answer some of the questions, pre-survey modelling is very useful.

Since the available algorithm for Kirchhoff migration operates in the two-dimensional space, it is important to acquire the data in such a way that all sources, receivers, and raypaths are coplanar. This is possible if rapid changes in the structural picture are confined to a single azimuth, i.e., dip direction. In the context of the present study a road coinciding with the regional dip direction provided proper location for the deployment of sources.

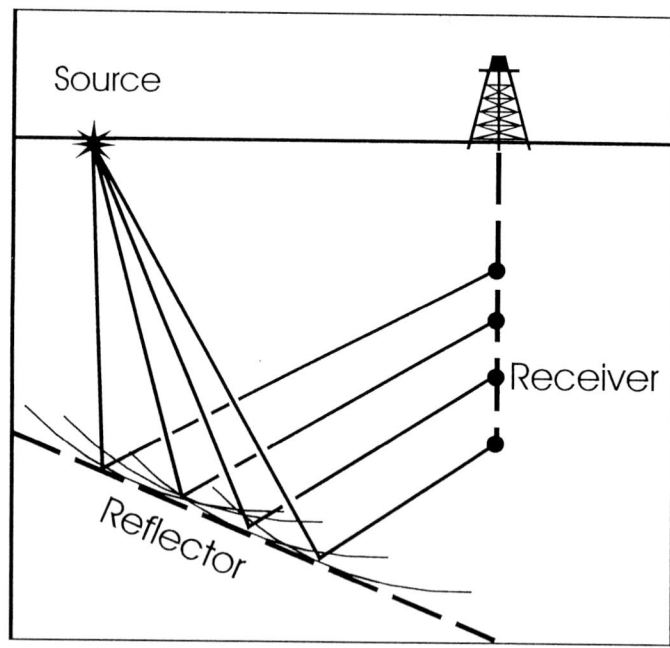

Fig. 3a. Schematic diagram illustrating rays reflected from a planar dipping interface.

The axis of the anticline inferred from the surface seismic data was expected to be perpendicular to the lease road allowing the data acquisition in the dip direction. Based on the geological knowledge of the area, the northeastern flank of the anticline was expected to be much steeper than the southwestern flank. The wellbore was believed to be located to the northeast of the steep flank of the anticline. Therefore,

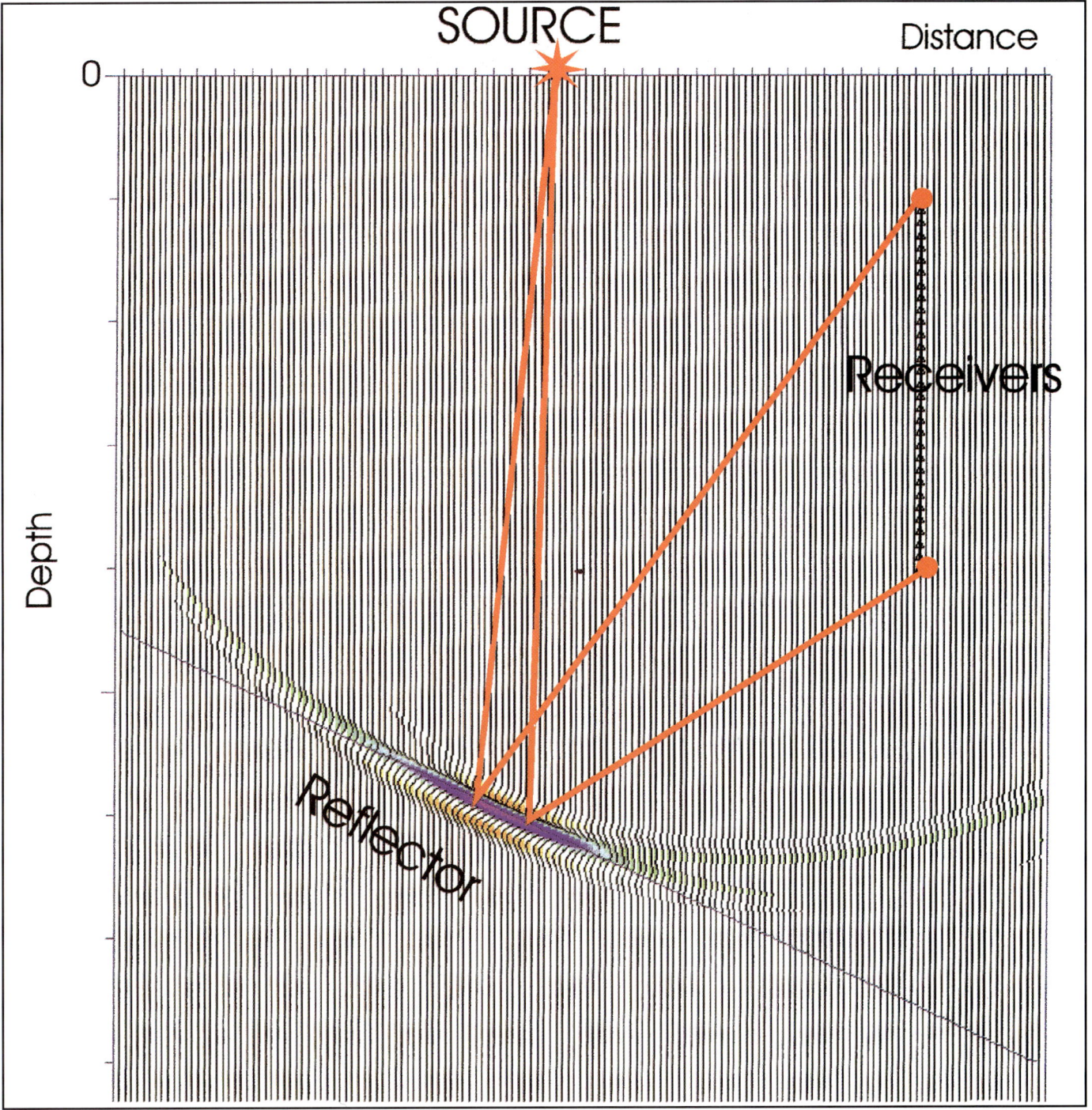

Fig. 3b. Migration of synthetic data obtained from a planar dipping interface. Note the excellent correspondence of high amplitude with the dipping interface. Also note the "elliptical smile" resulting from the construction of the image.

in view of those factors, the sources were placed closer together above the SW limb of the anticline, while a wider aperture was applied over the NE limb to capture, at a larger distance, the rays from a steeply dipping reflector. Also, by raytracing a simple model based on the current estimate of the geologic structure, the optimal locations of receivers in the wellbore were established. This raytracing process indicated that a range of receivers spaced 20 metres apart over the bottom 500 metres of the well would adequately image the target.

Vibrators were selected as the energy source due to the excellent repeatability of the signal and a relative ease and rapidity of operation. A linear "up-sweep" from 8 Hz to 60 Hz was chosen as previous VSP studies in this area indicated a significant loss of higher frequencies in the signal, and it was inferred that signals with frequencies higher than about

60 Hz would not be recorded. To increase the energy generated by the source, a pair of synchronized vibrators was required at each source location.

Data Acquisition

Based on the pre-survey planning the receivers were deployed in the bottom portion of the well from 1940 metres to 1400 metres at a 20 m spacing. This interval was considered the minimum interval necessary to obtain a useful image.

Three pairs of vibrators were used to minimize the acquisition time. One pair was vibrating while the other two were moving to another source location and preparing to generate the signal. Twenty offset locations were acquired. Nineteen were used for imaging and one was used for velocity information. Seven offsets were located NE of the wellbore and spaced about 200 m apart to image the steeply dipping flank of the anticline. The remaining twelve offsets were located SW of the wellbore and spaced about 100 m apart to image the more gently dipping flank (Figure 1). For each source-geophone combination, four sweeps were used. The signal, as observed during the initial stages of acquisition, appeared to have a good signal-to-noise ratio; hence, four sweeps per receiver level were considered sufficient in view of the overall time constraint.

In spite of the large scale of the acquisition, the entire operation took less than 36 hours. The data were acquired by Schlumberger of Canada. Upon completion of the acquisition the data were transferred to Calgary for rush processing in the offices of Western Atlas Logging.

Data Processing

The data from each offset were edited, summed and the first arrivals picked to produce nineteen total wavefield datasets. Editing and summing of the data indicated that four sweeps provided a marginally adequate signal-to-noise ratio. Typical borehole surveys in this area use six to eight sweeps when time is not such a severe constraint.

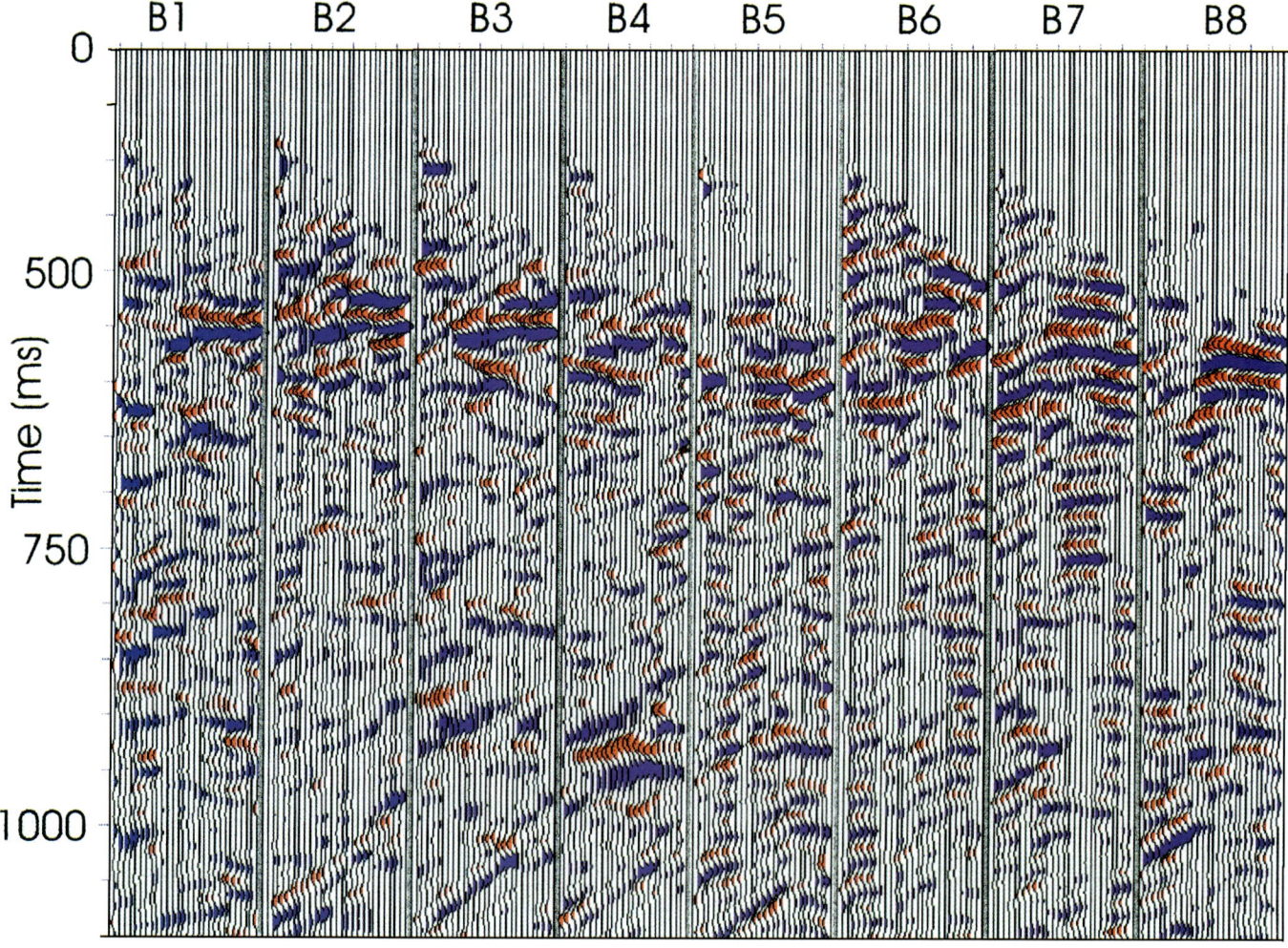

Fig. 4. Selected deconvolved reflected wavefields from the SW source array constituting the input to migration.

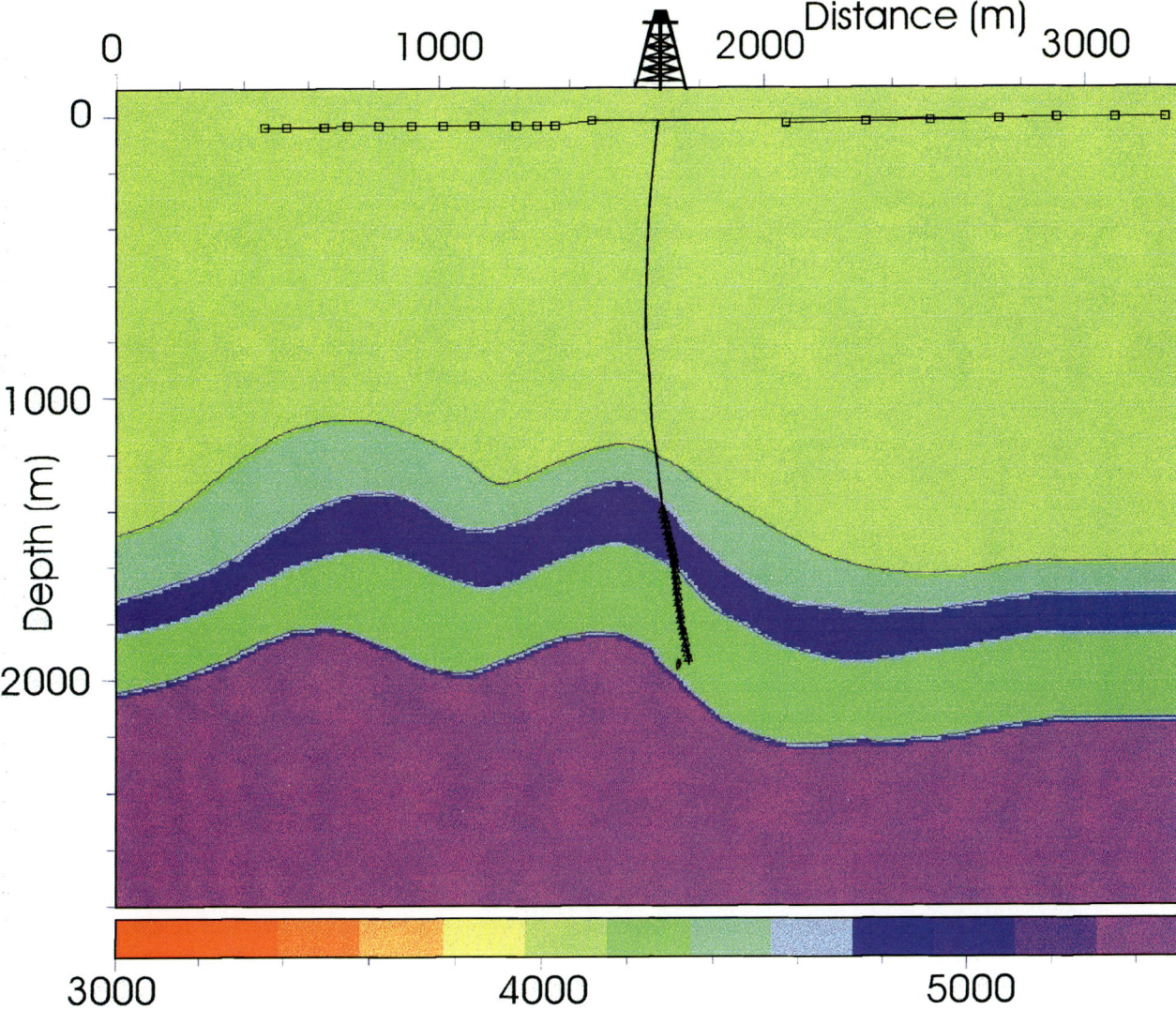

Fig. 5. Structural velocity model for migration. Velocities are derived from checkshots and regional information. Note the deployment of sources along the surface and receivers near the bottom of the wellbore.

Velocity information was obtained from the checkshot VSP data. In order to minimize processing time, the data from the NE offsets were merged in common-shot gathers in order to form a single dataset. Similarly the data from the SW offsets were merged together. The two merged datasets were processed independently using identical processing flows.

The appearance of the total wavefields was complicated by the presence of converted waves as well as reflected energy with moveouts similar to those of downgoing wavefield. The latter effect was due to the steeply dipping reflectors. Separation of the downgoing (direct) and upgoing (reflected) wavefields with opposite moveouts was accomplished by median filtering. The downgoing converted waves were attenuated by a second application of median filtering. Removal of reflected converted waves and reflections with downgoing moveouts could not be adequately treated without severely degrading the image.

Deconvolution yielding a zero-phase wavelet was performed. In VSP data such deconvolution consists of designing an operator from the direct arrival and applying it to the reflected waves. Since a direct measurement of the downgoing wavefield is recorded, the source signature and multiples from it are known and may be deconvolved to produce a zero-phase reflected wavefield.

Deconvolved data sets from the SW array and NE array were migrated independently using Kirchhoff depth migration. A partial display of the input data from the SW offset is shown in Figure 4.

The domain in which a migrated VSP image is displayed is equivalent to that of a normal geological cross-section spanned by horizontal and vertical distances, whereas the domain of the pre-migration display, spanned by depth and traveltime, does not have a simple pictorial geological equivalent. Thus, for VSP data, the process of migration consists of changing the domain in which the data is displayed. Prior

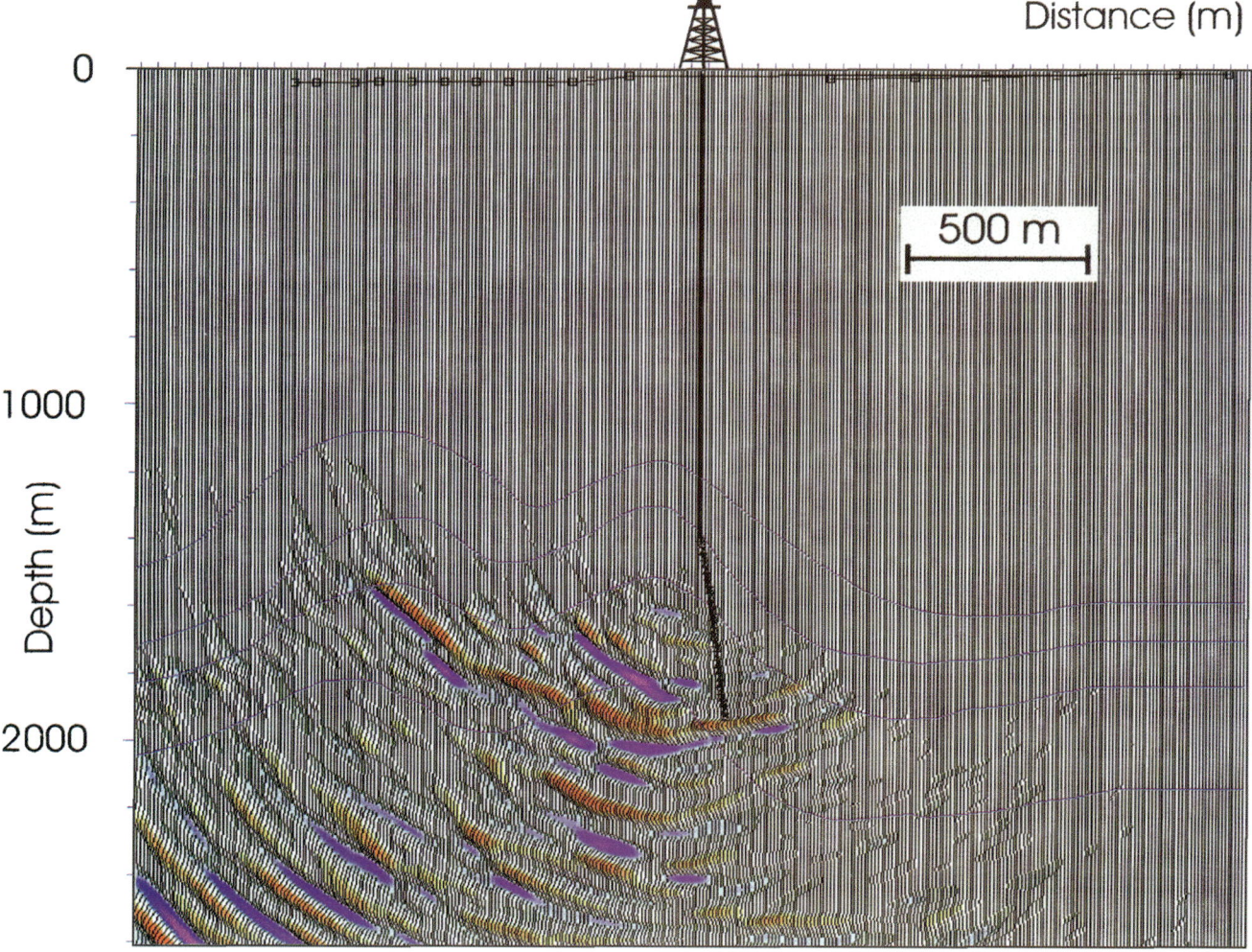

Fig. 6a. Migrated image with velocity model. High-amplitude zones indicate likely positions of reflectors illuminated by source-receiver configurations. Positions of sources near the surface and receivers near the bottom of the wellbore are shown.

to migration, the depth to a given receiver in the wellbore is placed on the horizontal axis, while traveltime is placed on the vertical axis. After migration, horizontal distance is placed on the horizontal axis, while depth is placed on the vertical axis.

A 2D-input model is required for VSP migration. The model defines the structure of the velocity field for traveltime calculations. The model, based on interpretation of the surface seismic, velocity information from the checkshot survey and regional velocity information, is shown in Figure 5. The sources and receivers are then projected onto the model with the wellhead being located at 1600 m on horizontal axis. The resulting migrated images from each array were analyzed and corrected for static differences and merged to create a single image as shown in Figure 6a. The entire processing sequence was accomplished in 18 hours so that a timely interpretation could be made in order to whipstock the well and minimize rig costs.

INTERPRETATION

The interpretation of VSP data suffers, like all seismic data, from the problem of non-uniqueness. The very construction of the image through the Kirchhoff algorithm generates many artifacts; most notably, the elliptical shapes which contribute to the construction of the image are still visible and might render the interpretation more difficult. The interpreter will, almost inevitably, be faced with ambiguous results. A relatively reliable interpretation can be achieved by applying geological constraints, as well as all available information, e.g., dipmeter, sonic logs, etc. Since the likely position of the reflector, or a scattering point, is indicated by a high amplitude on the image, it has been found that a colour-coded amplitude display provides more information and allows a more reliable picture of the subsurface to be deduced than when using a black and white, variable-area display.

A simple interpretation of the anticlinal structure is illustrated in Figure 6b. Note the high amplitude denoted by

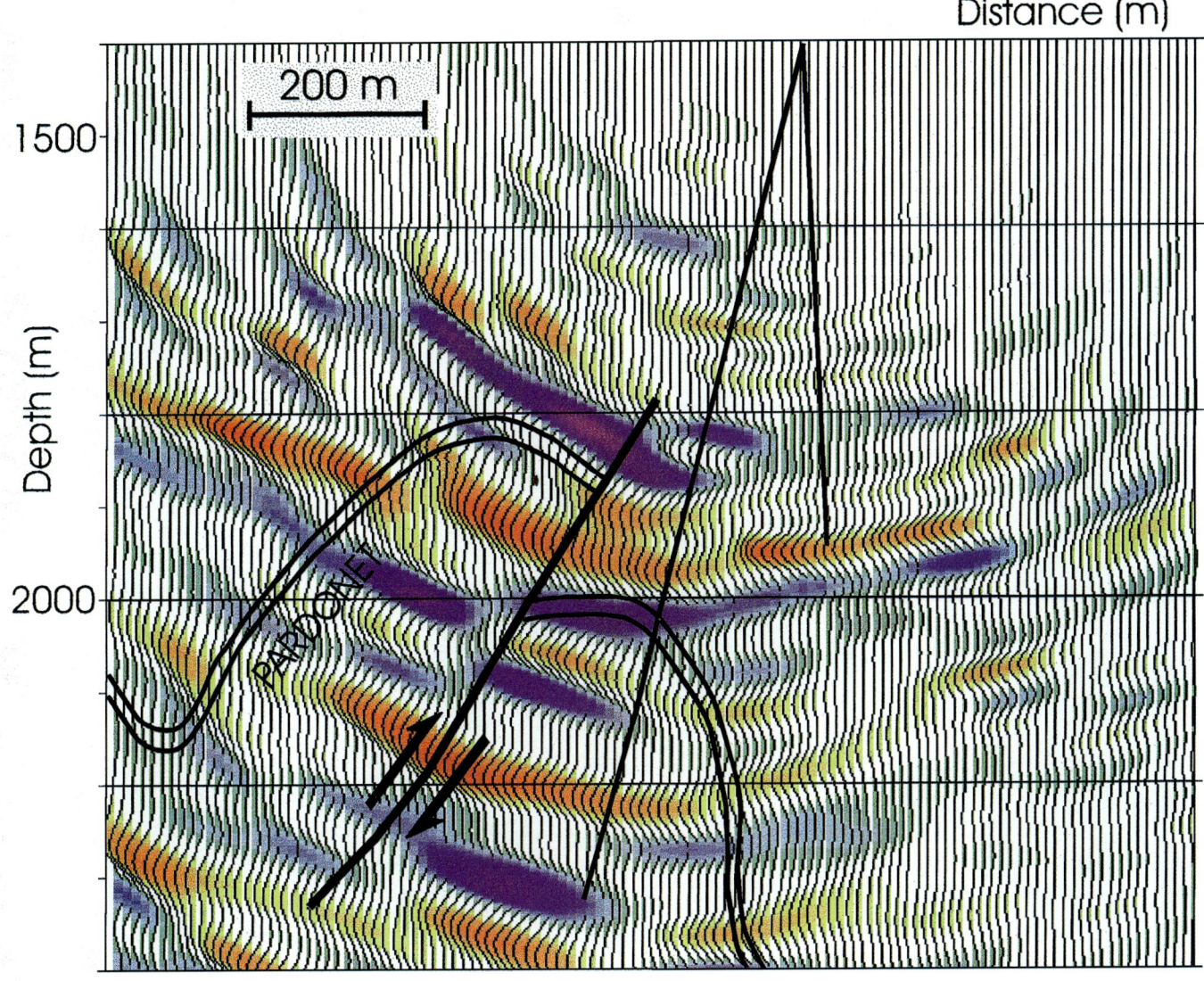

Fig. 6b. Magnification of the part of the image shown in Figure 6a. The interpretation suggests a Triassic anticline disturbed by a reverse fault whose existence is also inferred from well-logs. The position of the downthrown block was confirmed by drilling.

purple interpreted as the location of the reflector which can be compared with the image illustrated in Figure 3b. The "smiling" shape of the image is not a function of the geometry of reflectors but an inherent result of image construction by superposition of ellipses. One must learn to interpret the migrated VSP image in spite of this misleading artifact. The presence of the reverse fault is inferred from both the VSP image and the interception of the fault plane by the wellbore as indicated by well logs (see Figure 7).

In the Kirchhoff migrated image from all the offset locations, the zones of high amplitude were located behind the wellbore. This feature indicated that the crest of the anticline was situated to the southwest of the well trajectory. The presence of several zones of high amplitude suggested that the structure was not a simple anticline but rather an anticlinal uplift which consisted of several imbrications separated by faults. The lack of a clear location corresponding to the single apex entailed several complications as well as opened several options. The final decision concerning the selection of a rather moderate deviation with respect to trajectory of the original well was based on several factors. Firstly, the position of the adjacent well (c-45-J) had to be considered in view of the spacing unit. Secondly, by drilling further away from the aforementioned well, the likelihood of encountering and draining new reserves was increased.

DISCUSSION AND CONCLUSIONS

The modified well trajectory penetrated the structure and economic hydrocarbons were found. After all information was gathered a final interpretation was performed. New information, derived mainly from the dipmeter log in the new wellbore was, in its general outline, consistent with the original interpretation. Certain adjustments, referring to the exact spatial position of reflectors, were necessary as a result of the discrepancy between the actual and modelled velocity

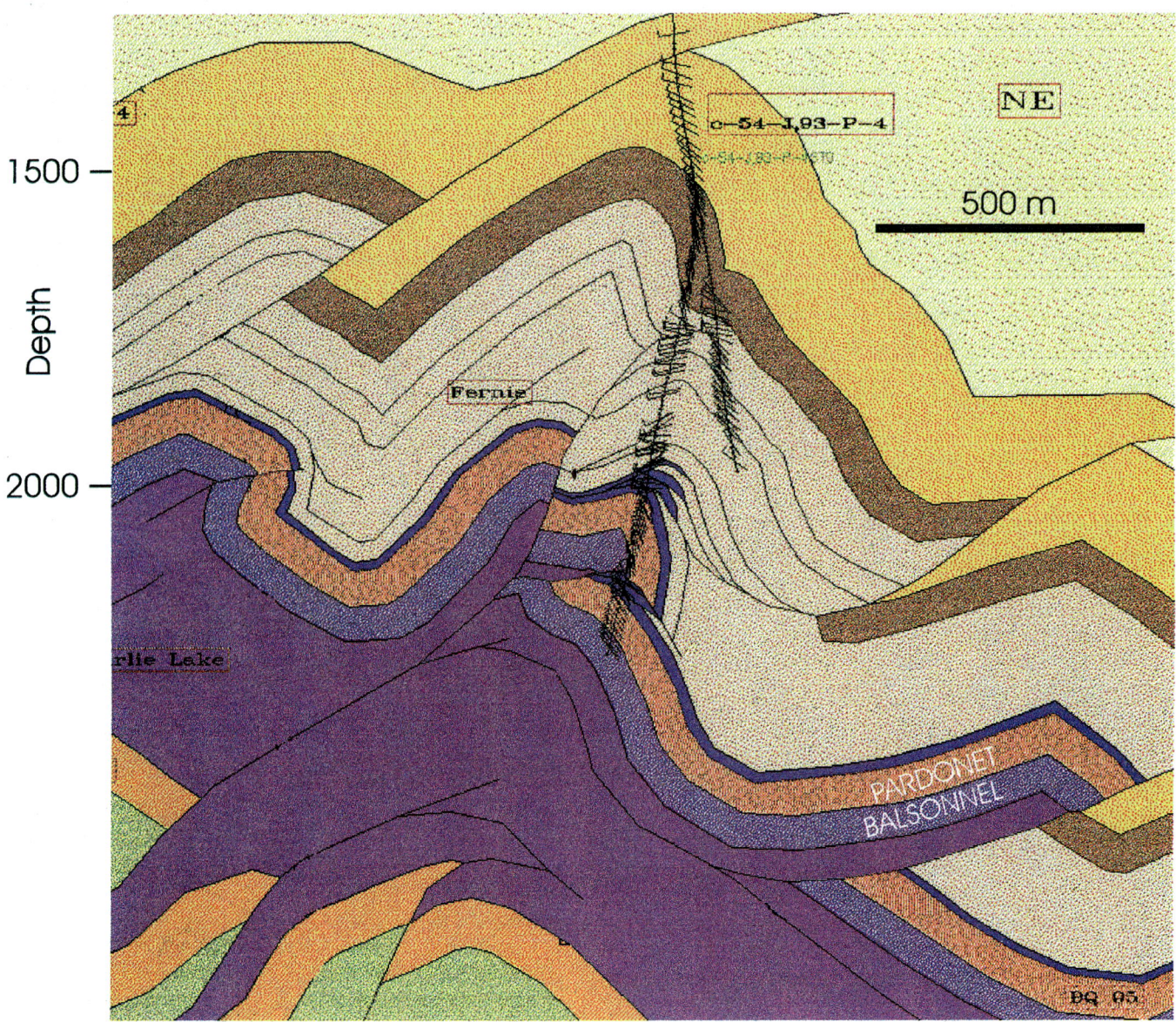

Fig. 7. Final interpretation of the subsurface based on all available information. The original and final (deeper) trajectories are shown. The adjacent well c-45-J (not shown) penetrates the anticlinal structure to the southwest of c-54-J.

fields. The high-amplitude zone visible in Figure 6b at approximately 1800 metres in depth, most likely corresponds to the highest point on the Triassic structure, while the high-amplitude zone at about 2000 metres in depth corresponds to the downthrown block.

The VSP proved to be a useful tool; it provided a timely result which was used in combination with other geological and geophysical information, to constrain the solution and make a decision on the deviation of the well. The image derived from the VSP has been subjected to only partial verification by drilling, which tested one zone marked by a high amplitude interpreted as a downthrown block. Another high-amplitude event, interpreted to be the crest of the Triassic anticline remains untested.

The importance of the VSP study performed on the Talisman/Ocelot c-54-J well goes beyond its use for this particular well. It contributed to the understanding of several important aspects of pre-survey planning, acquisition, processing and interpretation. This study is an example of the VSP as an exploration tool and shows the importance of using multiple sources in imaging complex structures.

REFERENCES

Payne, M.A., Eriksen, E.A., and Rape, T.D., 1994, Considerations for high-resolution VSP imaging: The Leading Edge, 13(3), 173-180.

Zhu, J., and Lines L., 1994, Imaging of complex subsurface structures by VSP migration: Can. J. Expl. Geophys., 30, 73-83.

Prestack depth migration of an Alberta Foothills data set—The Husky experience

W.-J. Wu[*], L. Lines[‡], A. Burton[**], H.-X. Lu[‡], J. Zhu[§], W. Jamison[§§], and R. P. Bording[‡‡]

ABSTRACT

We produce depth images for an Alberta Foothills line by iteratively using a number of migration and velocity analysis techniques. In imaging steeply dipping layers of a foothills data set, it is apparent that thrust belt geology can violate the conventional assumptions of elevation datum corrections and common midpoint (CMP) stacking. To circumvent these problems, we use migration from topography in which we perform prestack depth migration on the data using correct source and receiver elevations. Migration from topography produces enhanced images of steep shallow reflectors when compared to conventional processing. In addition to migration from topography, we couple prestack depth migration with the continuous adjustment of velocity depth models. A number of criteria are used in doing this. These criteria require that our velocity estimates produce a focused image and that migrated depths in common image gathers be independent of source-receiver offset. Velocity models are estimated by a series of iterative and interpretive steps involving prestack migration velocity analysis and structural interpretation. Overlays of velocity models on depth migrations should generally show consistency between velocity boundaries and reflection depths. Our preferred seismic depth section has been produced by using prestack reverse-time depth migration coupled with careful geological interpretation.

INTRODUCTION

In recent years, industrial research groups participated in prestack migration comparisons using the Marmousi data, a synthetic data set designed at the Institut Français du Petrole (Versteeg, 1994). In conducting similar experiments for a structurally complex Alberta foothills data set, we use a line recorded by Husky Oil and Talisman Energy. The data set, known as the Husky Structural Data Set, was provided for the 1995 SEG and 1996 CSEG Convention Workshops on Structural Imaging by Larry Mewhort of Husky Oil and by Christof Stork of Advance Geophysical. It is a data set used in testing current algorithms for imaging complex structures. The line is from the Benjamin Creek area of southern Alberta, and many of the preliminary results of the studies were compiled by Stork et al. (1995). It is anticipated that this real data set will serve as a standard for many future processing tests in the same way that the Marmousi data has provided a means for testing seismic imaging algorithms on model data. In this area of the foothills, the topography has large elevation changes (300 m over the length of the line), and the thrust belt has considerable structural complexity. For this structural imaging problem, major efforts involved the construction of a realistic velocity model and the migration from topography, a technique introduced by Wiggins (1984) and described in Gray and Marfurt (1995) and Lines et al. (1996). For the acquisition of this foothills line, there were 143 shots. The shot spacing was variable because of the rough mountainous terrain. Typical shot spacing was 100 m. There were generally 300 traces per shot with geophone spacing of 20 m, giving a common depth point (CDP) spacing of 10 m.

Along with the migration from topography concept, a major issue involved with effective foothills imaging is the

---

Presented at the 66th Annual International Meeting, Society of Exploration Geophysicists. Manuscript received by the Editor January 24, 1997; revised manuscript received July 31, 1997.
*Geo-X Systems Ltd., 425 1st Street S.W., Calgary, Alberta T2P 3L8, Canada.
‡Formerly Dept. of Earth Sciences, Memorial University of Newfoundland, St. John's, Newfoundland A1B 3X5; presently University of Calgary, Dept. of Geology and Geophysics, 2500 University Dr. N.W., Calgary, Alberta T2N, N4, Canada.
**Dept. of Earth Sciences, Memorial University of Newfoundland, St. John's, Newfoundland A1B 3X5, Canada.
§Formerly Dept. of Earth Sciences, Memorial University of Newfoundland, St. John's, Newfoundland A1B 3X5, Canada; presently G-X Technologies, 5847 San Felipe, Houston, Texas, 77057.
§§The Upper Crust Inc., 1430, 700-4th Ave. S.W., Calgary, Alberta, T2P 3J4, Canada.
‡‡Institute for Geophysics, The University of Texas at Austin, Texas 78759-8397.
© 1998 Society of Exploration Geophysicists. All rights reserved.

development of an accurate seismic velocity model. Based on the model studies of Lines et al. (1993), a preferred method of velocity analysis in such areas of structural complexity uses iterative prestack depth migration. Prestack migration velocity analysis appears to have the fewest restrictive assumptions involving the structural nature of the velocity model. Stork (1992) and Whitmore and Garing (1993) describe methods for optimizing velocity based on the fact that common reflection-point (CRP) depth images should be independent of source-receiver offset or incident plane-wave angle. Methods that optimize the focusing of prestack depth migrations by estimation of velocity have been discussed in MacKay and Abma (1992). In this study, we use combinations of these prestack migration velocity analysis methods, coupled with interpretation to optimize the velocity models.

METHODOLOGIES

Three methodologies were crucial to the success of imaging complex foothills geology:

1) Prestack migration from topography,
2) Iterative and interpretive adjustment of velocity for optimizing prestack migration,
3) Structural interpretation of the prestack migrations during the iterative process.

Poststack reverse-time migration has proven to be a useful and general migration method for imaging both 2-D and 3-D data as evidenced by the many papers including those of McMechan (1983), Whitmore (1983), Baysal et al. (1983), Mufti et al. (1996), and Wu et al. (1996). Reverse-time migration methods have also been used successfully for prestack depth migration as exhibited by the research of Chang and McMechan (1986), Whitmore and Lines (1986), Chang and McMechan (1990), Wu et al. (1995) and Lines et al. (1996). Reverse-time migration is a general method that is able to successfully define complex structures such as overhanging salt domes and overthrust folded layers. We shall show a comparison between the poststack and prestack forms of reverse-time migration for steeply dipping beds and variable topography.

However, major problems with the conventional seismic processing of Canadian foothills data occur because of the wide variation in surface elevations and the complex geology of thrust faulting and folding of sedimentary layers producing locally very steep dips. Thrust-belt geology can cause problems with conventional methods that attempt to apply statics corrections to a datum plane. Conventional elevation statics corrections (Dobrin, 1976) assume vertically traveling energy (raypaths) between the datum plane and the surface. If the surface layer velocity is considerably less than the velocity of deeper layers and reflections from deeper layers are traveling in a near vertical direction, then the vertical ray statics assumption is reasonably valid. This assumption is a good one for an area such as the Western Canadian Basin which is covered by a low-velocity unconsolidated glacial drift layer. However, this assumption is often violated in thrust-belt environments where steeply dipping layers of high seismic velocity outcrop at the surface. In foothills geology, seismic reflection energy travels at oblique angles through steeply dipping surface layers. Second, thrust-belt geology often causes a violation of the common reflection-point (CRP) assumptions. Source-receiver combinations having a common midpoint (CMP) will generally not share a CRP for a steeply dipping event. Therefore, the sorting and stacking of CMP traces will cause a smearing of reflection energy. This smearing effect will be particularly deleterious for shallow steeply dipping events that often occur in foothills geology. Because of the breakdown of these assumptions, it is not surprising that shallow steeply dipping events are not imaged clearly by conventional processing.

Both the statics correction problem and the CMP problem can be obviated by the use of prestack migration with sources and receivers at their correct elevation provided we have an accurate velocity model. This concept is illustrated with synthetic examples in McMechan and Chen (1990). Figure 1 shows the basic model for prestack migration where the source and receiver positions are located correctly. This model follows the discussion of Liner and Lines (1994) who describe prestack migration in terms of aplanatic surfaces. For a particular traveltime, the aplanatic surface defines the locus of possible reflection points for particular locations of source and receiver. For a constant velocity medium, the aplanatic surfaces would be ellipses. For variable velocity media, the aplanatic surface is a distortion of an ellipse that can be computed by solutions

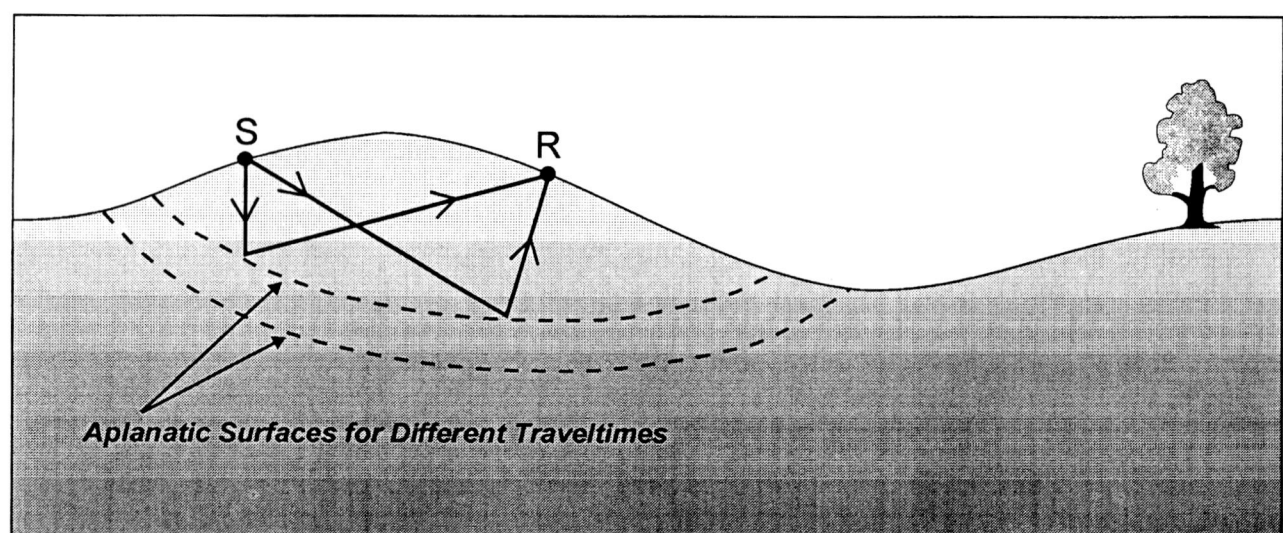

FIG. 1. Aplanatic surfaces define possible reflection points for arbitrary source-receiver elevations. This figure from Lines et al. (1996) used with the permission of the Cana. J. Expl. Geophys.

to the eikonal equation or by solutions to the wave equation. The use of eikonal solvers to define aplanatic surfaces is used in Kirchhoff migration (Gray and May, 1994; Zhu and Lines, 1995), whereas prestack reverse-time migration methods can use finite-difference wave equation solutions to define aplanatic surfaces effectively (Chang and McMechan, 1986, 1990; Botelho and Stoffa, 1988). For prestack Kirchhoff migration, seismic trace energy is spread over these aplanatic surfaces for a given trace. We then repeat this procedure for a multitude of traces and then sum the trace migrations to obtain the migrated image in depth.

Prestack migration from topography does not suffer from either the shortcomings of datum statics corrections or the assumptions that CMPs are the same as CRPs. As we shall demonstrate, these shortcomings are particularly problematic for foothills data. The key ingredient for the success of prestack migration is the availability of an accurate velocity model.

One prestack migration velocity analysis technique uses the criterion that the depth image at a CRP should be independent of offset. That is, the correct depth estimate of the reflector should not depend on the geometry of the seismic experiment. To test the validity of the migration result and the velocity model, we apply prestack depth migration to the data, and for a CRP we examine the offset variation of the depth migrations. A variation in the depth estimates versus offset leads to velocity adjustment of the model to eliminate "smiles" or "frowns" as a function of offset for migrated CRP depth gathers. The CRP depth gathers should be "flat," i.e., independent of offset. This approach is very similar to normal moveout analysis of CMP data. As explained in Whitmore and Garing (1993), the same kind of analysis can be performed using plane-wave angles instead of source-receiver offset distances.

A second velocity analysis technique attempts to improve the focusing of the prestack migration by velocity adjustment. This interpretive method is similar to the focusing of a camera in that the correct velocity model is assumed to be the one that provides the most focused image. This form of migration velocity analysis is analogous to velocity analysis by CMP stacking.

Iterative traveltime tomography (Bording et al., 1987) represents yet another general method for velocity model estimation. This method can use prestack traveltimes from direct arrivals, refracted arrivals, or reflected arrivals. Tomography sometimes experiences difficulties with complex geology since one has to be careful that the picked traveltimes are the traveltimes that one is modeling by ray tracing.

For the Husky data set, there is also additional information in the form of well logs and VSP data that can provide constraints on the velocity model. An inversion method developed by Lines (1993) attempts to match depth migrations to formation tops. Although this was helpful for this study, it had the shortcoming that many layers in the model do not intersect the logged portions of the wells. Therefore, this migration/inversion method will not provide the entire velocity model picture in this case. Iterative prestack depth migration, coupled with geological interpretation, was our main tool in our development of a velocity model. Geological interpretation of the complex structural velocity model also provided a key part of the velocity model definition.

Several iterative adjustment to the original velocity model were based upon the characteristics of prestack migration. The improved versions of the seismic depth image were reinterpreted by the structural geologist. This reinterpretation uses current concepts of structural geometries and interactions in fold-and-thrust systems and adhered to the basic concepts of structural balancing (but the new interpretation was not rigorously balanced). The improved version of the seismic record used for this reinterpretation allowed dilineation of some duplex faulting and suggested changes in the positions of thrust faults and the associated hanging wall structures.

RESULTS

The initial velocity model for iterative prestack depth migration is shown in Figure 2. This model, provided to the SEG

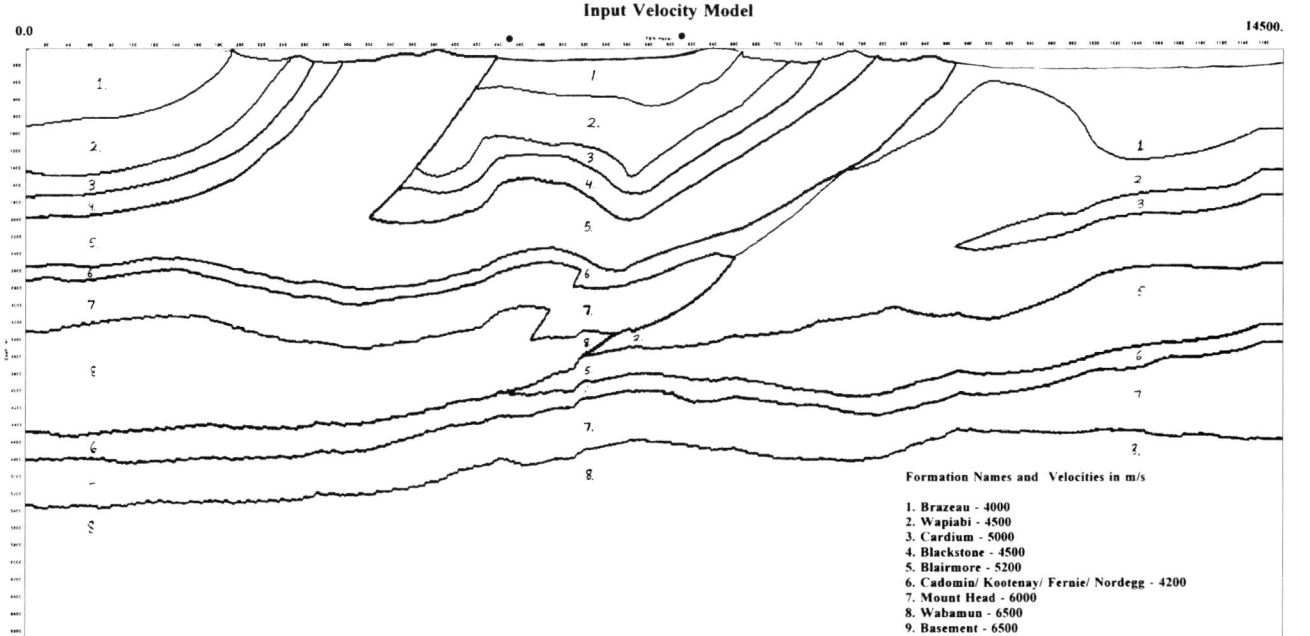

FIG. 2. Initial velocity model shows steeply dipping, high velocity layers of foothills formations. Model width = 14.7 km. Model depth = 7 km.

workshop participants, was based on surface outcrops, well information, and geological interpretation of seismic data. As shown by this model, there are steeply dipping events that outcrop at the surface. Such steeply dipping events are problematic for CMP stacking and poststack migration. These problems are evidenced by the lack of coherent reflectors near the surface, as shown in the poststack migration of Figure 3. As previously mentioned, these near-surface reflection imaging problems are related to the breakdown of the elevation statics and CMP stacking assumptions.

To obviate these difficulties, we used prestack "migration from topography." The resulting prestack migration of Figure 4 for the same initial velocity model indicates that "migration from topography" does not suffer from the same near-surface imaging problems as poststack migration. The steeply dipping layers that outcrop at the surface are imaged more clearly in Figure 4 than the blurred near-surface images of Figure 3. We see significant differences between these migrations clearly in the upper left portion of the sections (upper 3 km and westmost 4 km of the section) where there is a syncline containing dipping beds of the Brazeau, Wapiabi, Cardium, Blackstone, and Blairmore formations. The poststack migration also shows some reflection energy with conflicting dips; these conflicting dips do not appear in the prestack migration. In comparing the poststack and prestack migrations of Figures 3 and 4, we note that the depth migration differences become less obvious for deeper reflections. This is an expected result since as offset/reflection depth ratios decrease, raypath differences between prestack and poststack models become less pronounced. For shallow steeply dipping reflectors; however, the

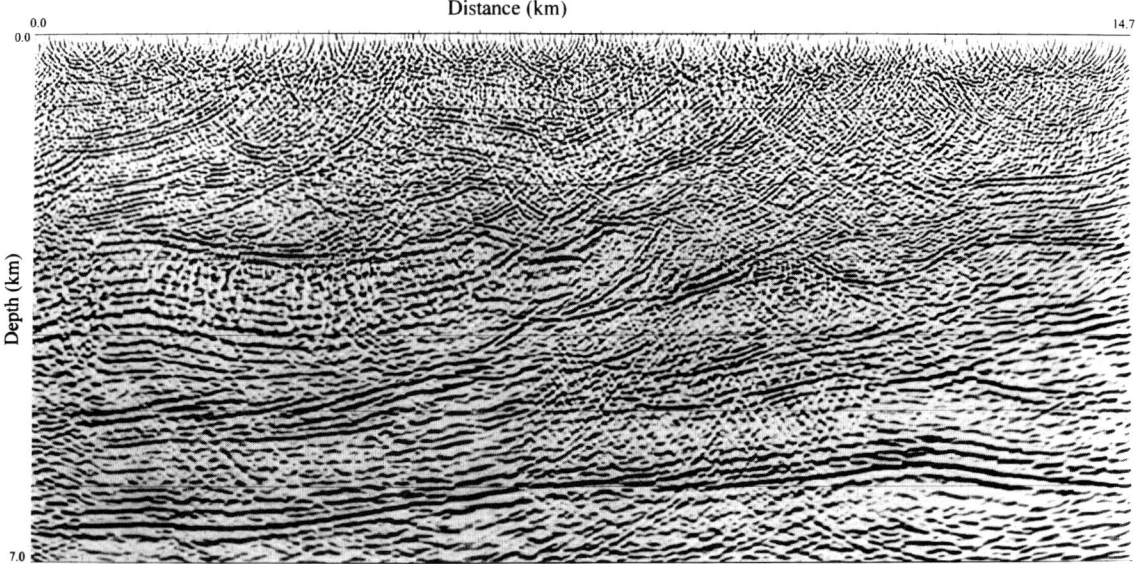

FIG. 3. Poststack migration of data following conventional elevation datum corrections from a flat datum plane. Section width = 14.7 km. Section depth = 7 km. This figure from Lines et al. (1996) used with the permission of the Cana. J. Expl. Geophys.

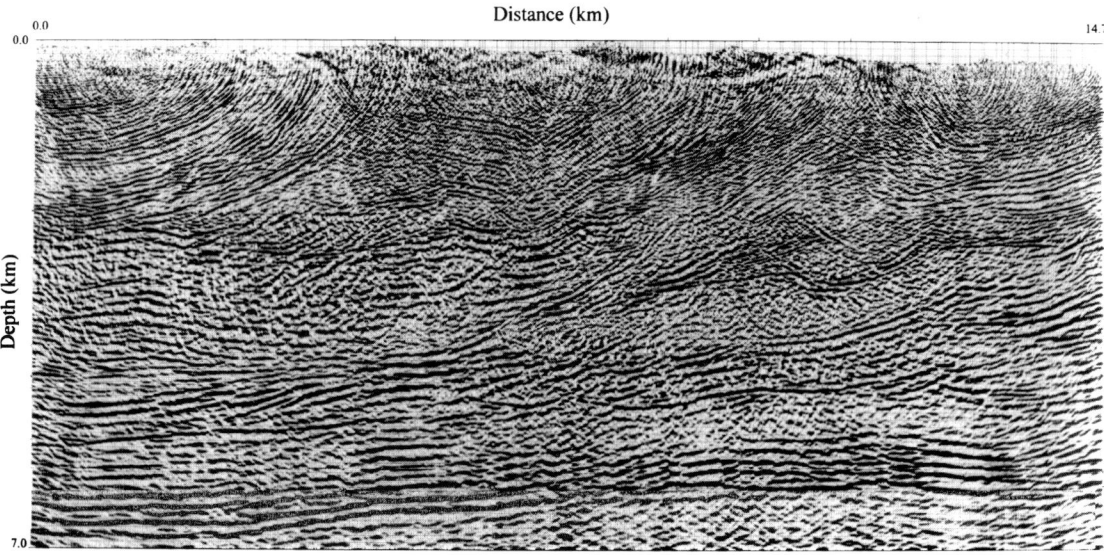

FIG. 4. Prestack migration from topography using the same velocity model as in Figure 3. Section width = 14.7 km. Section depth = 7 km. This figure from Lines et al. (1996) used with the permission of the Cana. J. Expl. Geophys.

prestack migration result is better than the poststack migration result.

Further improvements to the image are obtained through improvement of the velocity model. A series of iterative and interpretive steps was used to refine this model. For the near-surface layers, traveltime tomography proves to be a good technique for determining velocity from direct arrivals and head-wave arrivals. For the deeper reflected events, we used common-image gather analysis, focusing analysis, and a structural interpretation of the depth migration to estimate a layered velocity model. The resulting velocity model is derived resulted from an integration of interpretation and quantitative analysis, and it is difficult to describe an exact recipe.

Velocity models were generally derived by using isotropic analysis. According to Skuce (1995), velocity models should also take anisotropy into account, especially in shale layers. This appears to be a very difficult inversion problem for complicated structures unless anisotropic rock property measurements are available as constraints on the problem. The problem is made difficult by the fact that the original depositional layering (and hence the anisotropy) was likely horizontal. Following this depositional phase, thrust faulting caused the layers to be steeply dipping over a wide range of angles. Therefore, the layers would have a complex anisotropic nature, which would be very difficult to infer from seismic data alone. Anisotropic inversions for velocity will generally have a high degree of nonuniqueness and would seem to require additional rock property measurements as constraints. Since such measurements were not available to us, the anisotropic inversion problem was not pursued but is a prospect for future investigations.

One aspect of the velocity modeling/prestack migration process became abundantly clear through the imaging procedure. It was important to have the geological interpreter "in the loop" when defining the velocity model. The reinterpretation of the seismic depth migration in Figure 4 gave an adjusted velocity model from the original model in Figure 2. Figure 5 shows the result of velocity model adjustment after three iterations of prestack migration and velocity model adjustment. Finally, this new velocity model of Figure 5 is used to produce the prestack reverse-time depth migration shown in Figure 6. It is believed that this migration gave a better definition of the shallow steeply dipping events than did the prestack depth migration using the initial model. At each iteration, the layer boundaries from the velocity model are derived from the reflecting boundaries in the depth migration. One of the useful methods for deducing this model involves the use of color overlay plots of velocity models on the depth migration. There should be general consistency between the layer boundaries and the reflecting interfaces of the depth migration since most reflection energy is derived from velocity contrasts in these sedimentary environments. Figure 7 shows a color overlay of the final velocity model on the depth migration. It is believed that this velocity model/reflection image is accurate to a depth of about 5 km. Detailed velocity model adjustment was not continued below this depth since the velocity leveraging decreases as offset/depth is less than unity (Lines et al., 1993). In this case, the largest offset was about 6000 m, so that velocity analysis by "moveout methods" below a depth of 6000 m would be considered somewhat inaccurate. Nevertheless, a verification of the depth migration results by using two wells shows seismic depths to be within 30 m of the formation tops for each identifiable horizon.

CONCLUSIONS

Our conclusions from the study thus far are two-fold. First, prestack depth migration with correct source-receiver elevations is preferable to poststack migration from a datum in areas of large elevation change. Second, velocity model adjustments

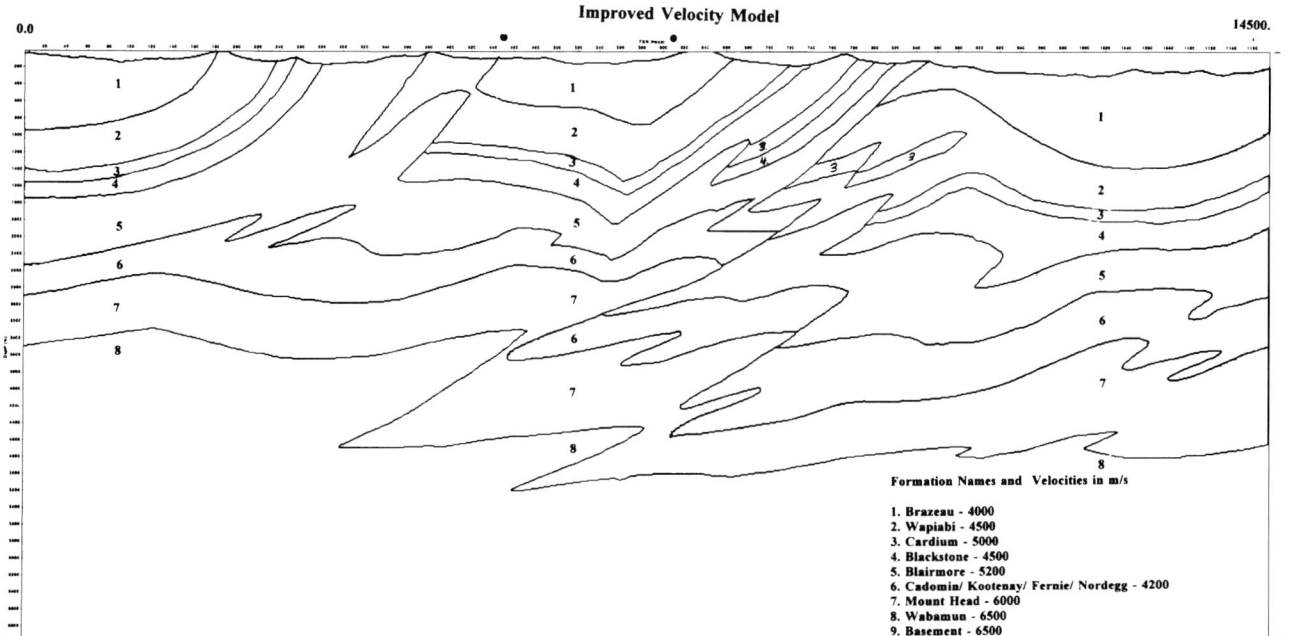

FIG. 5. Updated velocity model shows alteration of the velocity model in Figure 2 following three iterations of prestack migration velocity analysis and interpretation.

should be done in an interpretive setting with several tools when using prestack migration on foothills data.

Useful depth migrations of foothills data depend on accurate velocity models, general migration and velocity analysis algorithms, and detailed interpretation. The results of the Husky data set experiments for the 1995 SEG workshop will undoubtedly see future publications. Thus far, the largest improvements in processing that we have seen are a result of the use of prestack "migration from topography" and "iterative interactive interpretation." We conclude that for shallow, steeply dipping foothills reflections, prestack migration from topography will provide a substantial improvement over the conventional processing flow of elevation datuming, CMP stacking, and poststack migration. Furthermore, the improved migrations are made possible by the iterative interpretation modifications of the previous migration's velocity models during the imaging process.

ACKNOWLEDGMENTS

We wish to thank NSERC, Petro-Canada, and sponsors of the MUSIC project for their financial support of this research. We also acknowledge the support of Larry Mewhort of Husky Oil Inc. and Christof Stork of Advance Geophysical for making this Alberta foothills data set available to us.

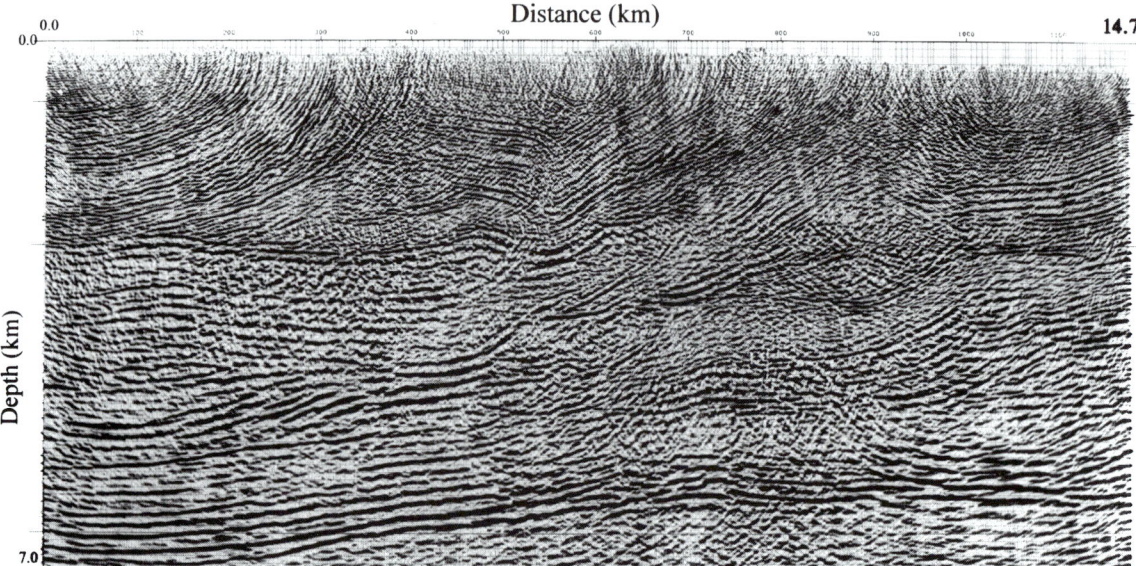

FIG. 6. Prestack depth migration using the improved velocity model of Figure 5. Section width = 14.7 km. Section depth = 7 km.

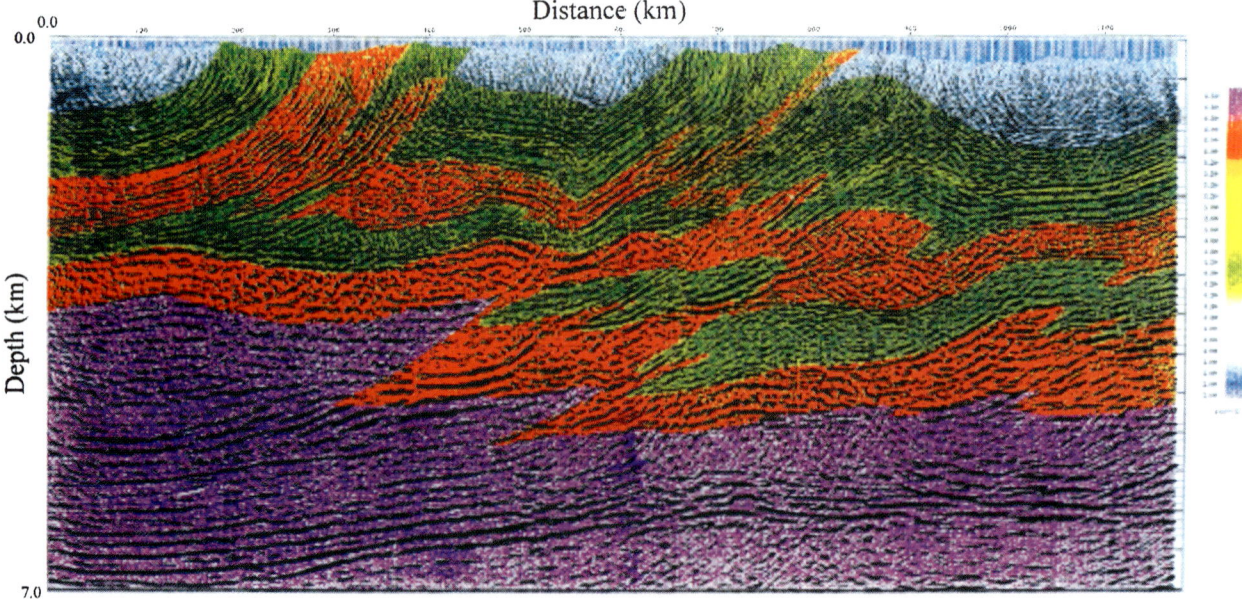

FIG. 7. Color overlay of velocity model on prestack depth migration.

REFERENCES

Baysal, E., Kosloff, D. D., and Sherwood, J. W. C., 1983, Reverse-time migration: Geophysics, **48**, 1514–1524.

Bording, R. P., Gersztenkorn, A., Lines, L. R., Scales, J. A., and Treitel, S., 1987, Applications of seismic traveltime tomography: Geophysical J. Roy. Astr. Soc., **90**, 285–303.

Botelho, M. A. B., and Stoffa, P. L., 1988, Velocity analysis using reverse-time migration, EOS, **49**, 1326.

Chang, W. F., and McMechan, G. A., 1986, Reverse-time migration of offset vertical seismic profiling data using the excitation-time imaging condition: Geophysics, **51**, 67–84.

Chang, W. F., and McMechan, G. A., 1990, 3D acoustic prestack reverse-time migration: Geophys. Prosp., **38**, 737–755.

Dobrin, M., 1976, Introduction to geophysical prospecting: McGraw-Hill Book Co.

Gray, S. H., and May, W. P., 1994, Kirchhoff migration using eikonal equation traveltimes: Geophysics, **59**, 810–187.

Gray, S. H., and Marfurt, K. J., 1995, Migration from topography: Improving the near-surface image: Can. J. Expl. Geophys., **31**, 18–24.

Liner, C. L., and Lines, L. R., 1994, Simple prestack migration of crosswell seismic data: J. Seis. Expl., **3**, 101–112.

Lines, L. R., Rahimian, F., and Kelly, K. R., 1993, A model-based comparison of modern velocity analysis methods: The Leading Edge, **12**, No. 7, 750–754.

Lines, L. R., 1993, Optimization of seismic migration through use of well information: Can. J. Expl. Geophys., **29**, 419–428.

Lines, L. R., Wu, W., Lu, H., Burton, A., and Zhu, J., 1996, Migration from topography: experience with an Alberta Foothills data set: Can. J. Expl. Geophys., **32**, 24–31.

MacKay, S., and Abma, R., 1992, Imaging and velocity estimation with depth-focusing analysis: Geophysics, **57**, 1608–1622.

McMechan, G. A., 1983, Migration by extrapolation of time-dependent boundary values: Geophys. Prosp., **31**, 413–430.

McMechan, G. A., and Chen, H. W., 1990, Implicit static corrections in prestack migration of common-source data: Geophysics, **55**, 757–760.

Mufti, I. R., Pita, J. A., and Huntley, R. W., 1996, Finite-difference depth migration of exploration-scale 3-D seismic data: Geophysics, **61**, 776–794.

Skuce, A., 1995, Seismic imaging in the Canadian Rocky Mountain foothills: Presented at the 1995 Ann. Mtg., Soc. Expl. Geophys. workshop #6.

Stork, C., 1992, Reflection tomography in the post-migrated domain: Geophysics, **57**, 680–692.

Stork, C., Welsh, C., and Skuce, A., 1995, Demonstration of processing and model building methods on a real complex structure data set: Presented at the Ann. Mtg. workshop #6, Soc. Expl. Geophys.

Versteeg, R., 1994, The Marmousi experience: velocity model determination on a synthetic complex data set: The Leading Edge, **13**, No. 927–936.

Wiggins, W., 1984, Extrapolation and migration of nonplanar data: Geophysics, **49**, 1239–1248.

Whitmore, N. D., 1983, Iterative depth migration by backward time propagation: 53rd Ann. Internat. Mtg. Soc. Expl. Geophys., Expanded Abstracts, 382–385.

Whitmore, N. D., and Lines, L. R., 1986, Vertical seismic profiling depth migration of a salt dome flank: Geophysics, **51**, 1087–1109.

Whitmore, N. D., and Garing, J. D., 1993, Interval velocity estimation using iterative prestack depth migration in the constant angle domain: The Leading Edge, **12**, No. 757–762.

Wu, W.-J., Lines, L. R., and Lu, H.-X., 1995, Reverse-time migration with finite-difference solutions of the acoustic wave equation: 65th Ann. Internat. Mtg. Soc. Expl. Geophys., Expanded Abstracts, 1149–1152.

Wu, W.-J., Lines, L. R., and Lu, H.-X., 1996, Analysis of higher-order finite-difference schemes in 3-D reverse-time migration: Geophysics, **61**, 845–856.

Zhu, J., and Lines, L. R., 1995, Practical subsurface imaging by prestack depth migration: CSEG Ann. Mtg., Expanded Abstracts, 5–6.

An imaging comparison of three depth migration algorithms on Alberta Foothills datasets

Lanlan Yan and Larry, R. Lines, FRP, Dept. of Geology and Geophysics, Univ. of Calgary

Summary

In performing seismic migration from the Alberta foothills, we used three popular imaging methods, Kirchhoff, reverse-time, and f-x depth migration. Migration results from real data demonstrated that despite inaccurate velocity models from the initial interpretation, once a reliable structural stack was available, each method could provide us with a reasonable post-stack depth migration result. Through relative comparisons among these three algorithms, the method based on the Kirchhoff integral proved to be the most viable of the three methods for seismic imaging of complex geologic structures - with respect to imaging robustness, computational efficiency, as well as target-oriented capability. The post-stack depth migration obtained will be an intermediate result, which would be subsequently used to get a relatively accurate velocity model for pre-stack depth migration. With an accurate velocity field available and with advanced computer technology, these migration methods will make a new improvement in subsequent 3-D pre-stack depth migration.

Introduction

Seismic migration is usually the last step of a series of processing sequences. Since it can provide the image of the Earth's subsurface for geophysicists and geologists to decide on prospective locations for hydrocarbon traps, many advanced migration methods implemented in different domains (such as x-t, x-f, f-k, t-k, etc.) have come into use in the past two decades. Kirchhoff integral migration and reverse-time migration are two popular migration methods performed in the x-t domain. The advantages of both methods lie in the fact that they have no limitations of imaging steeply dipping reflections and can handle both vertical and lateral velocity variations, which is typical of the complex structure of foothills. F-x migration is a fairly recently developed migration technique, since it is implemented in the hybrid domain of frequency-space (f-x); it is also called f-x or omega-x migration. Just like Kirchhoff and reverse-time migration methods, the strengths of f-x migration are in handling steeply dipping structures (up to 90 degrees) and its exceptional ability to deal with velocity variations, whether vertical or lateral.

Implementation of Kirchhoff migration basically involves two steps: 1) calculation of the travel-times from the reflector point to the shot point and to the receiver point, respectively. 2) Weighted summation of the amplitude of the derivative traces along the presumed diffraction trajectory. Due to its characteristics of computational efficiency, good imaging, as well as target-oriented capability, Kirchhoff migration has proven to be one of the more versatile and robust migration methods (ref. Gray, 1997, 1998).

In contrast to the Kirchhoff migration, reverse-time migration is a relatively recent seismic imaging technique. Through using the two-way acoustic wave equation, the reverse-time method performs its migration work in steps of wave field extrapolation backward in time followed by seismic imaging. Since reverse-time depth migration is performed in the t-x domain which can handle vertical and lateral velocity variations and dips up to and beyond 90 degrees; such an algorithm will be extremely computer-intensive (ref. Zhu et. al., 1998; Wu et. al, 1998)

Unlike Kirchhoff and reverse-time migrations, explicit finite difference migration is a newer method performed in f-x domain (Balcquiere, et. al, 1991; Han, 1998). Although it is grouped with finite-difference techniques, no mathematical finite-differences are employed in the whole procedure. The f-x migration performs its post-stack depth migration using explicit spatially variant extrapolators (also called convolution filters applied in the f-x domain). With a precomputed table of migration convolutional operators being frequently employed, this method has the characteristics of lower running time and fairly stable imaging result. Especially, compared with traditional implicit finite-difference techniques (15 degree, 45 degree and 65 degree finite-difference methods), the f-x migration approach is more accurate and suffers much less from dispersion effects.

Comparisons of post-stack depth migration on a real dataset

Commonly, seismic data acquired in very complicated geological areas such as the steeply dipping thrust belts in the Alberta foothills areas are often not adequately imaged by the conventional process of using NMO, DMO, stack, and migration, because the structurally complex geology often has both vertical and lateral velocity variations there. In this case, pre-stack depth migration should undoubtedly be applied. However, under the circumstance that the true subsurface velocity is unknown, the sensitivity of pre-stack depth migration to the accuracy of the velocity model is a disadvantage. Actually, estimation of the interval velocity model is much harder than migration itself since velocity analysis and migration are closely related to each other in the whole migration procedure. As such, post-stack depth migration, an intermediate image result, can still serve in the iterative modification of the velocities until a relatively accurate velocity field is obtained. A successful application has been presented in sub-salt imaging in the Gulf of Mexico by Ratcliff et al. (1994). From a practical viewpoint, the investigation of post-stack depth migration is important for refining our imaging result before computing the final pre-stack depth migrations. The seismic data used in migration is a 2-D exploration line acquired in the Alberta foothills. The structural stacked section is shown in Figure 1. This line extends in the NE-SW direction over a triangular back-thrust environment where deformation is characterized by reverse-thrusts over a duplex core structure, accompanied by ramping and sole detachment in the deeper section. The topography along the seismic line has a significant variation. Since the structural stack was processed well and its inherent geological structure is already evident, we sequentially used three velocity models to perform migrations and compare their imaging results by three different migration algorithms.

The simplest model used in the migration comparisons, shown in Figure 2, was a model with vertical velocity variation V(z) only. All of the three methods, reverse-time migration (Figure 3), Kirchhoff migration (Figure 4), and f-x migration (Figure 5) give an acceptable imaging result for this model. Figure 6 was the V(x,z) model given to us by Mobil and MUSIC (Memorial University Seismic Imaging Consortium). It was obtained from the interpretation of the industry processed pre-stack Kirchhoff diffraction migration. For this model, Kirchhoff migration (Figure 7) gave the best result, with f-x migration being slightly blurred, and reverse-time migration being inferior. With a hybrid model which averaged the V(z) and V(x,z) models, Kirchhoff migration algorithm again proved to give the superior result (shown in Figure 8).

Conclusions

The imaging comparisons of reverse-time, Kirchhoff and f-x post-stack depth migration made us arrive at the following conclusions:
(1) All of the three migration methods can provide us with acceptable and well-imaged intermediate results on the simplest model. Post-stack depth migration was not very sensitive to the accuracy of the velocity field once a reliable structural stack was available.
(2) With respect to imaging accuracy and computational cost, the Kirchhoff migration method seemed to be superior to reverse-time and f-x migration algorithms because its results on different kinds of velocity cases were basically kept coherent and consistent with the velocity model given by Mobil.
(3) Due to lack of an accurate velocity field, reverse-time migration was less stable than the other two imaging methods, especially in the case of the original Mobil velocity model.
(4) Improved pre-stack depth migration can be achieved through the combination of the post-stack depth migration and the iterative, interpretive adjustment of velocity fields by eliminating smiles and frowns on common image gathers

Acknowledgements

We gratefully acknowledge the financial support for this work by the Foothills Research Project and the CSEG Superfund. Special thanks go to Dr. Helen Isaac, Jennifer Cunningham and Han-xing Lu for their help in use of ProMax System. We also deeply appreciate the data set provided to us for this research by Gary Taylor of Mobil Oil Canada.

References

Balcquiere, G., Duijndam, A., and Romijn, R., 1991, Efficient f-x depth migration of shot record practical aspects. First Break, Vol. 9, No.1, 9-23.

Gray, S. H., 1997, Seismic imaging: Use the right tool for the job: The Leading Edge. Vol. 16, 1585-1588.

Gray, S. H., 1998, Speed and accuracy of seismic migration methods: unpublished manuscript.

Han, B., 1998, A comparison of four depth-migration methods, 68th Ann. Internat. Mtg., Soc. Expl. Geophys. Expanded Abstract, 1104-1107.

Ratcliff, D. W., Jacewitz, C. A. and Gray, S. H., 1994, Subsalt imaging via target-oriented 3-D pre-stack depth migration: The Leading Edge, Vol. 13, 163-169.

Wu, W., Lines, L. R., Burton, A.J., Lu, H., Zhu, J., Jamison, W., and Bording, R.P., 1998, Pre-stack depth migration of Alberta foothills datasets- The Husky Experience: Geophysics, Vol. 63, 392-398.

Zhu, J. and Lines, L. R., 1998, Comparison of Kirchhoff and reverse-time migration methods with applications to pre-stack depth imaging of complex structure: Geophysics, Vol. 63, No. 4,   1166-1176.

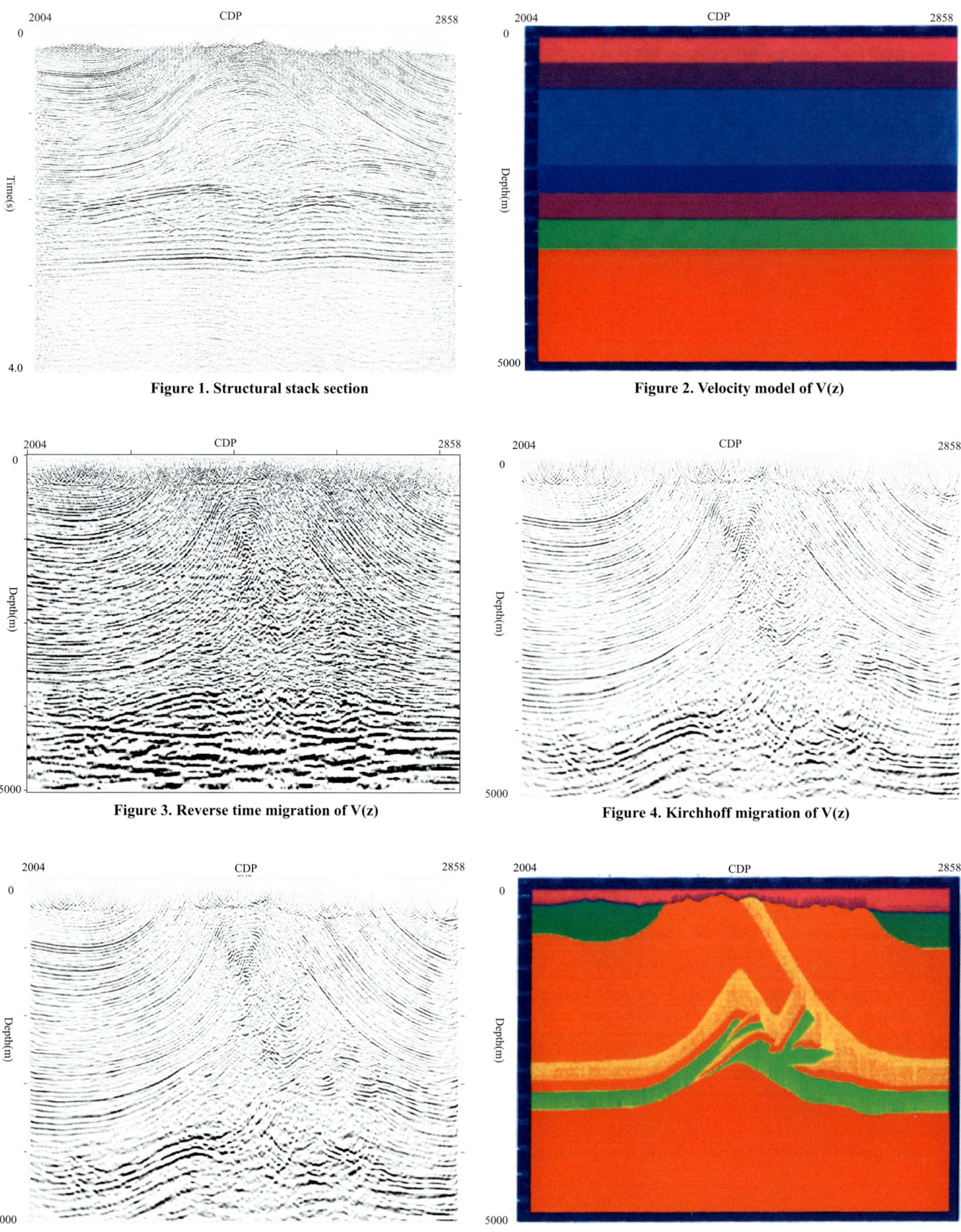

Figure 1. Structural stack section

Figure 2. Velocity model of V(z)

Figure 3. Reverse time migration of V(z)

Figure 4. Kirchhoff migration of V(z)

Figure 5. F-x migration of V(z)

Figure 6. Original velocity model of V(x,z)

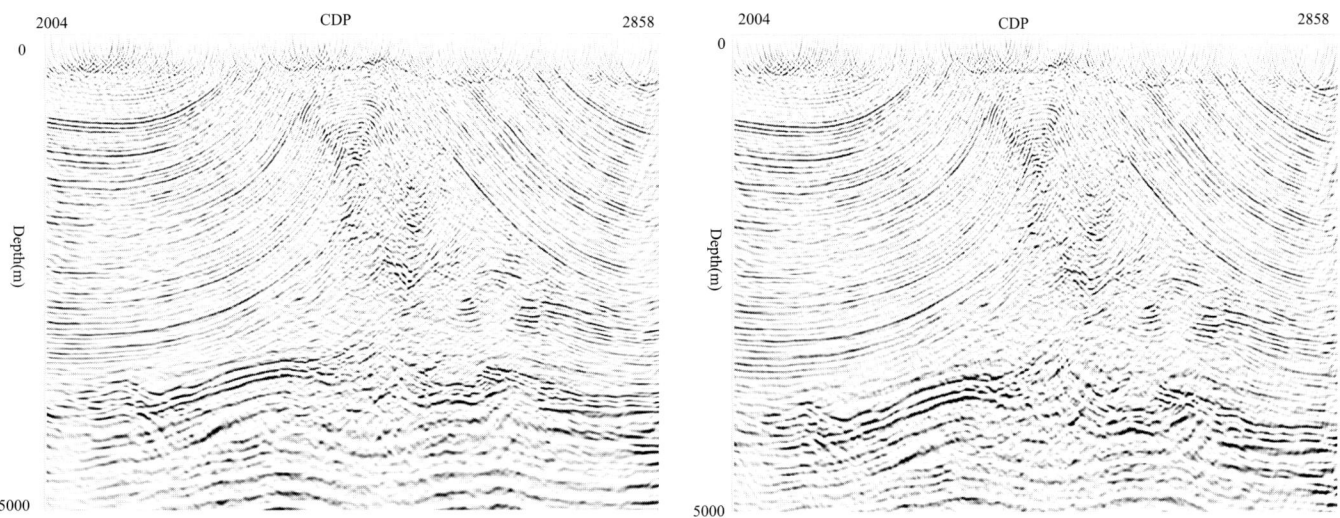

Figure 7. Kirchhoff migration of V(x,z)

Figure 8. Kirchhoff migration of hybrid velocity model

Crooked line, rough topography: advancing toward the correct seismic image
Gary Maclean*, Amoco Canada; Samuel H. Gray, Amoco Canada; Kurt J. Marfurt, Amoco EPTG

Summary

Seismic exploration in mountainous areas imposes serious compromises on both acquisition and processing. Access restrictions usually result in profiles that are not straight and are not recorded along the true dip direction (if there is a true dip direction!). Processing constraints often result in very poor approximate corrections for elevations and for deviations from a straight line. Most fundamentally, 2-D acquisition and processing assumes that the Earth is 2-D; this assumption is often seriously violated in mountainous areas. While we cannot efficiently correct 2-D seismic data for the effects of a fully 3-D subsurface, we can improve the data quality in thrust areas where the 2-D assumption is reasonable. We do this in a series of small steps, which improve the accuracy of several approximations made in processing 2-D land data.

Introduction

Because of acquisition and processing constraints, seismic data acquired in mountainous areas often produce poor subsurface images. The economics and logistics of acquiring seismic data in remote areas, as well as the topographic relief and the very rocks on which the geophones are planted, usually result in suboptimal data acquisition. Typically, isolated lines are shot along traverses chosen more for convenience than for the ultimate purpose of structural imaging. These lines are often crooked and only approximately dip to the geologic structures of interest. Additionally, the hard rocks which outcrop often permit only poor geophone coupling. In processing, many commonly-used procedures, adapted from processing flows where only gentle topography is encountered, lose their validity when applied to data from mountainous areas.

The end result of these problems is an inaccurate seismic image, especially in the near-surface. This is harmful for two completely different reasons. First, although the objective horizons are often deep, uncertainty in estimating the near-surface velocity leads to inaccurate deeper imaging and depth conversion. Second, a poorly-imaged near-surface decreases our ability to place wells optimally and guide the drill bit through the shallow formations, leading to unnecessarily high drilling costs.

This paper addresses the processing side of the data quality issue. Given seismic data acquired along a crooked line which is approximately dip to the geologic structure, we propose the following steps to improve the seismic image:
(1) Process the data directly from topography rather than from a flat or floating datum.
(2) In imaging, pay close attention to the treatment of amplitudes.
(3) Pay close attention to the effect of local topographic variations along the recording spread.
(4) In imaging, account for the crookedness of the acquisition line.

Methods for improving the seismic image

(1) Process the data directly from topography rather than from a flat or floating datum. While recognizing theoretical problems associated with refraction and reflection static corrections, as well as uphole corrections for buried shots, we perform these corrections anyway. However, we avoid the largest component of the error by removing the elevation statics before migration, so that traces can be migrated from the Earth's surface. Alternatively, to accomplish the same goal, we could retain the elevation statics, and migrate using a zero-velocity layer.

(2) In imaging, pay close attention to the treatment of amplitudes. Even carefully processed seismic data are surprisingly sensitive to errors in the migration weights, which should be calculated from topography, and not from a datum surface which might lie well above topography (Gray and Marfurt, 1995). Figures 1-2 illustrate the improvement obtained by performing this step.

(3) Pay close attention to the effect of local topographic variations along the recording spread. As the topography varies along the recording spread, changing from flat to steep, reflection energy from a dipping bed changes its apparent dip. By considering subarrays of the recording spread, we can condition the migration program to reject undesired information, such as certain types of refracted energy, while passing desired information, such as steep-dip reflection energy striking a steeply-sloping part of the spread. We do this by orienting the migration operator, not vertically (which is valid for migrating from a flat surface), but relative to the varying, non-horizontal acquisition surface (Figures 2-3).

(4) In imaging, account for the crookedness of the acquisition line. In principle, this requires a 3-D migration algorithm, but by assuming a single dip direction (i.e., no Earth variations along strike), we can approximate a 3-D migration by a 2-D migration that incorporates out-of-plane wave propagation. For example, we can approximate the traveltime from the surface location (x_0, y_0, z_0)

Crooked line, rough topography

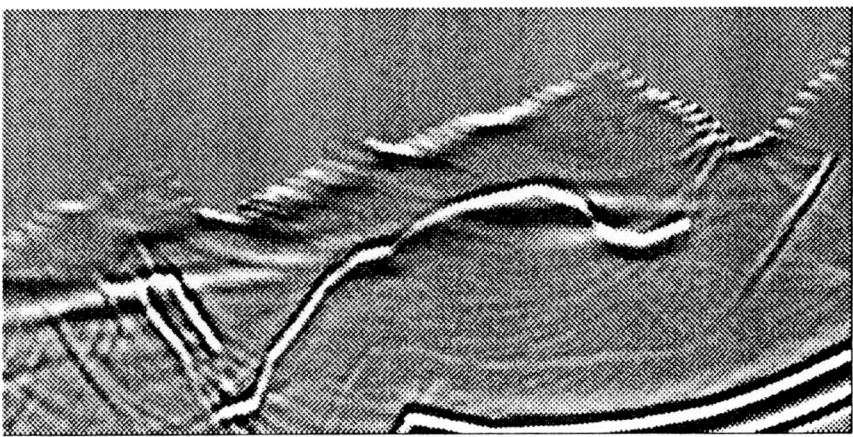

Figure 1. Synthetic data: prestack depth migration (detail) from topography, with migration amplitudes computed from a datum above topography. Note the blurring of the near-surface image.

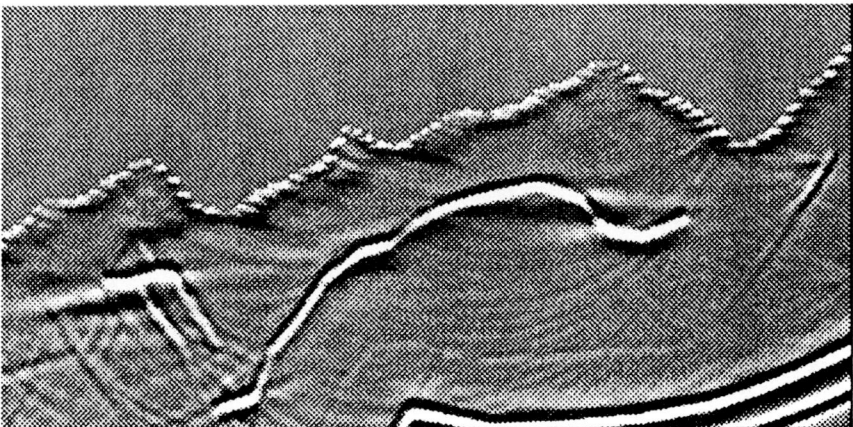

Figure 2. Prestack depth migration from topography, with migration amplitudes computed from topography. The near-surface image is improved over that in Figure 1.

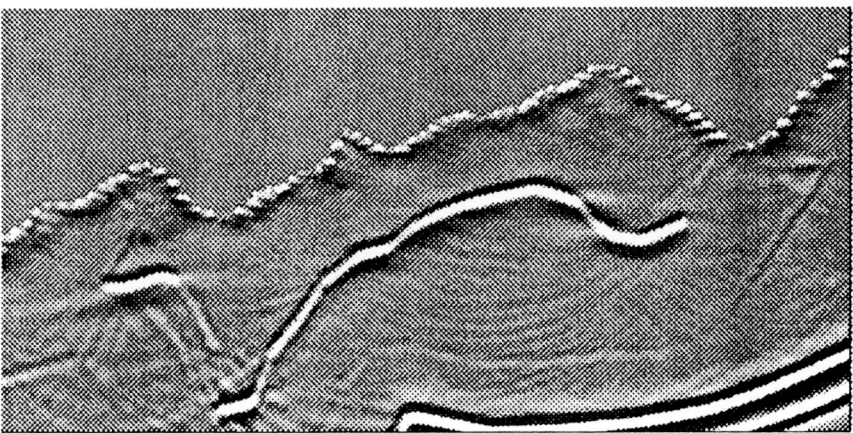

Figure 3. Prestack depth migration from topography, with the near-surface migration operator oriented perpendicular to the acquisition surface, and not vertically as in Figures 1-2. Although the steeply-dipping event on the right is less visible than on Figure 2, the near-surface image has been further clarified.

Crooked line, rough topography

to the subsurface location $(x, y_0 + \Delta y, z)$ by performing a ray-based NMO correction of the traveltime from (x_0, y_0, z_0) to (x, y_0, z). Figure 4 shows CRP gathers from two prestack migrations of synthetic data acquired along a (flat) crooked line of sinusoidal shape; the line is 9 km long in the dip (x) direction with 4 km total variation in the strike (y) direction. Reflection data from two line scatterers placed at depths of 1 km and 2 km in the center of the line were migrated. In Figure 4(a), the data were migrated as if there were no y-variation along the line, and in Figure 4(b), the migration accounted for the crookedness of the line. The residual moveout in the gathers in Figure 4(a) is due not to a velocity error, but to the error of processing the crooked line as a straight line. Although the error decreases with depth, it is serious enough in the near-surface to result in a poor seismic image and a poor estimate of seismic velocities.

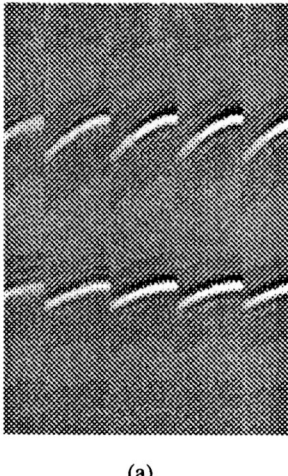 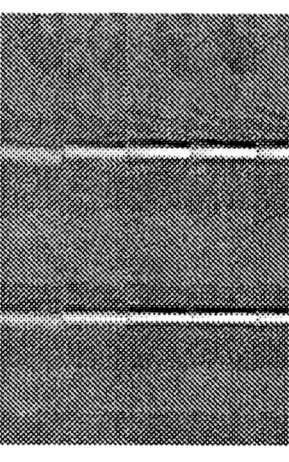

(a) (b)

Figure 4. Migrated CDP gathers from a crooked line. (a) The data have been migrated as if there were no y-variation along the line, resulting in residual moveout in the gathers. (b) The migration has accounted for the crookedness of the line.

These processing recommendations have implications for data acquisition. For example, receiver locations separated by a constant horizontal distance have greater station spacing on steep slopes than receiver locations separated by the same distance measured along the sloping acquisition surface. In order to ensure that different subarrays along the recording spread have the same spatial characteristics for noise rejection and migration purposes, it is desirable to acquire data using constant receiver spacing, not constant horizontal spacing, along the acquisition surface.

Conclusions

A cascade of topographically-related corrections will improve the interpretability of the final seismic image. These corrections are elementary in nature but, aside from the first of them, are not routinely applied. To apply them successfully will require modifications of both the acquisition (i.e., in the surveying of station locations) and the processing (i.e., in the imaging step) of seismic data collected along an irregular traverse. The final seismic image, especially in the near-surface, can benefit from their application by providing added clarity and an improved velocity analysis. These improvements in the near-surface image can have an additional positive effect on the drilling process.

Reference

Gray, S. H. and Marfurt, K. J., Migration from topography: Improving the near-surface image: Can. J. Expl. Geophys., **31**, 18-24.

NUMERICAL SEISMIC MODELING AND IMAGING OF COMPLEX FAULT-FOLD STRUCTURES IN A MOUNTAINOUS SETTING

M. Graziella Kirtland Grech, Deborah A. Spratt and Don C. Lawton, University of Calgary

ABSTRACT

Numerical seismic modeling experiments were performed to investigate optimum migration algorithms for imaging complex fault-fold structures. Two models were created, both based on structures in carbonate rocks of Paleozoic age, that outcrop in the Rocky Mountain Front Ranges of Western Canada. In the first model, these structures were buried under a lower-velocity layer representing Lower Cretaceous and Jurassic rocks to provide a flat surface on which the sources and receivers were placed during ray tracing. In the second model, the present day erosional surface was superimposed on the structures and ray tracing was done from topography. Synthetic seismic data were acquired by simulating a multi-offset acquisition geometry, and were then processed and migrated. The best images were obtained using prestack depth migration incorporating the exact velocity model. When migration velocities were varied randomly within 20% of the true values, the general shapes of the structures were retained, but events were mispositioned and had incorrect dimensions. Small-scale features (of the order of 3 km) at depths between 4 and 7 km were the least well imaged, even when exact velocities were used.

INTRODUCTION

In areas of complex geology, with folds, thrusts and imbricate structures typical of mountain belts and foothills around the word, the assumption that the common midpoint (CMP) stack represents a zero-offset section is no longer valid. This is attributed to the strong lateral changes in velocity and conflicting events that have different stacking velocities. Similar problems are also encountered in areas of salt mobilization and extensional faulting (Fagin, 1991). Under such circumstances, migration after stack will no longer produce a true image of the subsurface, and depth or time migration before stack becomes necessary (Yilmaz, 1987; Gray, 1998). Furthermore, the conventional assumptions of elevation statics corrections are often violated in areas with large topographic variations, typical of fold-thrust belts. In such environments, prestack migration from topography was found to produce the best image of steeply-dipping, shallow reflectors (Wu et al., 1998).

The objectives of this paper are to demonstrate which migration algorithms perform best for imaging these complex fault-fold environments, evaluate the sensitivity of different migrations to variations in the velocity model, examine the effect of rugged topography on the migration result and to determine which structural styles are the most difficult to image.

These objectives are of importance to the petroleum industry, as many hydrocarbon traps occur in complex fold-thrust environments.

The Numerical Models

The numerical models were based on a two-dimensional (2D) cross-section along the line SS' through the Front Ranges of the Canadian Rocky Mountains (Figure 1). Outcrops on either side of the line were projected onto SS' (Figure 2). The resulting cross-section with the different formations is shown in Figure 3 (Model 1) and Figure 4 (Model 2), and the corresponding velocities and lithologies are given in Table 1. Model 1 is 33 km long and 16 km thick. A wide range of dips is encompassed, starting from a low dip of 5° on the westerly dipping basement and with increasingly steeper dips (40°-60°) shallower in the section, average dip being 30°. The velocities were taken from sonic logs from several wells in the Foothills and Front Ranges. The structures enclosed in dashed boxes and marked A, B, C and D are of particular interest and were the main imaging targets for migration.

Heart Mountain (Figure 3, A) is made up of multiple carbonate sheets, with velocities ranging between 5500 m/s and 6400 m/s. These sheets are carried by the Lac des Arcs and Exshaw Thrusts and form hanging wall (HW) anticlines and footwall (FW) synclines; the limbs of these anticlines are relatively steep with dips up to 60°.

Structure (B) is Mount Yamnuska and is a fault-bend fold, the fault being the McConnell Thrust. The west half of the structure is made up of a HW flat and FW ramp, whereas the right half has a HW flat and FW flat. The youngest rocks in the whole model are included in the FW of this structure. They are Upper Cretaceous sandstones and shales with a velocity of 3900 m/s.

The Panther River Culmination (Figure 3, C) consists of three repeats of Devonian and Cambrian carbonate rocks with velocities of 6400 m/s and 6250 m/s, respectively and a dip of 10°-20°. Some younger, lower-velocity rocks (Triassic to Upper Mississippian age) are also included. This structure is underlain by granitic basement rocks

Structure (D) is a broad (6 km) syncline in the FW ramp of the Lac des Arcs thrust, underlying a HW flat.

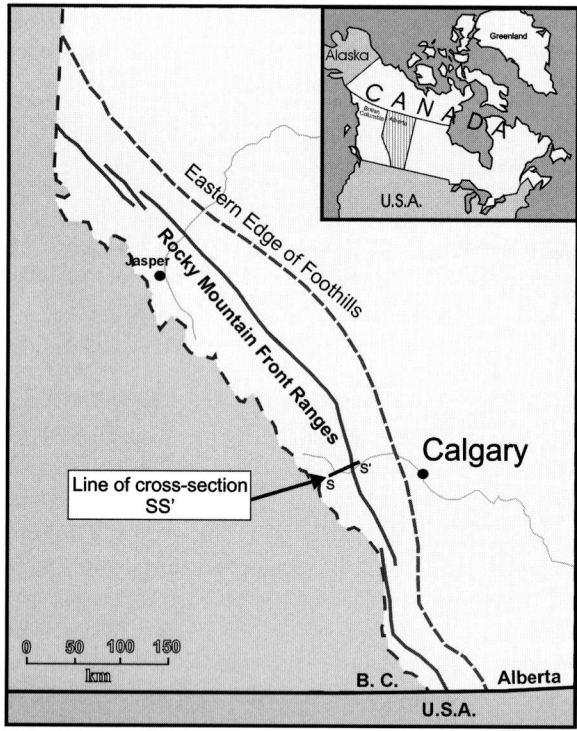

Figure 1. Map of Alberta, showing the eastern limit of the Foothills and the Rocky Mtn. Front Ranges. The models used in this paper are based on the cross-section SS'.

The model includes several major thrust faults (Figure 3) that carry Paleozoic carbonates to the present day erosional surface. For modeling purposes, the entire section was buried under Lower Cretaceous and Jurassic shales and sandstones, with a velocity of 4250 m/s, to give a flat surface.

Model 2 (Figure 4) is similar to Model 1, but in this case the shallower material has been removed, showing present day topographical variations of 1250 m. The model is now only 11 km deep. Structures (C) and (D) described above were not affected by the topography and are identical to those of Model 1, but only part of the Heart Mountain structure is now visible (Figure 4, A') and Mount Yamnuska (Figure 3, B) is not included as only the FW flat can be seen on this model. In these environments, the CMP stacking and elevation statics assumptions no longer hold, and prestack migration from topography was found to produce a better image in the near surface (Gray and Marfurt, 1995; Lines et al., 1996).

METHOD

Data Acquisition

Synthetic seismic data were acquired over the models using multi-offset ray tracing (GX II) with split-spread geometry. The spread used was 6000m long on each side of the shot point, with a total of 240 channels recorded for each shot. The geophone separation was 50 m. The shot interval used was 200 m and the capture radius was half the distance between receivers. Reflections only were simulated, generated from all horizons, as well as faults. Diffractions were not included. Ray tracing was carried out from the top boundary on Model 1 and from topography on Model 2.

After ray tracing, synthetic seismic traces were generated, with a record length of 6500 ms and a sampling interval of 2 ms. Transmission, spreading and attenuation losses were not included. A Ricker wavelet was used, with a dominant frequency of 30 Hz, which is typical for such geological environments.

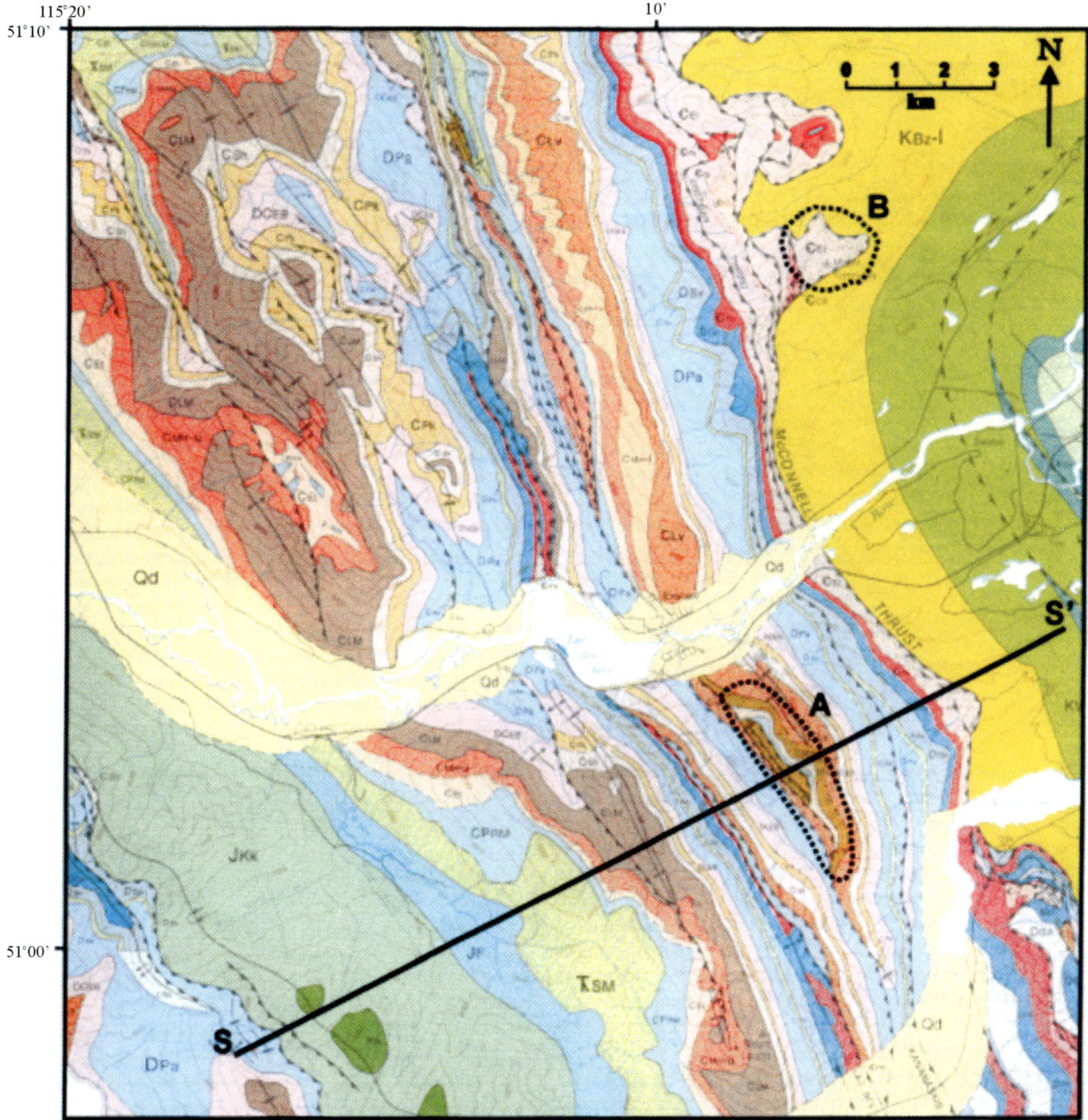

Figure 2. Geological map of the study area and its surroundings, showing the line of cross-section SS' and some of the major thrusts and outcrops used in the models. The Heart Mtn. structure (A on Model 1 and A' on Model 2) as well as Mt. Yamnuska (B on Model 1) are enclosed by a dashed line. This map is taken from Geological Survey of Canada map 1865 A (McMechan, 1995).

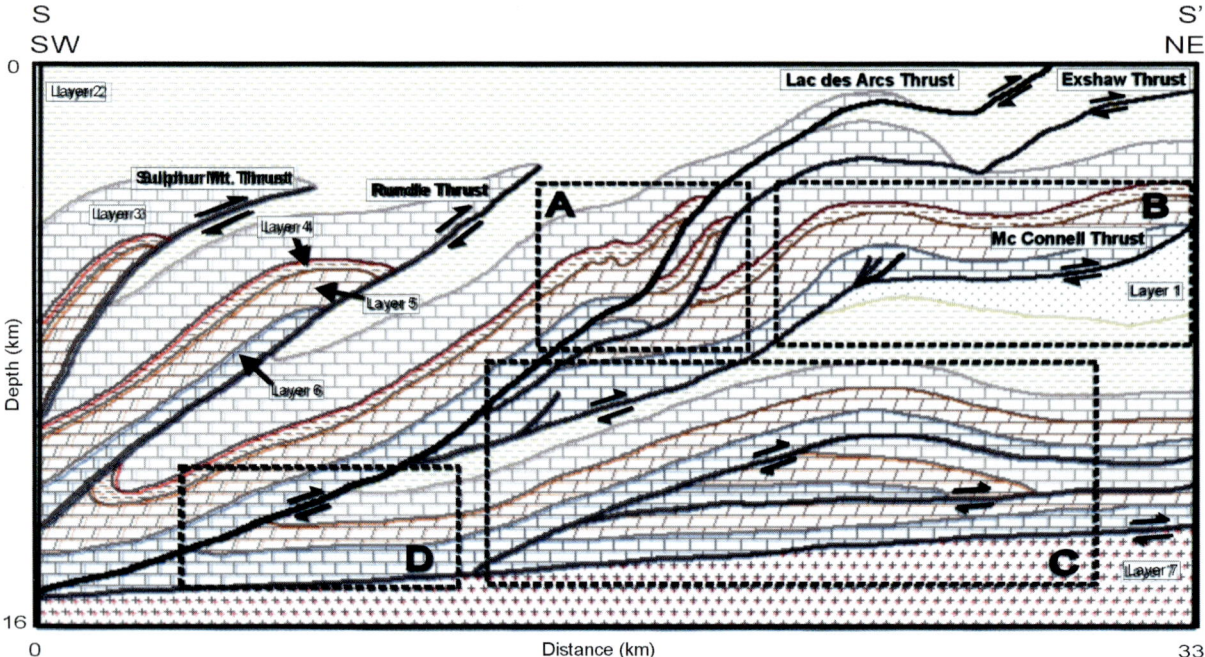

Figure 3. Model 1 – The geological cross-section based on line SS'. The structures are buried under a layer of Lower Cretaceous – Jurassic rocks. No vertical exaggeration.

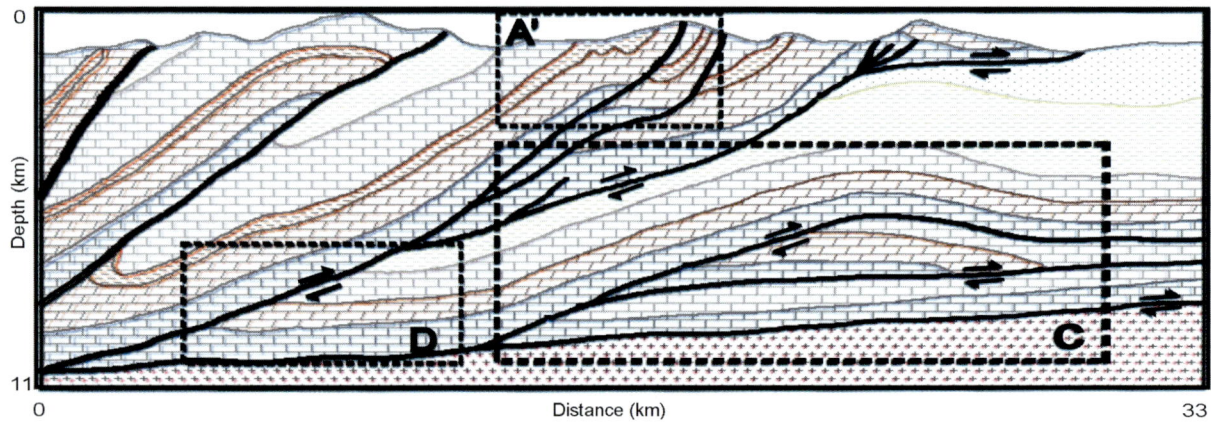

Figure 4. Model 2 – Similar to model 1 above, but with topographical variations of 1250 m superimposed. This model shows the present erosional surface. No vertical exaggeration.

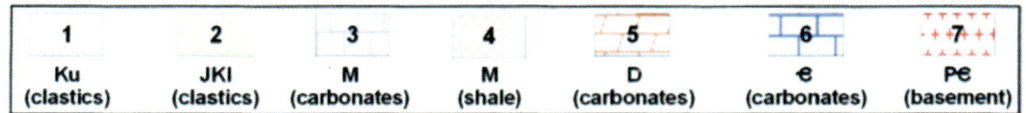

Table 1. Legend showing layer numbers, Period and dominant lithology for both models.

Data Processing

Figure 5 shows the resulting stacked section for the data set from Model 1. It is made up of conflicting events that are difficult to interpret. The section was obtained after standard velocity analysis to determine a velocity function which was later used for NMO correction. Gaps in some of the reflections are due to shadow zones in the ray tracing. This stack was used as input for the poststack migrations. The velocity models for the time migration in this paper were obtained by a depth to time conversion along vertical rays of the exact depth velocity model.

Since the model is structurally complex, Kirchhoff and Finite Difference shot migration algorithms were used (Claerbout and Doherty, 1972; Schneider, 1978). Both pre- and poststack migrations were tested on the data set from Model 1, migrating from a flat datum in every case. With the exception of one prestack time migration from datum, only prestack depth migration algorithms from topography were tested on the data set from Model 2. In the case of migration from topography, elevation corrections were not required.

ANALYSIS

MODEL 1:

Kirchhoff migration with the exact velocity model

Results of a post-stack time migration are shown in Figure 6. The reflectors in the upper part of the section are well resolved, but imaging is poorer later in the section. The small structures within Heart Mtn. (A) are indiscernible, as is the anticline of Mt. Yamnuska (B). The HWF and FWF in (B) can be identified clearly. The Panther culmination (C) has an overall antiformal shape, but the crests of each folded layer are masked by diffractions, and the data appear to be over-migrated. The presence of higher velocity rocks in the thrust sheets above result in pull-up of the basement event. The FW cutoffs in (D) are not well imaged and the reflectors appear to be delayed in time forming a syncline.

The post-stack depth migration is shown in Figure 7. Only the shallower events are imaged well. Although the apparent structure in the basement seen on the time migration is not evident after depth migration, it is difficult to identify events above basement. The HW anticlines and FW

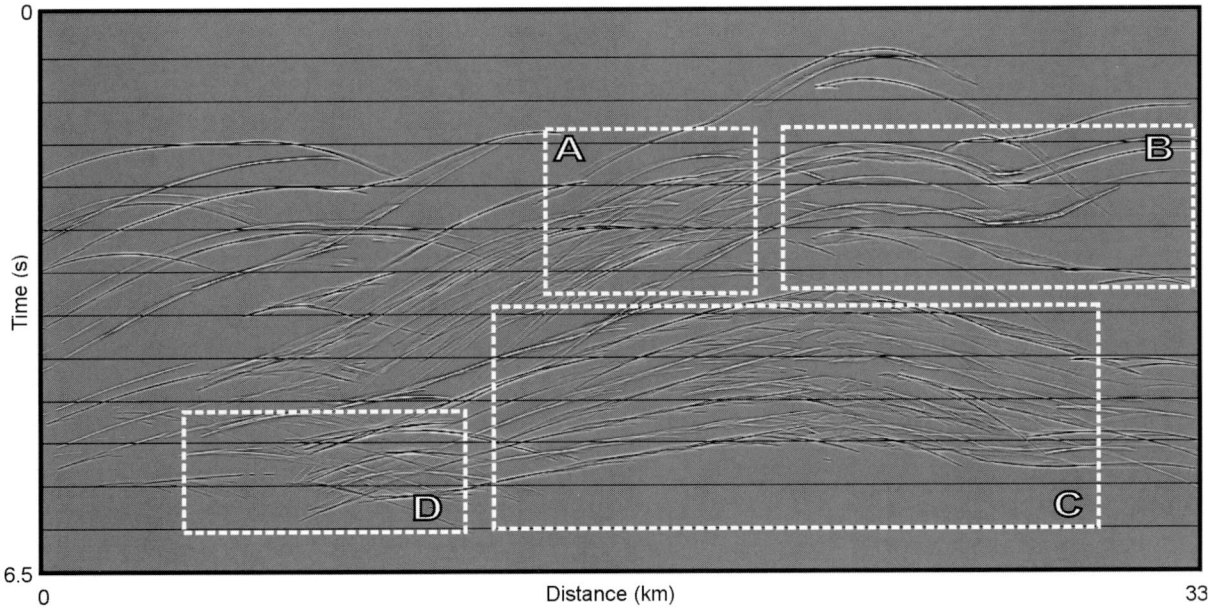

Figure 5. The stacked section from Model 1, obtained after standard velocity analysis and NMO correction.

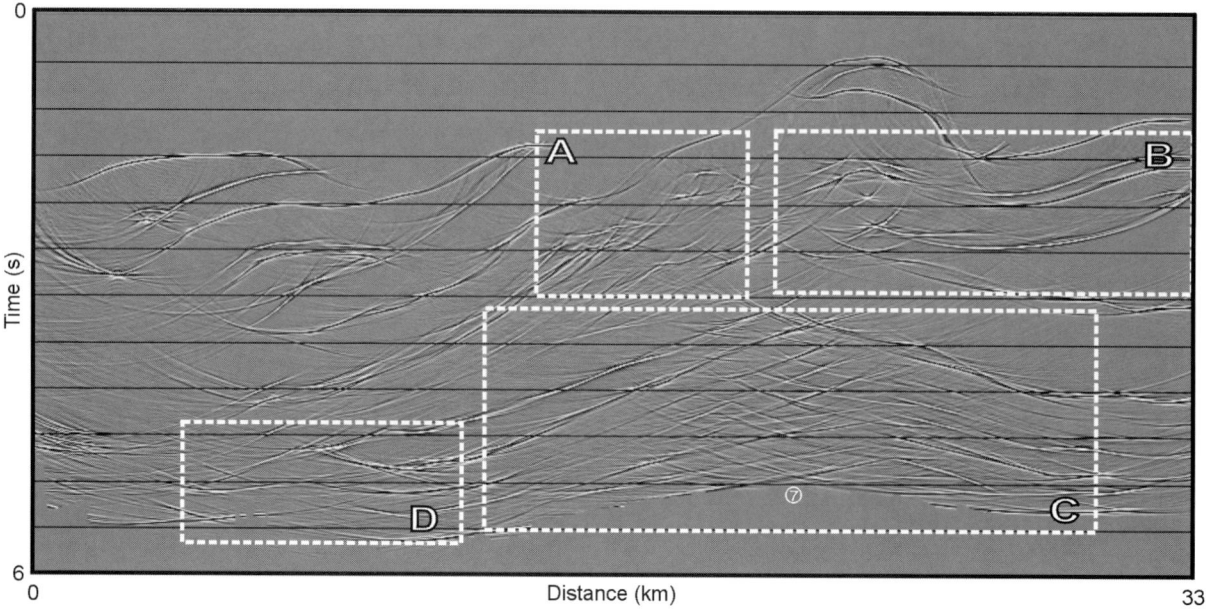

Figure 6. Kirchhoff poststack time-migrated section from Model 1 using the exact velocity model.

synclines of Heart Mtn. (A) are slightly better imaged then on the time migration. Mt. Yamnuska (B) is also better resolved, with the slightly flattened anticline in layer 4 being properly imaged. However, only the crest of layer 6 is seen. The FW ramp can also be identified. Although the general shape of the Panther Culmination (C) is evident, it is difficult to resolve the internal thrust geometry, and only the top three sheets can be interpreted. Structure (D) is least-well imaged, and the migration overall noisy.

A prestack time-migrated section is displayed in Figure 8. This result is not significantly better than the poststack time migration (Figure 6). The main improvements are the details of the Heart Mtn. structure and the east flank of the Panther Culmination (C).

Prestack depth migration (Figure 9) achieved the best result, with all four structures of interest being imaged well. The HW anticlines of the Heart Mtn. structure (A) can be identified, although it is still difficult to interpret the footwall syncline with certainty. The Exshaw Thrust, which marks the interface between the repeats of layer 6 (identified by 'x' and 'y' in Figure 9) is not discernible. The steep limbs of the sheets carried by the Lac des Arcs Thrust can be interpreted, but not those marked by 'z' (Figure 9). Mt. Yamnsuka (B) is also well imaged, including the contrast between layers 5 and 6. On the poststack migrated sections, only the crest was imaged, but the entire horizon can be interpreted in Figure 9. The FW ramp is very clearly imaged. The Panther Culmination (C) is also very well imaged, with the three repeats of carbonate sheets being fully discernible. The FW cutoffs (D) are easy to identify.

Kirchhoff migration with a perturbed velocity model

Randomly varied velocities are listed in Table 2 and the corresponding Kirchhoff prestack migration with these velocities is shown in Figure 10. How well events were imaged (including correct depth and dimensions) depended on how close the new velocities were to the true values. The greatest variation in velocity is in layers 5 and 6; the velocity error in layer 5 is +759 m/s, and in layer 6 is -1010 m/s. This resulted in poor imaging of the Panther Culmination (C), where several repeats of layers 5 and 6 occur. The anticline is irregular on the hinterland side and is more sharply peaked, but the different layers can still generally be distinguished. Mt. Yamnuska (B) is imaged reasonably well, although the anticline has a smaller amplitude and wavelength than in the model. In this case, layers 5 and 6 are involved only once. Despite the larger velocity contrast between layers 5 and 6, the boundary is hardly evident. Once again, the events in the HW of the Lac des Arcs Thrust can be distinguished, but not the events below it, and not all the FW cutoffs in (D) can be interpreted. The basement is no longer planar, but starts off flat, ramps up in the middle of the section and flattens again to the foreland.

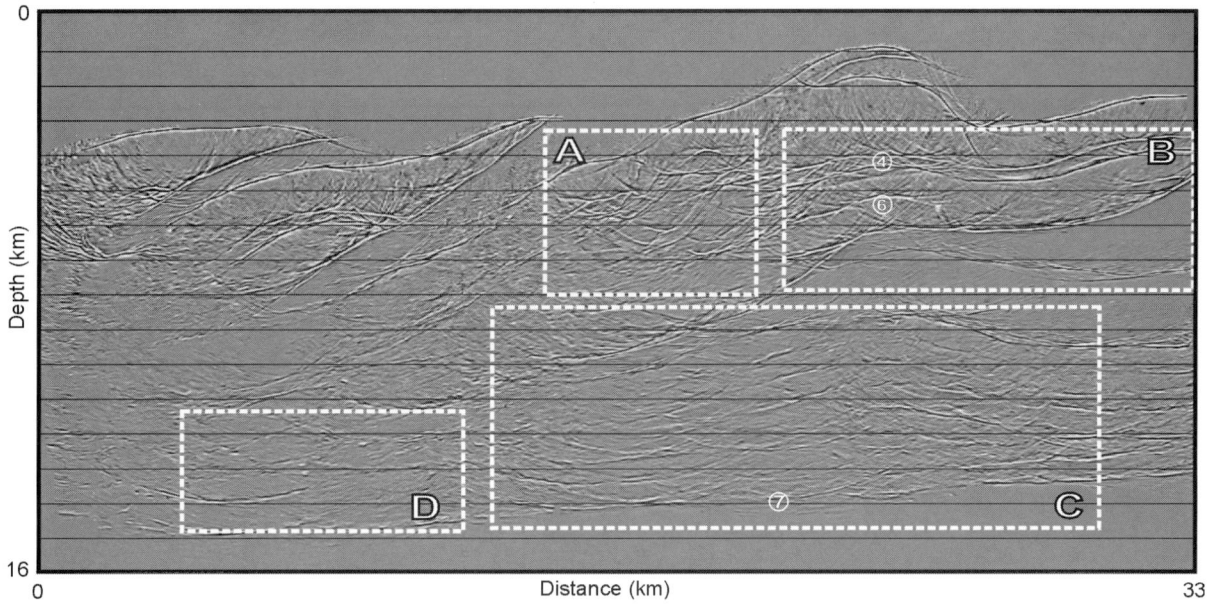

Figure 7. Kirchhoff poststack depth-migrated section from Model 1 with the known velocity model.

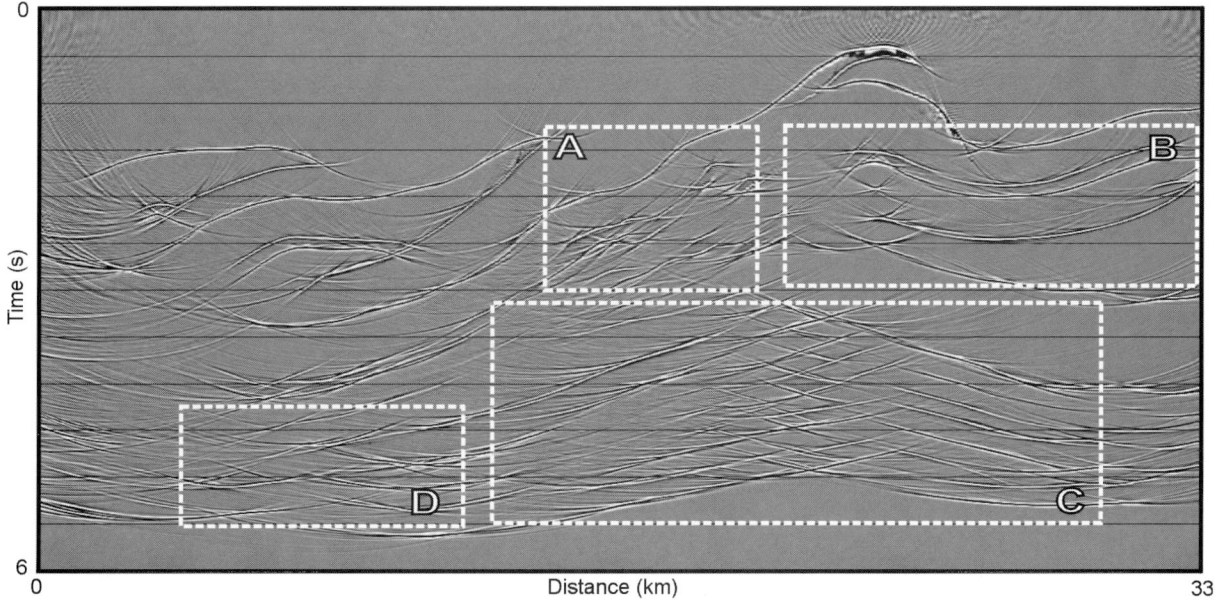

Figure 8. Kirchhoff prestack time-migrated section from Model 1 with the known velocity model.

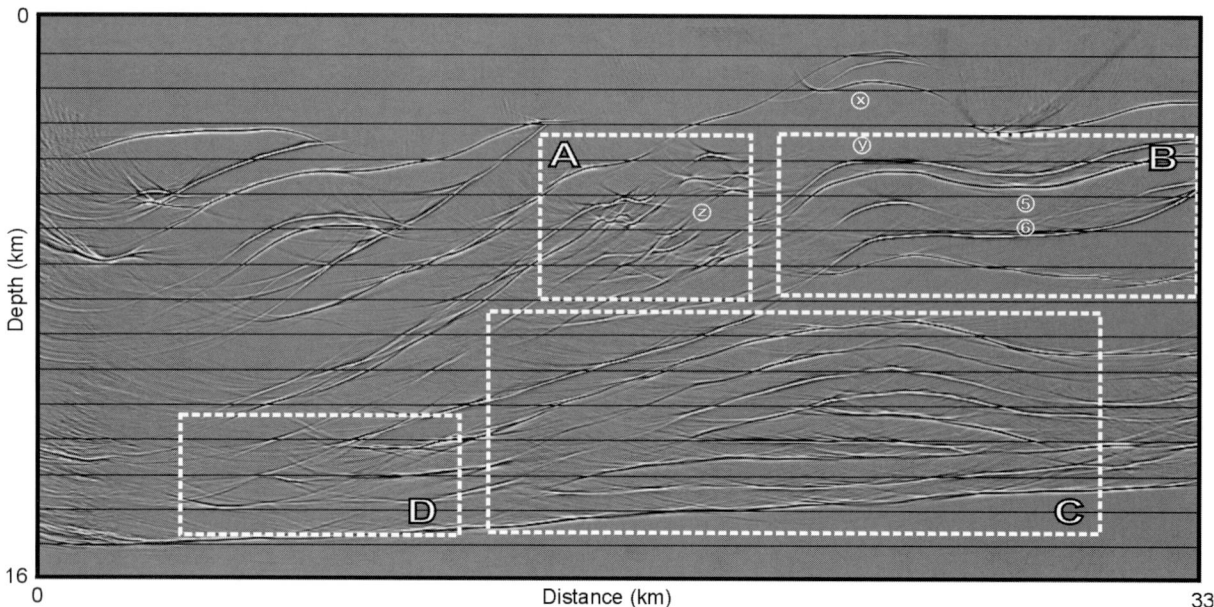

Figure 9. Kirchhoff prestack depth-migrated section from Model 1 with the exact velocity model.

Layer No.	1	2	3	4	5	6	7
Exact Velocity (m/s)	3900	4250	6100	5500	6400	6250	6500
Random Velocity (m/s)	4472	4573	5707	5917	7159	5240	6946

Table 2. The exact and randomly generated velocities for each layer.

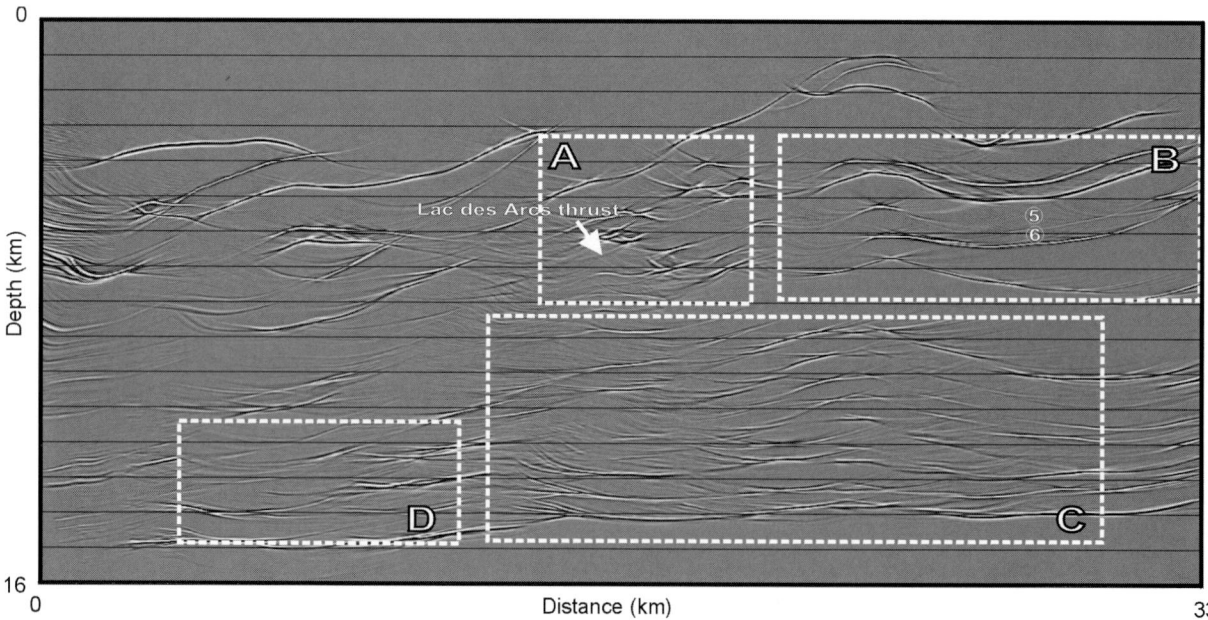

Figure 10. Kirchhoff prestack depth-migrated section from Model 1 with interval velocities varied randomly by +/- 20% from true values.

Model 2:

Kirchhoff prestack migrations

The Kirchhoff prestack time-migrated section (Figure 11) images the general shape of the structures, but the events are not well focussed, making thinner layers more difficult to interpret. The crests of Target C are discontinuous and appear to be over-migrated. There is also a velocity pull-up on the underlying basement, due to the high velocity rocks above. The FW syncline (D) also shows false structure and the cutoffs in the FW ramp extend into the HW flat. Some events are broken up and difficult to interpret. Target A' is the least well imaged, with the events becoming more indiscernible in the lower right hand corner. The section is also relatively noisy, particularly at early times.

Figure 12 is a Kirchhoff prestack depth migration from topography, with the exact velocity model. Targets C and D are well imaged, with the exception of cutoffs, due to the lack of diffractions during ray tracing. The section is relatively noisy in the shallower parts, making the interpretation of Target A' very difficult. Other events, like those marked by 'x' and 'y' appear to be discontinuous.

Kirchhoff prestack migration from topography with an inexact velocity model (Table 2) is shown in Figure 13. Targets C and D can still be interpreted, although they have an inaccurate structural shape and size. The interfaces between different layers are not well focussed overall. It is difficult to interpret thin layers like Layer 4 (marked by an arrow) as the top of the layer is not imaged in places. The section is also noisy in the shallower parts, making interpretation of Target A' impossible.

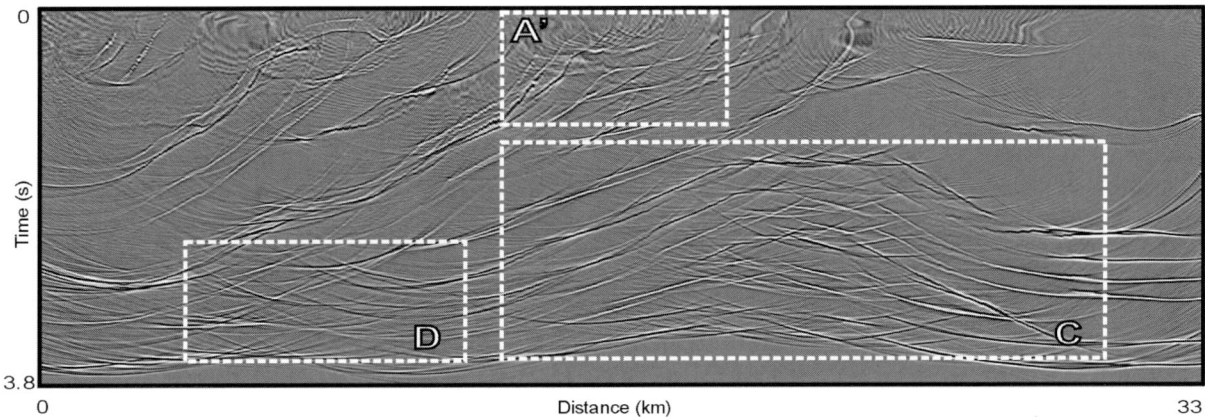

Figure 11. Kirchhoff prestack time-migrated section of Model 2 (from datum) with known velocity model.

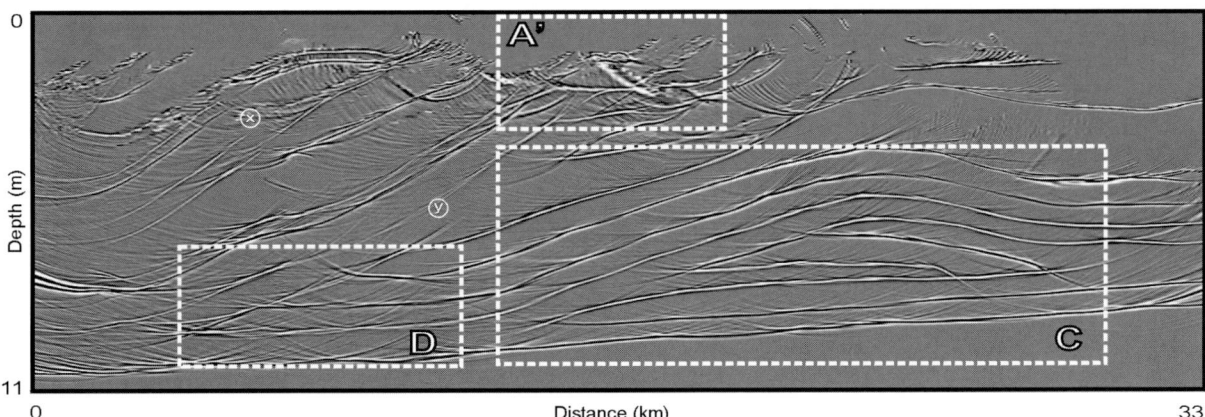

Figure 12. Kirchhoff prestack depth-migrated section from topography with the known velocity model (Model 2).

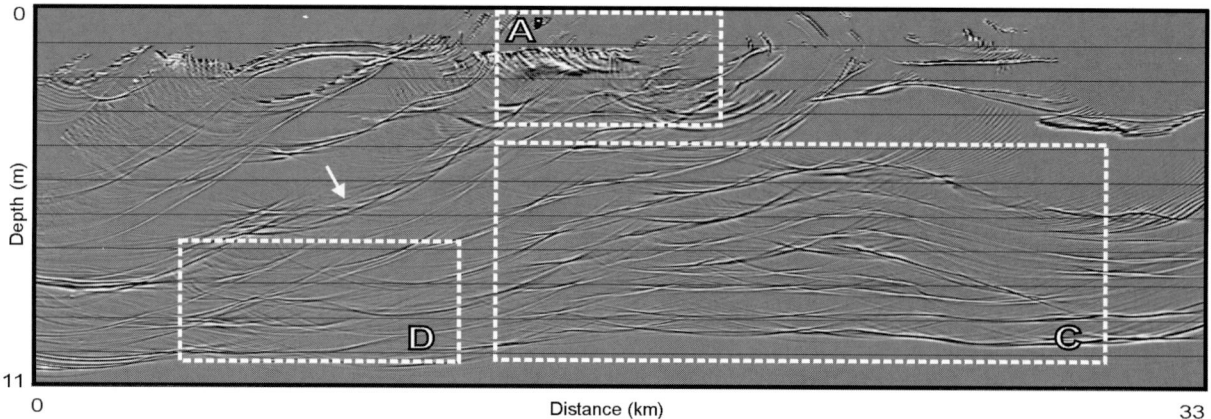

Figure 13. Kirchhoff prestack depth-migrated section from topography (Model 2) with an inexact velocity model.

Finite Difference prestack migrations

The optimum migrated section was obtained using FD shot migration in depth, from topography and with the exact velocity model (Figure 14). All the three structures can be interpreted with ease, false time structures have been removed and the events are well focussed. Fault planes are not imaged where there is no velocity contrast across them. Cutoffs of the FW ramp in (D) still tend to extend slightly in the HW flat. This effect could be minimized by including diffractions in ray tracing. Increasing the offset range used in migration improved the image of the steeply dipping thrusts.

It is interesting to compare this result (Figure 14) with Kirchhoff prestack migration from topography with the exact velocity model (Figure 12). Although the exact velocity model was used for both migrations, the FD result is superior to the Kirchhoff one – imaging and focussing of events is better, and the shallower events are not masked by noise. Some of this noise was eliminated by applying a mute on CDP gathers after migration, however it was difficult to find a compromise between the amount of noise to be removed and without destroying part of the reflected energy. A significant amount of noise thus still remains.

Figure 15 was obtained using the same algorithm as Figure 14, however the velocities used were again varied randomly within +/- 20% of the true values (Table 2) with the greatest velocity variations in layer 5 and layer 6. (The velocity error in layer 5 is +759 m/s, and in layer 6 is -1010 m/s). Target C was again the most severely affected, as it consists of several repeats of these two layers. The anticline has an irregular shape on the left side, but the different layers can be still distinguished. The basement again shows false structure. The FW syncline (D) is pushed down, showing a similar structural shape as on the time migration (Figure 11). The FW cutoffs, although they can be identified, extend into the HW ramp. The small-scale features of Target A' can still be identified. The fact that they are shallow in the section minimizes the accumulation of velocity errors and hence the small-scale features are imaged better. In general, the section becomes noisier and focussing of events becomes poorer deeper down in the section.

Comparing this image (Figure 15) with Kirchhoff prestack migration with the inexact velocity model (Figure 13), shows that the FD migration again yields a better image than the Kirchhoff migration. Events marked 'x', 'y' and 'z' are more continuous on the FD section and the events marked by 'w' on the FD section (Figure 15) are not imaged on the Kirchhoff section (Figure 13). The most striking difference is Target A'; it is well imaged on the FD section but it cannot be interpreted on the Kirchhoff result.

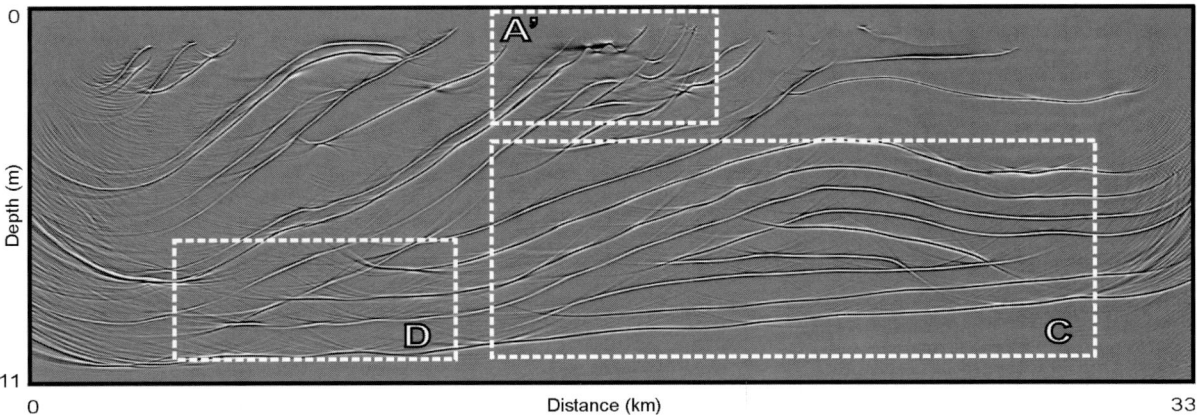

Figure 14: FD prestack depth-migrated section from topography (Model 2) with the known velocity model.

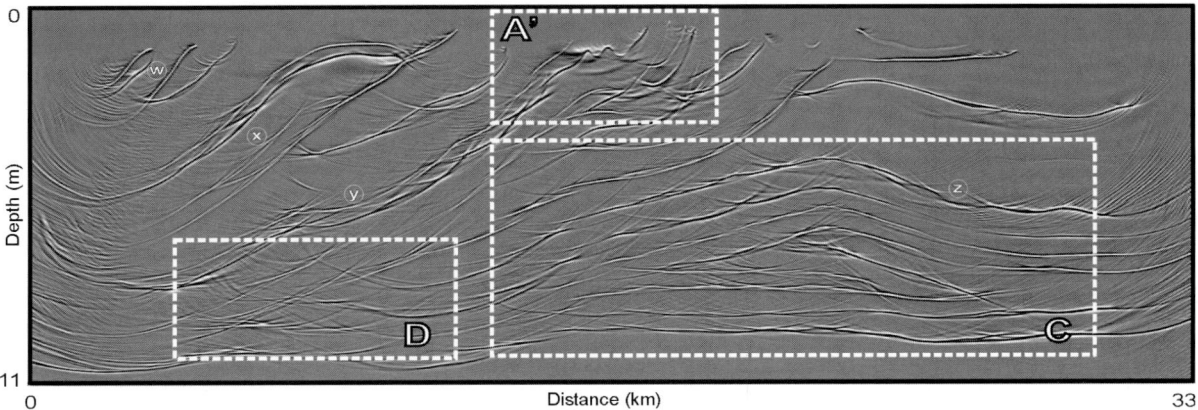

Figure 15. FD prestack depth-migrated section (Model 2) with inexact velocity model.

Figure 16 shows a close up of the three main areas of interest from Model 2 (a, b and c), with the corresponding images (d, e and f) taken from the best migration (Figure 14). The best migrated structure was Target C (Figure 2 c, f), having the best focussed and most continuous events. The arrows (Figure 2f) indicate fault planes that were not imaged due to lack of velocity contrast. Some cutoffs of Target A' (Figure 2 a, d), indicated by a double-headed arrow, are better imaged than those of the FW syncline (Figure 2 b, e). Amplitude variations can be observed along the events in both cases.

CONCLUSIONS

The following observations can be made when all the migration results are compared:

- The best migrated section is FD prestack depth migration from topography on Model 2 (Figure 14) with the exact velocity model. This image is superior to the Kirchhoff prestack migration from topography (Figure 12) of the same model and also to the Kirchhoff prestack migration result of Model 1 (Figure 9), where ray tracing and migration were from a flat datum. Focussing is better on Figure 14, cutoffs are sharper and the three targets (A', C and D) easier to interpret. Some of this may be attributed to the fact that the targets of Model 2 are buried less deep then those of Model 1 and differences in the way the two algorithms handle steep dips and variable velocities.

- As the velocity model deviated from the true velocity values, the small scale features, like those of Heart Mtn. (A and A'), became less well imaged. Identification of the large scale features particularly those of Mt. Yamnuska (B) and to a certain extent those of the Panther Culmination (C), was still possible when the velocities were in error within +/- 20%.

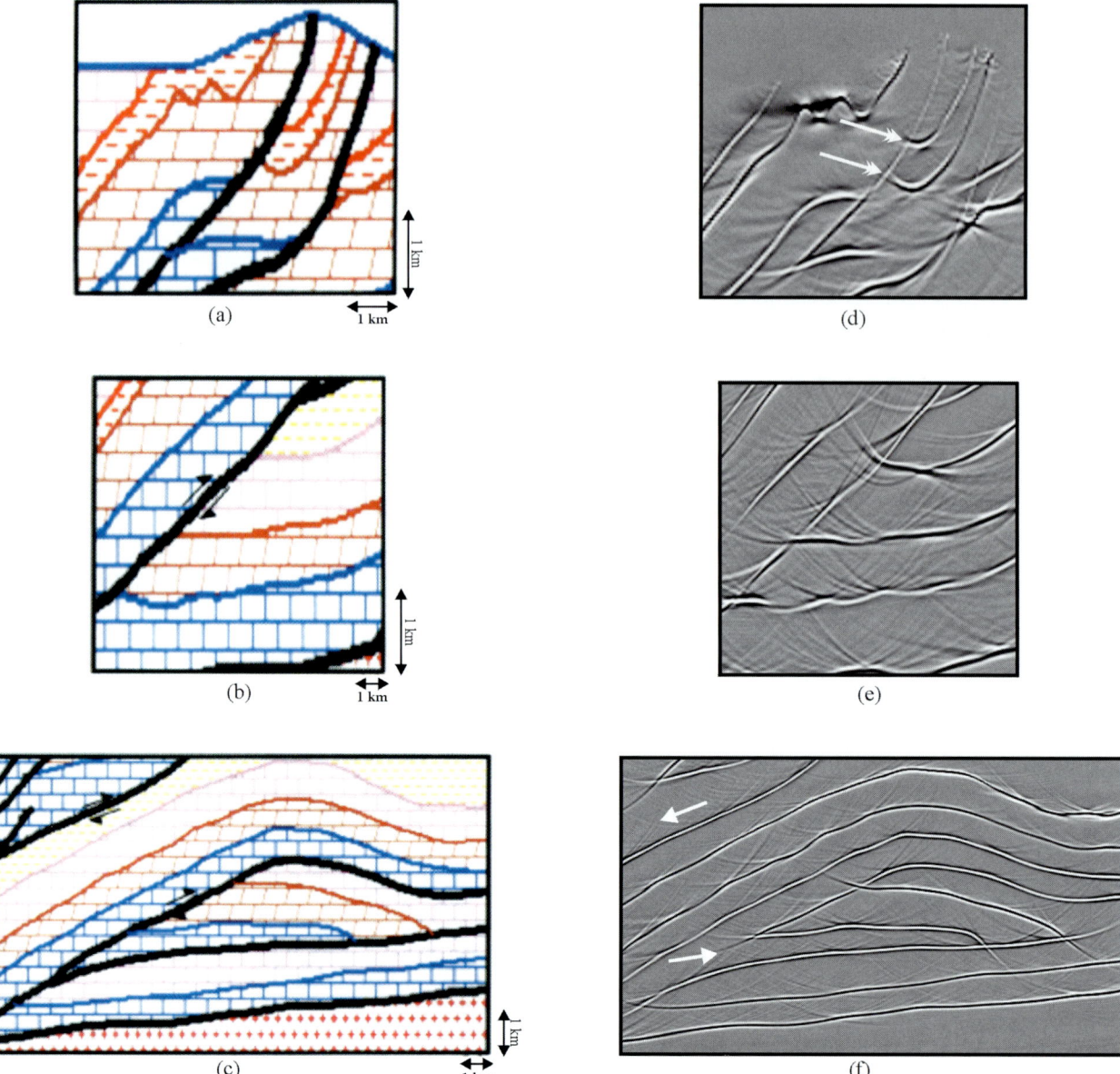

Figure 16. Detailed images of the three areas of interest. Figures a, b and c show the geological model, and d, e and f the respective images. The latter are taken from the FD shot migration from topography (Model 2) with the known velocity model. The arrows in (d) show sharp cutoffs, whereas the arrows in (f) indicate fault planes that were not imaged due to a lack of velocity contrast.

Structure (D), although a large scale feature, was more sensitive to velocity variations than were (B) and (C).

- The noise in the shallower parts of the the Kirchhoff migrations from topography can be attributed to inaccurate choice of migration weights in the Kirchhoff integral. Modifications of these weights may yield a better image (Gray and Marfurt, 1995).

- Smiles at the tips of fault planes and cutoff points are a result of the omission of diffractions during the ray tracing.

- Out of the four structures of interest, C and D were the best imaged in most cases. The interface between layers 5 and 6 on the flanks of the anticline on the Mt. Yamnuska structure (B, Model 1) was either not imaged at all or was very weakly imaged on the best migrated sections. The steep limbs of the Heart Mtn.

structure from Model 1 (A) were not well imaged, whereas the same structures from Model 2 were well imaged on the FD results (Figures 13 and 14).

- It is quite difficult to determine the exact details of the core of the Heart Mtn. Structure (A) even on the best migrated section from Model 1 (Figure 9), however the whole structure (A') can be easily interpreted on the best migrated section based on Model 2 (Figure 13).
- Fault planes were not imaged in most cases, making some cutoffs difficult to pick. There is also no indication of the Lac des Arcs or Exshaw Thrusts within layer 3 in the shallow, northeast part of Model 1 on any of the migrated sections pertaining to that model. The splays off the McConnell Thrust are not imaged in any of the sections.
- The Sulphur Mtn. and Rundle Thrusts and the strata they carry in their hanging walls are reasonably well imaged even where the velocities are not exact. One interesting thing to note is that the cutoffs of layer 4 in the hanging walls of both faults seem to extend into the footwall, beyond the bounding thrusts on most sections based on Model 1. This may also be caused by lack of diffractions in modeling. This effect is however not evident on Figure 12, the best migrated section of Model 2.
- The southwestern portion of Model 1 is never imaged due to the acquisition geometry resulting in a limited migration aperture. The first source location was at 0 m, but there was no energy recorded in any of the 120 channels to the west of this location, so no energy was available for migration. It is however imaged on results from Model 2, the model being 5 km less deep.

DISCUSSION

The best migrated section was obtained with FD shot migration in depth and from topography, using the known velocity model. This result was superior to Kirchhoff depth migration from topography and from a flat datum. As the velocities deviated from their true value, the deeper events were most affected, due to the accumulation of velocity errors with depth. The small-scale features of Target A' of Model 2, in the shallower part of the section, could still be interpreted when FD shot migration was used, even when the velocities were inaccurate. In the presence of velocity errors, the different structures could still be interpreted, but were at the wrong depth and had the wrong structural shape and size. Layers that were of the order of 1km in width and 500 m in thickness, such as Layer 4, were the most poorly imaged and focussed, both with time migration and with depth migration when the velocity model was inexact. Fault planes were not imaged in the absence of a velocity contrast across them, making some FW cutoffs difficult to interpret.

These experiments show that good depth images can be obtained in complex structural environments if the velocities are well known, the structural style is well understood (for building a good velocity model) and the S/N ratio is high.

Acknowledgments

We gratefully acknowledge the financial support for this work by sponsors of the Foothills Research Project and the Natural Science and Engineering Research Council of Canada. The synthetic data set was generated using *GXII 2D Geophysical Modeling Software*, courtesy of GX Technology Corporation. The data was processed in *ProMAX 2D*, courtesy of Landmark Graphics Corporation.

REFERENCES

Claerbout, J. and Doherty, S. M., 1972, Downward continuation of moveout corrected seismograms: Geophysics, **37**, 741-768.

Fagin, S.W., 1991, Seismic Modeling of Geological Structures - Application to Exploration Problems, Geophysical Development Series No. 2: Society of Exploration Geophysicists.

Gray, S. H., 1998, Interpretive seismic imaging in structurally complex areas: Geo-Triad'98, June 15-18, Calgary, AB, Expanded Abstracts, 282-283.

Gray, S. H. and Marfurt, K. J., 1995, Migration from topography: Improving the near-surface image: Canadian Journal of Exploration Geophysics, **31**, 18-24.

Lines, L., Wu, W., Lu, H., Burton, A. and Zhu, J, 1996, Migration from topography: Experience with an Alberta Foothills data set: Canadian Journal of Exploration Geophysics, **32**, 24-30.

McMechan, M.E., 1995, Geology, Rocky Mountain Foothills and Front Ranges in Kananaskis country, Alberta. Geological Survey of Canada, Map 1865A, scale 1:100 000.

Schneider, W., 1978, Integral formulation of migration in two and three dimensions: Geophysics, 43, 49-76.

Wu, W., Lines, L., Burton, A., Zhu, J., Jamison, W. and Bording, R., P., 1998, Prestack depth migration of an Alberta foothills data set – The Husky Experience: Geophysics, **63**, 392-398.

Yilmaz, O., 1987, Seismic Data Processing, Investigations in Geophysics No. 2: Society of Exploration Geophysicists.

Depth Migration Experiments with the Spratt Foothills Model

Larry R. Lines, Department of Geology & Geophysics, University of Calgary

Many of the features of thrust-faulted structures in the Canadian Foothills of the Rocky Mountains are described in a model by Deborah Spratt of the University of Calgary. Figure 1 shows an outline of sedimentary layer boundaries for this model, shown earlier in the chapter by Kirtland Grech et al. (1999). The model typifies many of the thrust faults in the front ranges of the Canadian Rocky Mountains and includes the Sulphur, Rundle, Lac Des Arcs, and Exshaw thrusts. These front range structures typify oil and gas traps found in the foothills oil and gas fields which are east of the front ranges. The model is interesting from a seismic imaging perspective since it contains steep dips, edge diffractors, several thrust faults, and velocity inversions.

The velocity file for this model by Graziella Kirtland Grech is contained in a binary file, named *velocity.dat* on the compact disc found in this book. The velocity file is a matrix of 645 rows (depth) and 1330 columns (offset) and is readable in FORTRAN with the following statements. We assume that v is dimensioned as v(645,1330).

```
open(unit=1, file='velocity.dat' form='unformatted')
ncol = 1330
mrow = 645
do j=1,ncol
    read(1)(v(i,j), i=1,mrow)
end do
```

Readers may wish to use their favorite synthetic seismogram program to compute a seismic model response from the **velocity.dat** file.

The first synthetic seismograms for this model computed by Kirtland Grech were made using ray tracing software from a software package donated by GX Technology Inc. Ray tracing was chosen for its fast computational speed, and is adequate for initial tests of prestack migration algorithms. Unfortunately, the complex structures also cause shadow zones in the ray traced synthetic seismograms, noted by some "vanishing reflections" as seen in a typical source gather shown in Figure 2. Also, these seismograms contain no diffracted arrivals; this impacts the migration results. A Kirchhoff prestack depth migration of these data which uses a solution to the eikonal solution to compute traveltimes is shown in Figure 3. This depth migration uses an eikonal equation traveltime solver due to Fuhao Qin (formerly of University of Utah) combined with the prestack migration summation techniques described by Gray and May (1994) and by Liner and Lines (1994). The prestack depth migration has a few artifacts (mostly caused by a lack of diffractions in the input data). Nevertheless, the migrations do show a reasonably good representation of the reflector boundaries.

In order to obviate the problems with ray tracing, Pat Daley is computing finite-difference synthetic seismograms which solve the complete wave equation. This is taking considerable computation time, probably on the order of weeks for this data set.

As a quick and inexpensive interim solution to handle these problems, I computed an exploding reflector model using finite-difference codes developed by Phil Bording (ref. Bording and Lines, 1997). The exploding reflector model (Loewenthal et al., 1976) simulates the ideal stacked section by placing seismic sources at reflector boundaries and propagating energy upward to the surface at half the seismic velocity. This simulates the experiment of coincident surface sources and receivers with two-way wave propagation and is much less expensive and time consuming than computing hundreds of synthetic seismograms for the individual source gathers. This produced an idealized stacked section of 1330 traces and 2000 time samples, as shown in Figure 4.. (Trace spacing was 25 m and time sample interval was 2ms.) A low frequency wavelet of about 15 Hz was used to reduce run time. The file containing this seismogram is found in the unformatted binary file **seismic.dat**. This file can read by the following statement:

```
real*4 x(1330,2000)
open(unit=2,file='seismic.dat',form='unformatted')
nsamp=2000
ntrace=1330
do i=1,ntrace
        read(2)(x(i,j),j=1,nsamp)
end do
```

This "idealized stacked section" provided a data set for two of the most popular depth migration methods, the reverse-time and Kirchhoff depth migration techniques (ref., Zhu and Lines, 1998). Figure 5 uses a reverse-time depth migration code created by Phil Bording and this is the code **revtime.f**, which is also on the enclosed CD. The trace spacing and depth sampling interval was 25m, and there are 645 depth intervals, so the migration represents the same dimensions (33250m by 16125m) as the velocity model in Figure 1. The migration has successfully transformed the stacked section in Figure 4 to a good replica of the desired reflection positions, as shown by Figure 5.

This migration would be expected to do a good job since we are effectively running the wave equation backward in time on a seismic section produced by running the wave equation forward in time. However, Figure 6 shows that Kirchhoff depth migration also does a good job of producing an accurate depth image. This method uses the eikonal equation to form aplanatic surfaces and summation. Except for the early arrivals that may be caused by interbed multiples or grid dispersion, the reverse-time depth migration does a slightly better job of focusing the reflection events. Also, Kirchhoff migration does show problems with edge effects at the boundaries of the image due to the summation of fewer traces. For a more complete analysis of these two methods, refer to Zhu and Lines (1998).

Conclusions and Future Directions

The model and synthetic seismograms introduced by Deborah Spratt and Graziella Kirtland Grech will prove very useful in testing seismic imaging algorithms. Thus far, our test results have proved encouraging. Future numerical tests will involve prestack depth migration of finite-difference wave equation source records for this velocity model. We also propose to compute a model that includes anisotropy and elastic waves, and to use depth migration on these model data sets.

References

Bording, R.P., and Lines, L.R., 1997, Seismic modeling and imaging with the complete wave equation: SEG publication.

Gray, S.H., and May W.P., 1994, Kirchhoff migration using eikonal equation traveltimes: Geophysics, 59, 810-817.

Kirtland Grech, G., Lawton, D.C., and Spratt, D., 1999, Numerical seismic modelling of fault fold structures in a mountainous setting: CSEG Annual Meeting Abstracts.

Liner, C.L., and Lines, L.R., 1994, Simple prestack migration of crosswell seismic data: J. Seis. Expl., 3, 101-112.

Loewenthal, D., Lu, L., Roberson, R., and Sherwood, J., 1976, The wave equation applied to migration: Geophys. Prosp., 24, 380-399

Zhu, J., and Lines, L.R., 1998, Comparison of Kirchhoff and reverse-time migration methods with applications to prestack depth imaging of complex structures: Geophysics, 63, 1166-1176.

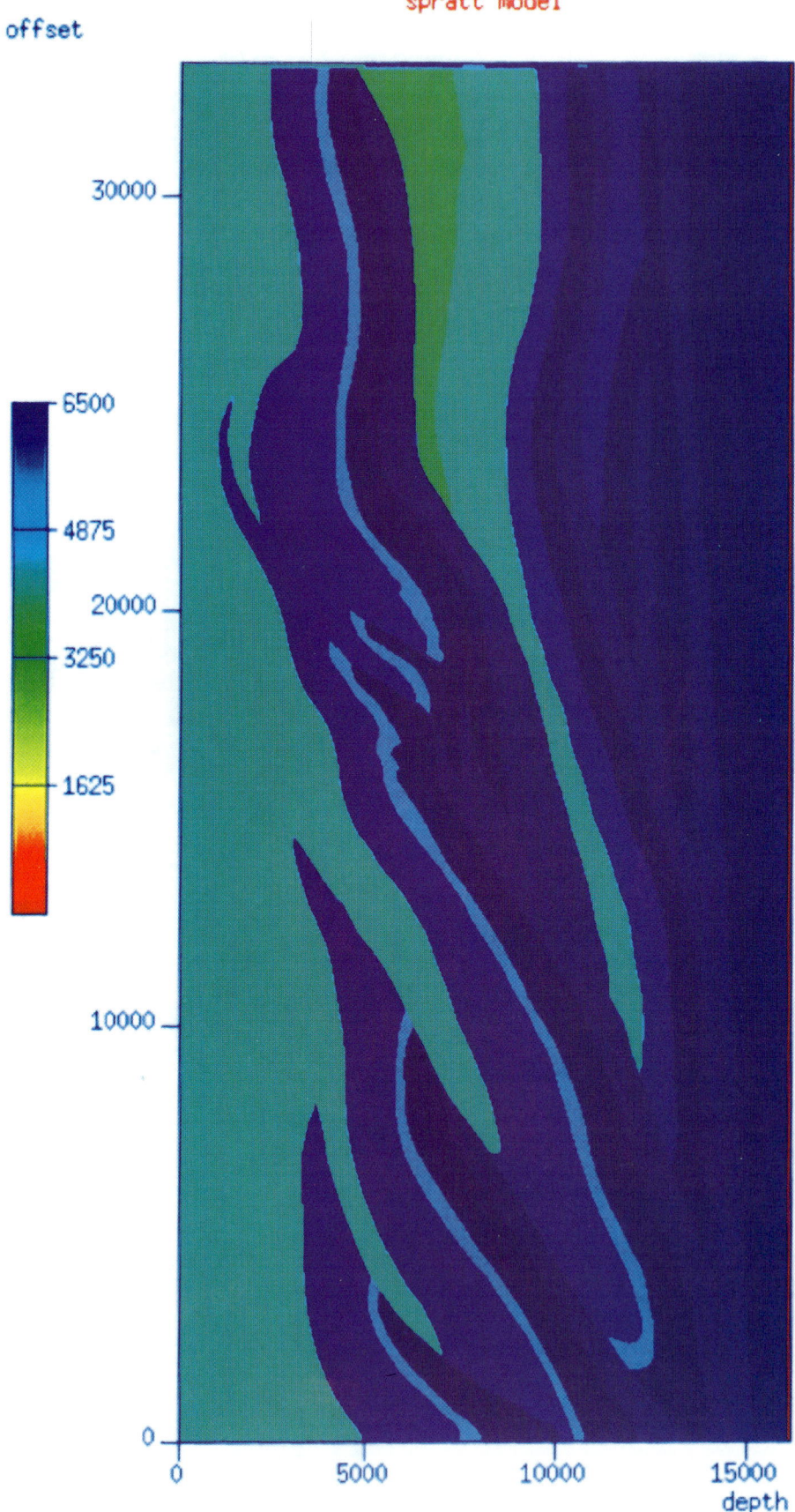

Figure 1. The Foothills model composed by Spratt contains a number of thrust faults including the Sulphur Mt., Rundle, Lac Des Arcs, and Exshaw thrust faults.

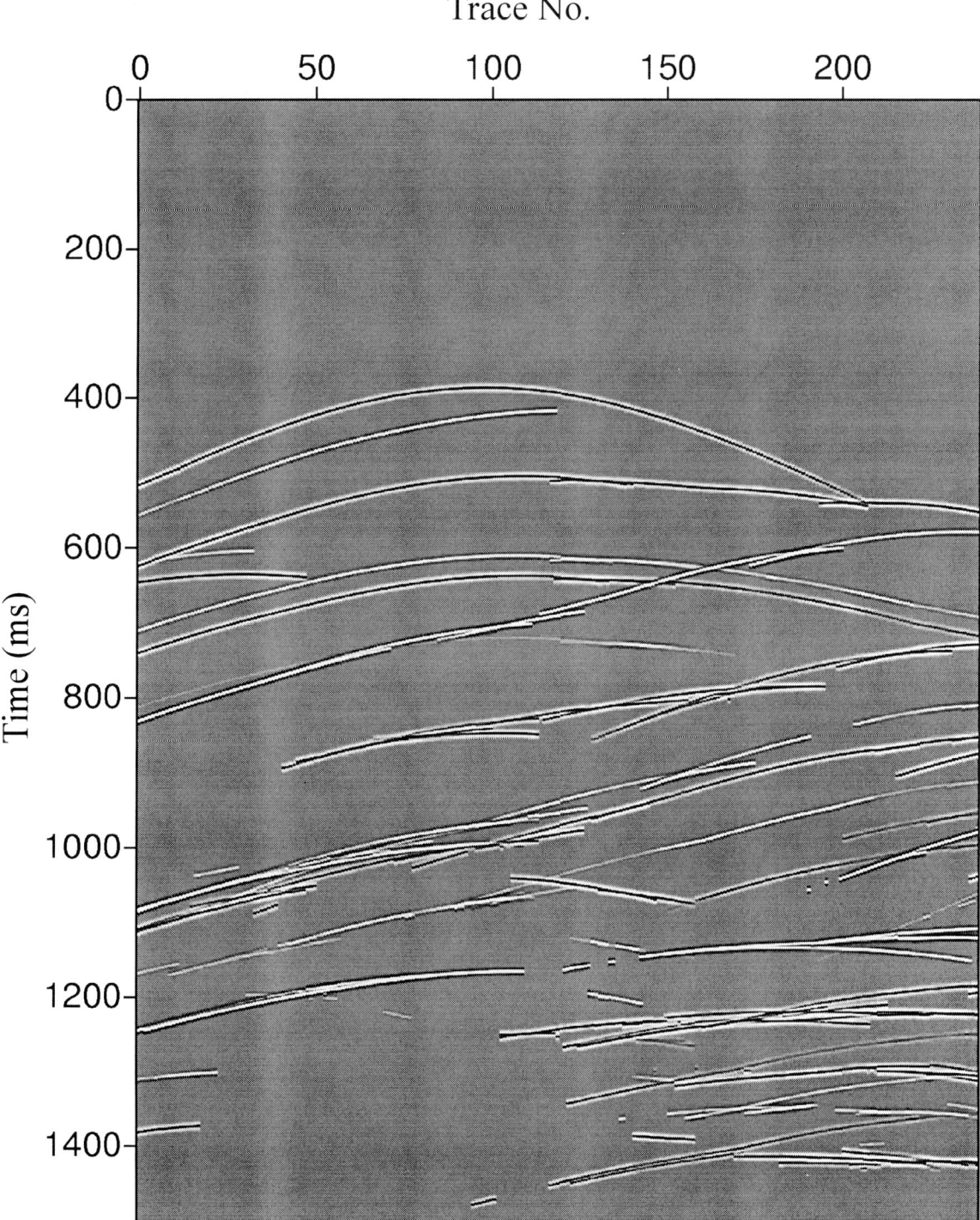

Figure 2. A typical source gather from the ray tracing calculations by Graziella Kirtland Grech.

Kirchhoff prestack depth migration

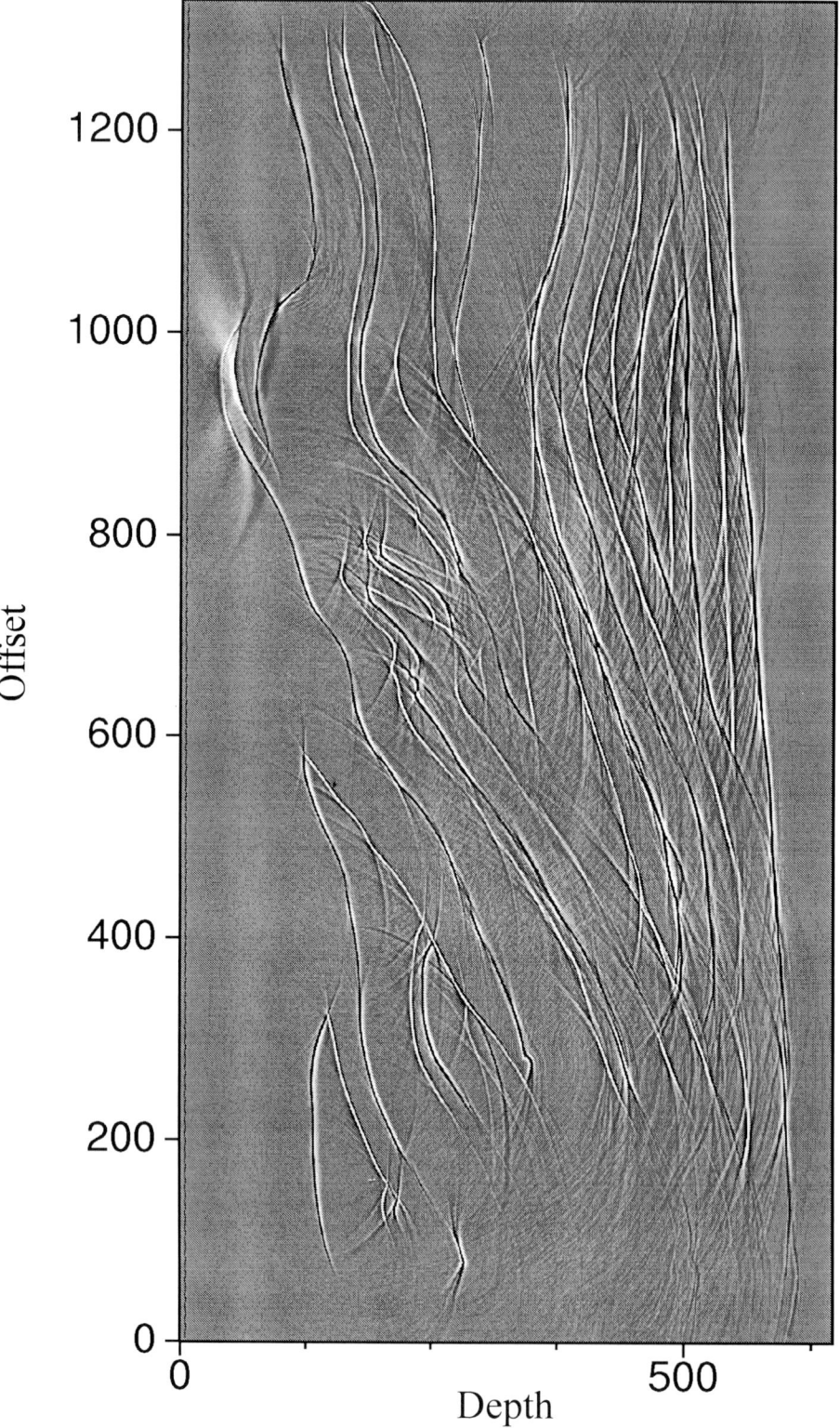

Figure 3. The Kirchhoff prestack depth migration of ray traced synthetic seismograms.

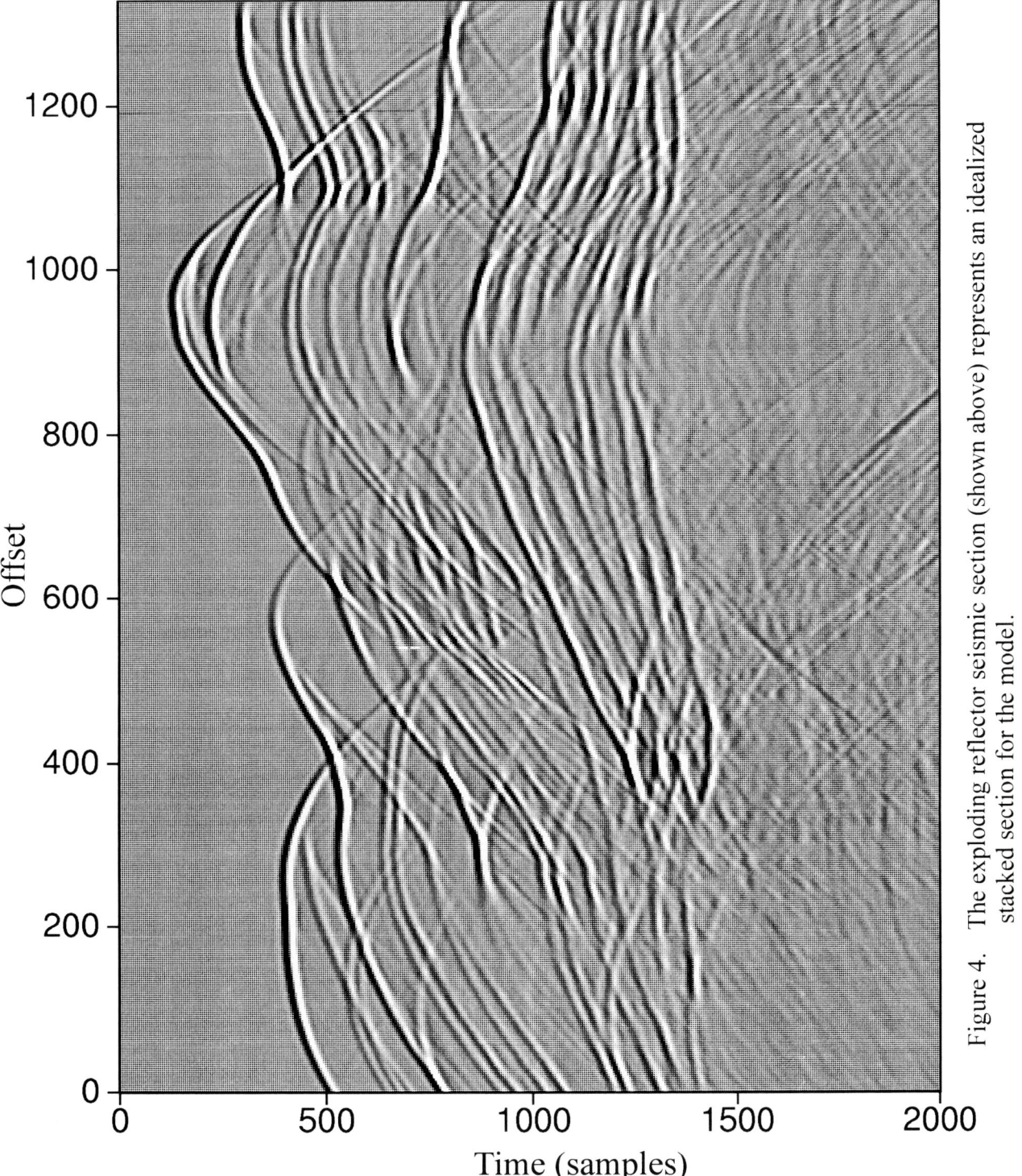

Figure 4. The exploding reflector seismic section (shown above) represents an idealized stacked section for the model.

reverse-time migration

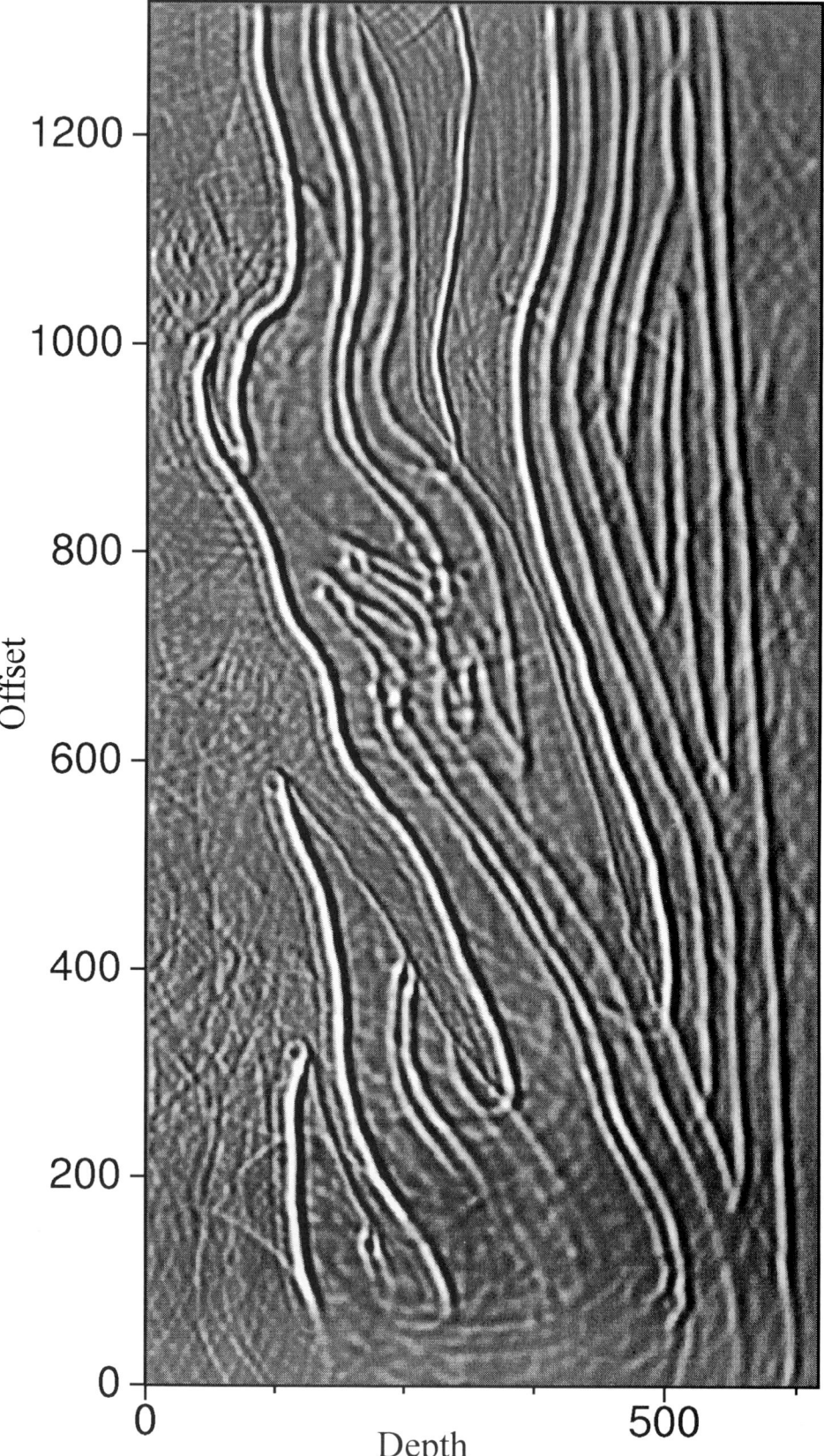

Figure 5. The reverse-time depth migration of the exploding reflector model.

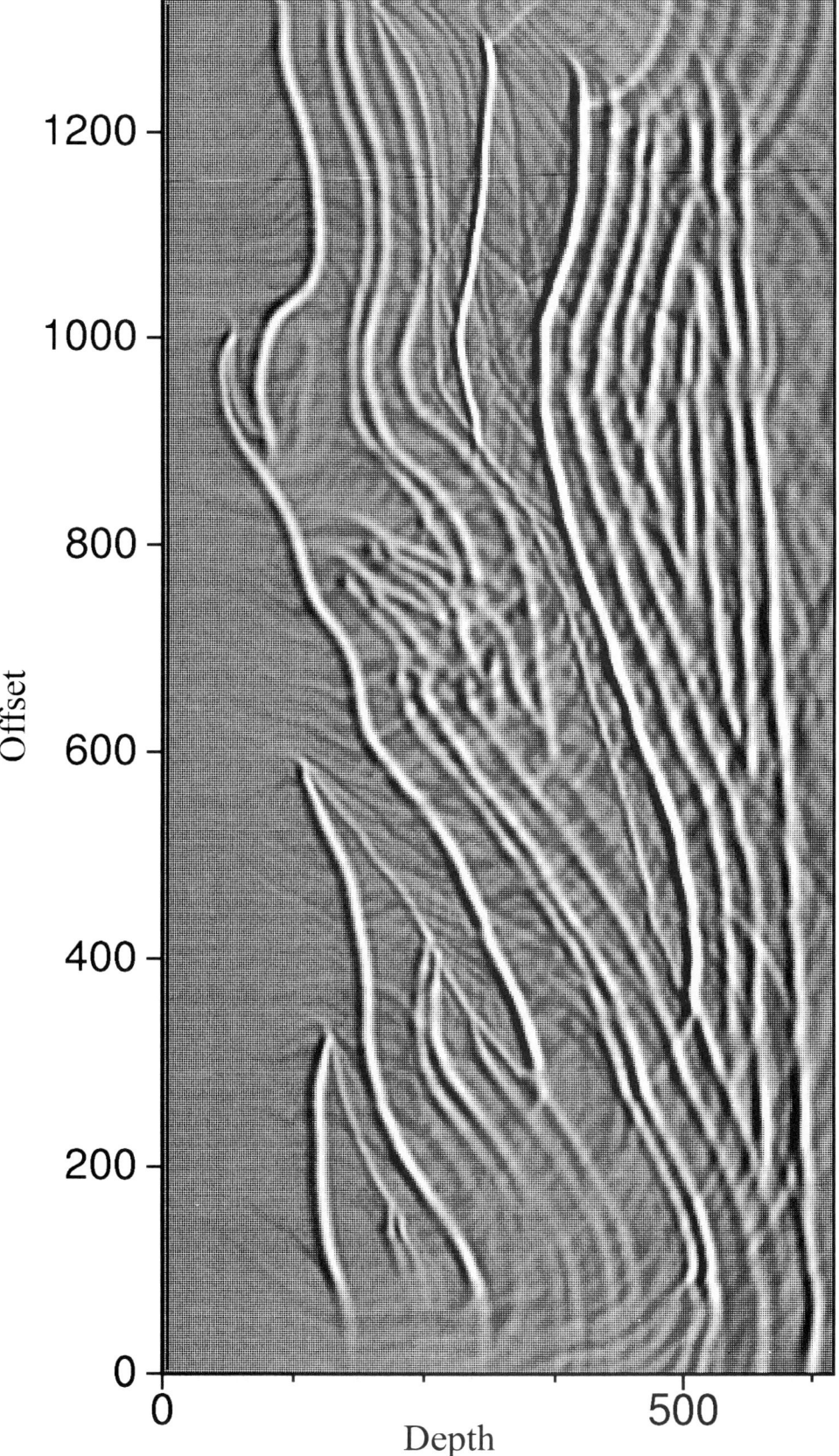

Figure 6. The Kirchhoff depth migration of the exploding reflector model.

Chapter 5 – Anisotropic Migration

As discussed in earlier chapters, one of the essential problems in prestack migration is the definition of a velocity model for depth imaging. However, as evidenced by the research of *Dai et al. (1998), Isaac and Lawton (1999), Leslie and Lawton (1998, 1999), and Vestrum et al. (1999)*, seismic velocity is a function of the direction of wave travel. That is, seismic waves exhibit anisotropic behavior and migration velocity models must be updated to include velocity anisotropy. This is particularly important in a foothills environments where sequences of anisotropic strata dip, often at high angles, and the lensing effects of these anisotropic strata cause position and depth errors in targets if isotropic velocity models are assumed (*Isaac and Lawton, 1999; Vestrum et al., 1999*). Algorithms that account for anisotropy in depth imaging include *Fei et al. (1998), Kirtland Grech et al. (1998), Vestrum et al. (1998), and Ferguson and Margrave (1997, 1998)*. In addition to the problem of algorithm design, it is important to determine anisotropic parameters (*Leslie and Lawton, 1999*). It is expected that more anisotropic migration algorithms will be developed in the future.

References:

Dai, N., Cheadle, S., and Isaac, J.H., 1998, Prestack depth migration in TI media: examples with numerical and physical modeling data: Geo-Triad '98, June 15-18, Calgary, Alberta, Expanded Abstracts, 83-84.

Fei, T., Dellinger, J., Murphy, G.E., Hensley, J., and Gray, S.H., 1998, Anisotropic true-amplitude migration: Geo-Triad '98, June 15-18, Calgary, Alberta, Expanded Abstracts, 85-86.

Ferguson, R.J., and Margrave, G.F., 1997, Nonstationary phase shift (NSPS) for TI media: CREWES Research Report, Volume 9, 32-1 - 32-27.

Ferguson, R.J., and Margrave, G.F., 1998, Examples of prestack depth migration in TI media: CREWES Research Report, Volume 10, 42-1 – 42-15.

Isaac, J.H., and Lawton, D.C., 1999, Image mispositioning due to dipping TI media: *Geophysics*, 64, 1230-1238.

Kirtland Grech, M.G., Isaac, J.H., and Lawton, D.C., 1998, Comparison of structural imaging in anisotropic media using *P*-wave and *S*-wave data: Foothills Research Project Research Report, Volume 4, 8-1 – 8-10.

Lawton, D.C., 1999, Slip-slidin' away – some practical implications of seismic velocity anisotropy on depth imaging: unpublished manuscript.

Leslie, J.M., and Lawton, D.C., 1998, Anisotropic prestack depth migration: *CSEG Recorder*, December 1998, 22-29.

Leslie, J.M., and Lawton, D.C., 1999, A refraction seismic field study to determine the anisotropic parameters of shales: *Geophysics,* 64, 1247-1252.

Vestrum, R.W., Lawton, D.C., and Schmid, R., 1999, Imaging structures below dipping TI media: *Geophysics*, 64, 1239-1246.

Slip-slidin' away – some practical implications of seismic velocity anisotropy on depth imaging

Don C. Lawton, Department of Geology & Geophysics, University of Calgary

In recent years, there has been increasing interest in imaging issues related to *P*-wave anisotropy, to the point that we can no longer ignore seismic velocity anisotropy in seismic data processing. Laboratory and field studies (e.g. Jones and Wang, 1981; Banik, 1984; Johnston and Christensen, 1995; Vernik and Liu, 1997; Leslie and Lawton, 1999) have provided compelling evidence that shales exhibit intrinsic transverse isotropy (TI). In this symmetry class, the seismic velocity parallel to the laminations is greater than that perpendicular to the layering, and the difference can be as high as 30%. In flat-layered clastic sequences, such as in the Western Canada sedimentary basin, the TI axis of symmetry is essentially vertical, and isotropic depth migration of reflection seismic data tends to overestimate the true depths of reflectors. This occurs because the imaging velocity, based on moveout analysis, is generally greater than the true vertical velocity.

Figure 1 is a photograph of flat-lying clastic strata in the Western Canada sedimentary basin, near Calgary, Alberta. These rocks are composed of alternating sandstones and shales of Cretaceous age, with individual strata being significantly less in thickness than a typical seismic wavelength ($\lambda = \sim 100$ m). Although individual strata may be isotropic, it has been shown that the entire sequence can be considered as being equivalent to a single TI layer (e.g. Postma, 1955; Backus, 1962; Schoenberg, 1994). The bedding-normal velocity is the slow velocity axis, and is defined as V_0. Conversely, the fast velocity direction is parallel to bedding, and is defined as V_{90} (Figure 1).

FIG. 1. Thinly layered clastic sediments from the Western Canada sedimentary basin, near Calgary, Alberta. These rocks are considered to be transversely isotropic (TI), with the fast velocity, V_{90}, parallel to bedding, and the slow velocity, V_0, perpendicular to bedding. Layering is at a much finer scale than a seismic wavelength, λ. Photograph courtesy of D. Spratt.

Figure 2 shows wavefronts propagating through a TI medium. In Figure 2a, the symmetry axis is vertical (VTI) and the wavefront propagates more quickly horizontally than vertically. This is the typical model of a sedimentary basin containing flat-lying, shale-prone sediments. In the case of elliptical anisotropy, for which the Thomsen (1986) parameters ε and δ are equal, the wavefront scribes an ellipse. In fold and thrust belts, such as the foothills of the Rocky Mountains, strata are carried in the hanging walls of thrust faults, and the axis of symmetry is rotated away from vertical. Figure 2b shows a schematic diagram of the extreme case of vertical dips, resulting in the axis of symmetry being horizontal. In this case, the wavefront propagates more rapidly in the vertical direction than horizontally. As a consequence, near-offset traveltimes to a deep reflector beneath an overburden with vertical dips (Figure 2b) will be less than the traveltimes to the equivalent reflector at the same depth beneath an overburden of flat-lying strata (Figure 2a). For any given seismic transect across a deformed belt it is common that there will changes in dip domains within the same stratigraphic unit, ranging from horizontal to vertical, and possibly also overturned. Hence, if this unit is anisotropic, there will exist traveltime anomalies in deep reflectors that are induced by anisotropic velocities in the variably dipping overburden (Leslie and Lawton, 1998).

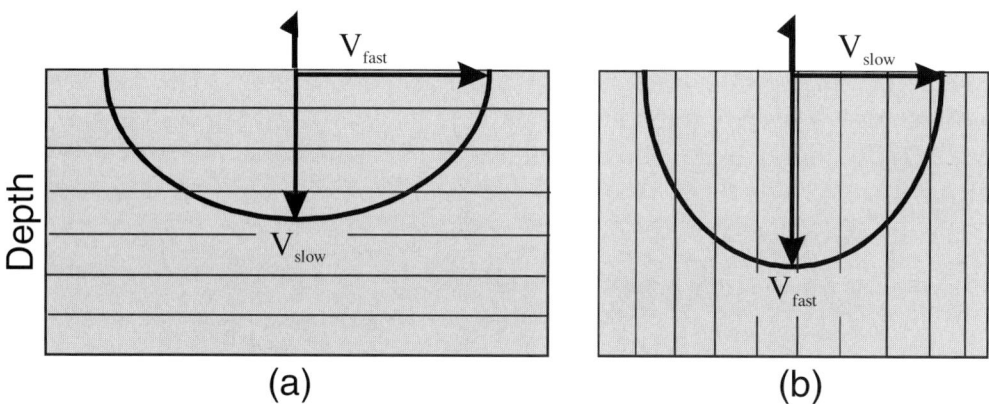

FIG. 2. Schematic diagram of a wavefront propagating in TI media with: (a) a vertical axis of symmetry, and (b) a horizontal axis of symmetry.

If anisotropic strata dip at intermediate angles, then the propagating wavefront is no longer symmetric about a vertical axis. We describe this geometry as the case of tilted transverse isotropy (TTI), and a schematic example for a dip of 45 degrees is shown in Figure 3. The wavefront propagates more rapidly in the down-dip direction than in the cross-dip direction and the point where the wavefront attains its deepest extent sideslips away from the source location, also in the downdip direction (Figure 3). The group velocity is the velocity of the energy transport in a direction radially away from the source (for homogeneous media) whereas the phase velocity is the local wavefront-normal velocity (Figure 3). The group and phase directions (and velocities) are equivalent only along the symmetry axes (i.e. the bedding-parallel and bedding-normal directions). Interactions of the wavefront with reflectors is a function of the phase angle and phase velocity rather than the group velocity and ray direction, although measured traveltimes are given by the ray length divided by the group (ray) velocity.

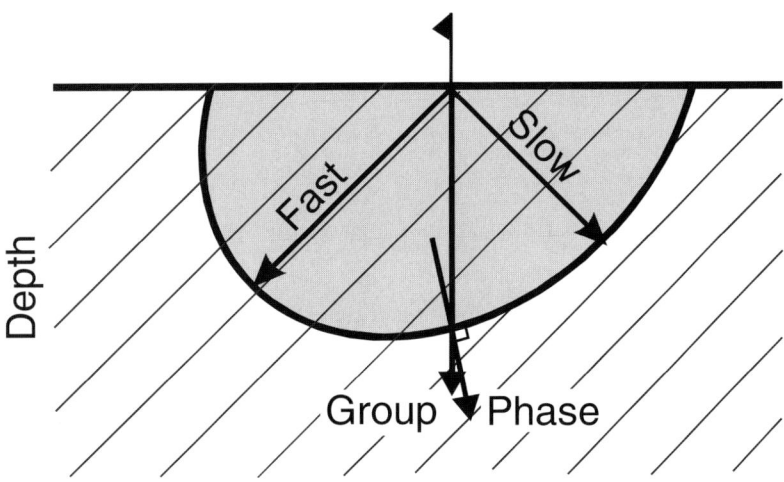

FIG. 3. Schematic diagram of a wavefront propagating in tilted transversely isotropic (TTI) media. The fast (bedding parallel) direction dips to the left at 45°.

Figure 4 shows a series of snapshots of a wavefront propagating through anisotropic strata that dips to the left at 45 degrees, and the TTI layer is underlain by a flat reflector. The fast velocity direction is assumed to be parallel to the dip of the strata. This geometry is common in fold-thrust belts, associated with footwall and hangingwall ramps. The downgoing wavefront sideslips in the down-dip direction (Figure 4a through Figure 4d), is reflected back toward the surface (Figure 4e) and intersects the surface (Figure 4f) where the minimum time (first arrival) occurs at the source location. Hence, in this example, the zero-offset ray is not vertical and the reflection point is displaced laterally in a down-dip direction from the source-point, even though the reflector is horizontal. The magnitude of the displacement can be significant, and may be as large as 20% of the reflector depth, depending on the dip and values of the anisotropic parameters of the overburden. A more detailed examination of the sideslip error is provided by Vestrum et al. (1999).

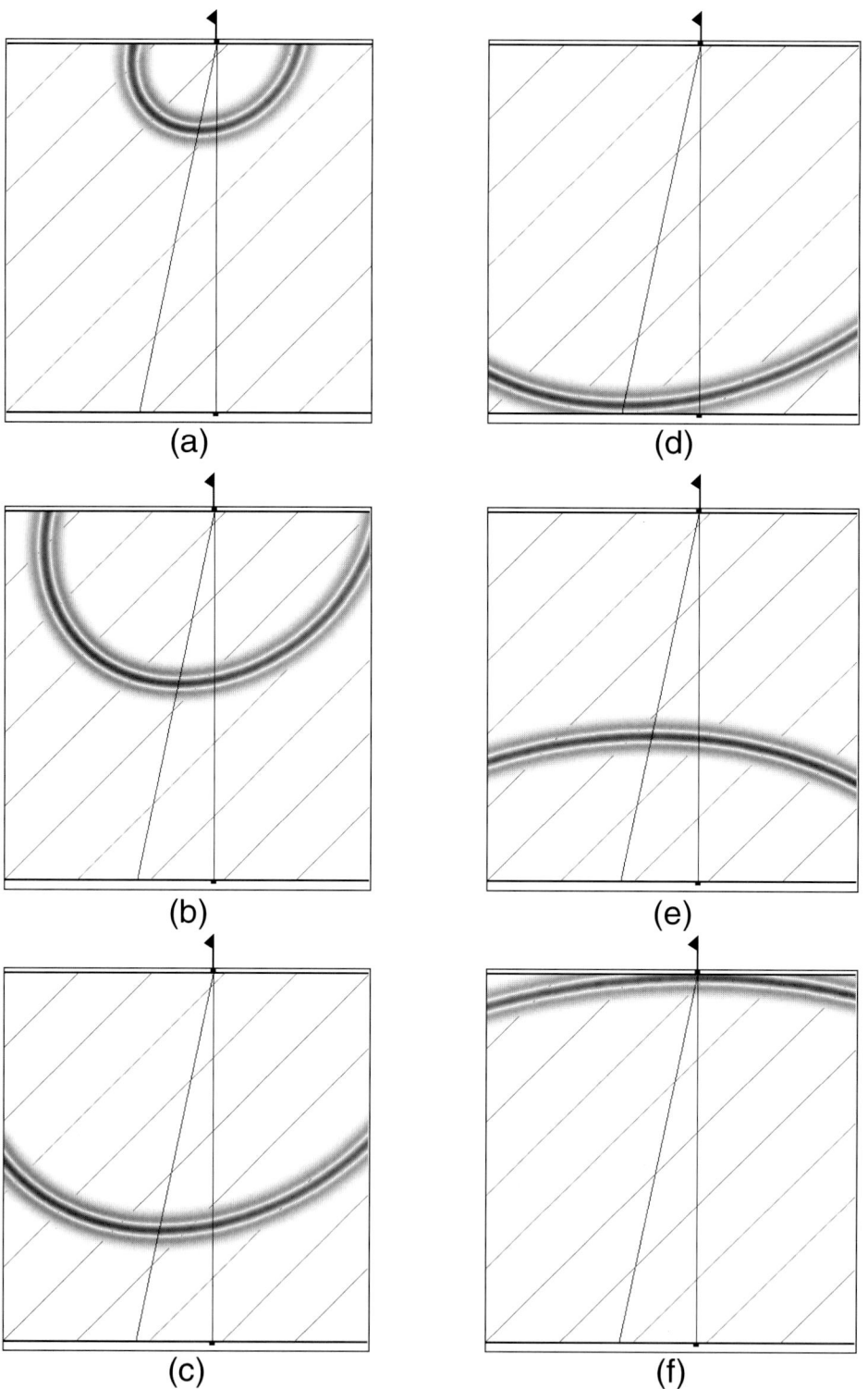

FIG. 4. Snapshots of a wavefront propagating through a TTI layer and reflecting from a flat underlying reflector: (a) through (c) show the downgoing wavefront, (d) the reflection point, (e) the upgoing wavefront, and (f) the wavefront impinging on the surface. The zero-offset reflection point is displaced downdip from the shotpoint. Images courtesy of R. Vestrum.

The 'side-slip' phenomenon, illustrated in Figure 4, has significant implications on the lateral positioning of targets imaged on reflection seismic sections. This is illustrated in Figure 5 which shows a schematic diagram of a surface seismic spread laid out over dipping, anisotropic strata. This dipping unit is underlain by an horizontal reflector (Figure 5a). If the layer above the reflector was isotropic, then the portion of the reflector that is illuminated by a shot located at the centre of the surface seismic spread would be distributed equally about the common mid-point (CMP). However, if this layer possesses TTI properties, then the segment of the reflector that is illuminated by the shot is displaced in the down-dip direction from the CMP, as shown in Figure 5a.

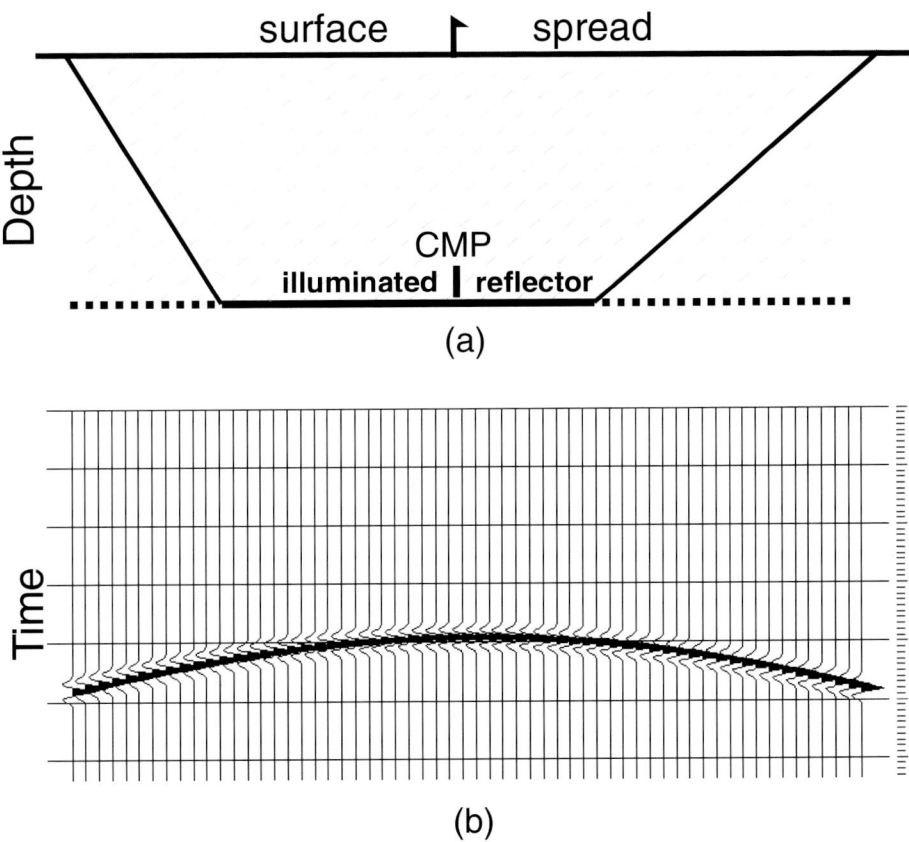

FIG. 5. (a) Diagram showing a reflector illuminated by a seismic spread laid out over a TTI layer. The subsurface reflector coverage is displaced in the downdip direction of the overburden from the CMP. (b) Multichannel seismogram from the model in (a), showing minimum traveltime at the shot location, and moveout that is close to being hyperbolic.

A synthetic shot gather from the model shown in Figure 5a is displayed in Figure 5b. It is noteworthy that the minimum traveltime is recorded at the zero-offset location, and that the reflection traveltimes recorded on the far traces are equal in both the updip and downdip directions. Hence, the pattern of reflection traveltimes (i.e. minimum time and moveout) could easily be interpreted in terms of a horizontal reflector underlying a homogeneous, isotropic layer. The only evidence that the layer is anisotropic is that the reflection traveltime moveout is slightly non-hyperbolic. Analysis of non-hyperbolic moveout can be used to estimate the anisotropic parameters (e.g. Tsvankin, 1997), but this approach becomes difficult in noisy foothills data, particularly if complex structures exist with wavelengths that are less than the seismic recording aperture.

If foothills seismic data are processed assuming that all velocities are isotropic, a well-imaged, migrated section may be obtained, but targets below dipping layers may be mispositioned. Migration error due to anisotropic velocities was investigated previously by Larner and Cohen (1993), and Alkhalifah and Larner (1994), and the magnitude of the lateral positioning error and its dependence on source-receiver offset is examined by Isaac and Lawton (1999). Figure 6 shows a diagram that illustrates how the lateral positioning error may impact a foothills explorationist. A common hydrocarbon target in fold-thrust environments is a hangingwall anticline with a steeply dipping forelimb (Figure 6); the objective typically is to drill reservoir rock in the forelimb. If TTI strata exist in the overburden, then the position and depth of the forelimb will be imaged incorrectly if the seismic data are processed and migrated using only isotropic velocities. The target will be located incorrectly in the updip direction of the overburden (Figure 6) which is usually toward the foreland. As a result, a well test may either miss the forelimb, or drill into it at a stratigraphic level higher than the target, and it will be required to be sidetracked toward the hinterland for a successful completion (Figure 6).

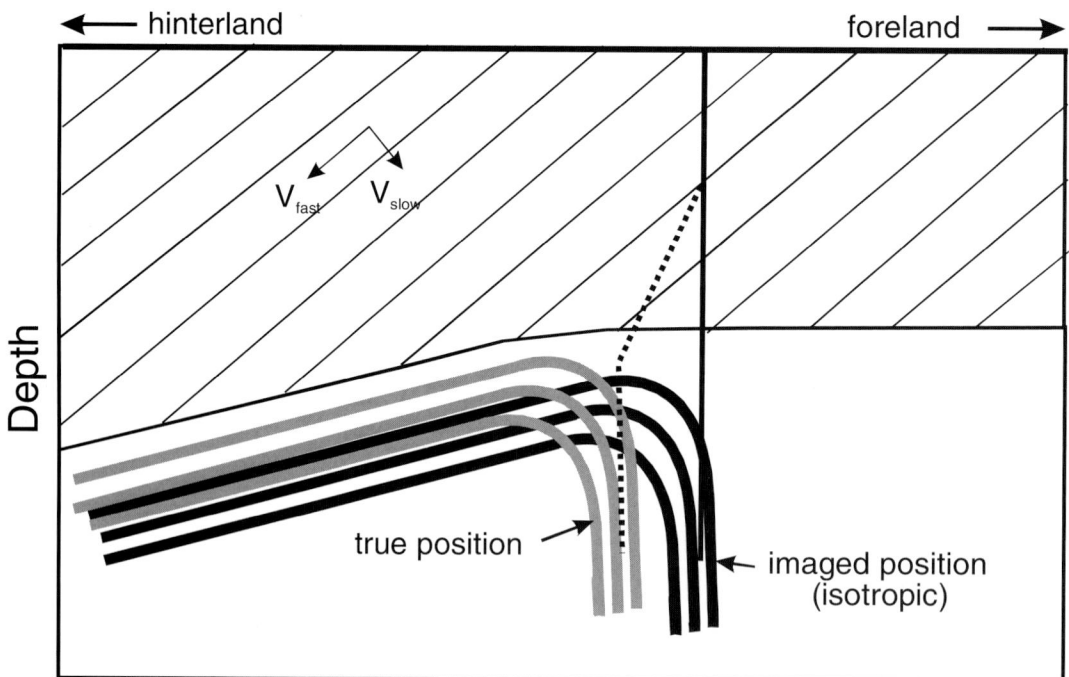

FIG. 6. The imaging problem likely to be encountered for structures below TTI overburden. If isotropic velocities are assumed during data processing, the target will be imaged incorrectly, and will appear to be displaced in the updip direction of the dipping overburden from its true location. Hence, a well drilled to test the forelimb of a fold, for example, will likely miss the structure in the foreland direction, or drill the structure stratigraphically higher than expected.

The solution to lateral positioning errors in processed seismic data is to incorporate anisotropic velocity models into the data processing flow, particularly depth migration. This parameterization is illustrated in Figure 7. Rather than an isotropic velocity field, the migration velocity model must include the stratigraphic dip, θ, the slow (i.e. bedding-normal) velocity, V_0, and the Thomsen anisotropy parameters ε and δ. In these models, θ is assessed from the interpretation of initial time or depth-migrated sections, V_0 is determined from regional sonic-log data and from migration velocity analysis, and ε and δ are assigned initial values, based on externally measured values of the stratigraphic units (e.g. Leslie and Lawton, 1999) or moveout analysis (e.g. Tsvankin, 1997). Implementation of anisotropic depth migration, and examples using physical model as well as field data, are presented by Vestrum et al., (1999).

FIG. 7. Velocity models for prestack depth migration in foothills areas with TTI strata must include the bedding-normal velocity, V_0, the stratal dip, θ, and the Thomsen (1986) parameters ε and δ. Photograph courtesy of J. Leslie

In summary, clastic strata in foothills environments are rotated in folds and in the hangingwalls of thrust faults. These strata often form coherent panels with high angles of dip, so that the TI axis of symmetry is no longer vertical. In these situations, significant depth and position errors of targets will occur on seismic sections if seismic velocities are assumed to be isotropic during data processing.

References

Alkhalifah, T., and Larner, K., 1994, Migration error in transversely isotropic media: Geophysics, 60, 1474-1484.

Backus, G.E., 1962, Long-wave elastic anisotropy produced by horizontal layering: J. Geophys. Res., 67, No. 11, 4427-4440.

Banik, N.C., 1984, Velocity anisotropy of shales and depth estimation in the North Sea basin: Geophysics, 49, 1411-1419.

Isaac, J.H., and Lawton, D.C., 1999, Image mispositioning due to dipping TI media: A physical seismic modeling study: Geophysics, 64, 1230-1238.

Johnston, J.E., and Christensen, N.I., 1995, Seismic anisotropy of shales: J. Geophys. Res.,100, 5991-6003.

Jones, L.E.A., and Wang, H.F., 1981, Ultrasonic velocities in Cretaceous shales from the Williston Basin: Geophysics, 46, 288-297.

Larner, K., and Cohen, J.K., 1993, Migration error in transversely isotropic media with linear velocity variation with depth: Geophysics, 58, 1454-1467.

Leslie, J.M, and Lawton, D.C., 1998, Anisotropic pre-stack depth migration: CSEG Recorder, 23, No. 10, 22-29.

Leslie, J.M., and Lawton, D.C., 1999, A refraction-seismic field study to determine the anisotropic parameters of shales: Geophysics, 64, 1247-1252).

Postma, G.W., 1995, Wave propagation in a stratified medium: Geophysics, 20, 294-392.

Schoenberg, M., 1994, Transversely isotropic media equivalent to thin isotropic layers: Geophys. Prosp., 42, 885-915.

Thomsen, L., 1986, Weak elastic anisotropy: Geophysics, 51, 1954-1966.

Tsvankin, I, 1997, Reflection moveout and parameter estimation for horizontal transverse isotropy: Geophysics, 62, 614-629.

Vernik, L., and Liu, X., 1997, Velocity anisotropy in shales: A petrophysical study: Geophysics, 62, 521-532.

Vestrum, R.W., Lawton, D.C., and Schmid, R., 1999, Imaging structures below dipping TI media: Geophysics, 64, 1239-1246.

ANISOTROPIC PRE-STACK DEPTH MIGRATION

Jennifer M. Leslie and Don C. Lawton
Department of Geology and Geophysics, University of Calgary
2500 University Drive N.W., Calgary, Alberta T2N 1N4

Seismic anisotropy is the variation of velocity with direction. This phenomenon may affect time-to-depth conversions of seismic data which, in turn, will result in incorrect images of subsurface structures. This misrepresentation can seriously alter the location of exploration targets. By studying seismic anisotropy, one can better assess a play target, thereby reducing the risks and costs involved. The main causes of seismic anisotropy are due to aligned mineral grains, aligned cracks, aligned crystals and periodic thin layering [Helbig, 1994]. The effects due to periodic thin layering are most pronounced in layered clastic sequences, specifically shales or interbedded shales and sandstones. It is known that the majority of sedimentary basins contain considerable amounts of shales and are also the main location of hydrocarbon reserves [Schoenberg, 1994; Sayers, 1994; Hornby et al., 1994]. Banik (1984) discovered a strong correlation between the presence of shales and a measured velocity anisotropy in North Sea sediments. It has been shown by many researchers that shales exhibit transverse isotropy (TI), with the symmetry axis perpendicular to bedding, when probed by seismic energy with wavelengths that are greater than the thickness of the layers [Johnston and Christensen, 1995; Backus, 1962; Postma, 1955; Levin, 1979; Schoenberg, 1994]. The difference between velocities perpendicular and parallel to bedding is significant, increasing with the degree of anisotropy and can be as much as 30% in shales [Backus, 1962; Berryman, 1979; Levin, 1979; Jones and Wang, 1981; Banik, 1984; Gaiser, 1990; Sayers, 1994; Hornby et al., 1994; Kebaili and Schmitt, 1996]. This can result in apparent, residual depth structures of several hundred metres for targets at several kilometres depth. Velocity anisotropy is also likely to exist in interbedded shales and sandstones in the Rocky Mountain Foothills, where folding and faulting can thrust these stratigraphic horizons, often at steep dip angles, to the surface. These dipping layers are expected to induce changes in velocity which result in traveltime anomalies in the seismic data [Crampin, 1977; Gaiser, 1990].

Physical models, representative of structures found in the Foothills of central Alberta and incorporating anisotropic material (Phenolic laminate), have been used to demonstrate the pull-up effects of anisotropy due to a steeply dipping shales. These physical models, initially described by Leslie and Lawton (1998a), are extremely useful for assessing effects of anisotropy on seismic images since the velocities, anisotropic parameters, dip angles, and the exact location and geometry of the thrust sheets are known *a priori*. This allowed us to determine if the effects of anisotropy, particularly transverse isotropy, significantly affect isotropic migration results. The results of the first model were conclusive [Leslie and Lawton, 1998a]: anisotropy is a significant effect, contributing to both lateral and depth positioning issues, and that isotropic migration is not able to properly account for these effects. A second model, made completely of Phenolic laminate, is considered to be a more geologically representative model as it better approximates the structures found in the Alberta Foothills. Numerical modelling performed on the second model indicates that the magnitude of the anisotropy of dipping strata affects traveltimes significantly, with distortions in the stacking velocity field greatest for intermediate dips [Leslie and Lawton, 1998a].

Pre-stack depth migration was performed on both physical models using both isotropic and anisotropic velocity models. The difference between isotropic and anisotropic migrations is in the code used (courtesy of Kelman Technologies Inc.), the traveltime generator for the anisotropic, Kirchhoff, pre-stack depth migration has been altered so that anisotropic group velocities are used to propagate the wavefronts. In the isotropic case, a point on the wavefront propagates normal to the wavefront using the local velocity. In the anisotropic case, a point on the wavefront propagates radially outwards from the sourcepoint at the group velocity, which may be oblique to the wavefront normal [Vestrum et al. 1998].

PHYSICAL MODEL 1

A 1:10 000 scale model of an anisotropic thrust sheet was constructed in the physical modelling laboratory of the Department of Geology and Geophysics at the University of Calgary (Figure 1).

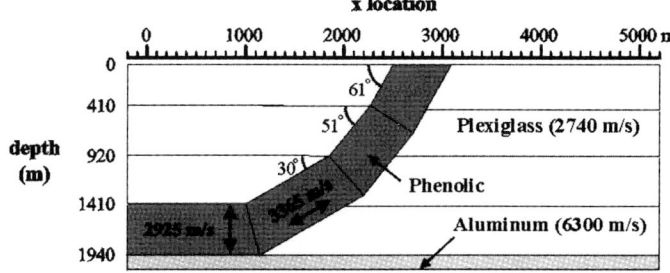

Figure 1. Diagram of first physical model built for the FRP. The reflector of interest is the aluminum base of the model at 1940 m (scaled).

The model mimics structures found in the central Alberta Foothills. Phenolic laminate, consisting of layers of cloth which have been glued together with epoxy, was used as the anisotropic material because it has well understood elastic properties [Cheadle et al., 1991]. A simulated thrust sheet was constructed such that the corresponding fast and slow velocities were V_{slow} = 2925 m/s and V_{fast} = 3365 m/s [Cheadle et al., 1991]. The Thomsen (1986) anisotropic parameters were determined to be $\delta = 0.081$ and $\varepsilon = 0.150$ [Cheadle et al., 1991]. The surrounding isotropic medium was made of Plexiglass, which has a measured velocity of 2740 m/s [Cheadle et al., 1991]. The marker of interest was the reflection from the horizontal, aluminum base of the model, at approximately 2 km depth (scaled), below the thrust sheet.

Results of the 2-D, zero-offset, ultrasonic survey across the model show an apparent structure in this basal reflector, with an

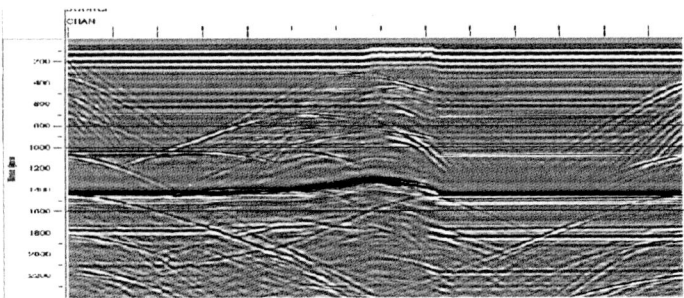

Figure 2. Zero-offset data from the ultrasonic survey performed on the first physical model. Note the pull-up (~100 ms) in the basal reflector located at approximately 1400 ms.

amplitude of approximately 100 ms (Figure 2). After eliminating the isotropic velocity pull-up from the data, due to the larger volume of faster material in the thrust sheet, the residual anomaly due to anisotropy is 50 ms. This translates into a residual depth structure of approximately 100 m, which is the magnitude of structures that are considered prospects in the Foothills. Hence the effects of anisotropy due to steeply dipping clastics are represented and sig-

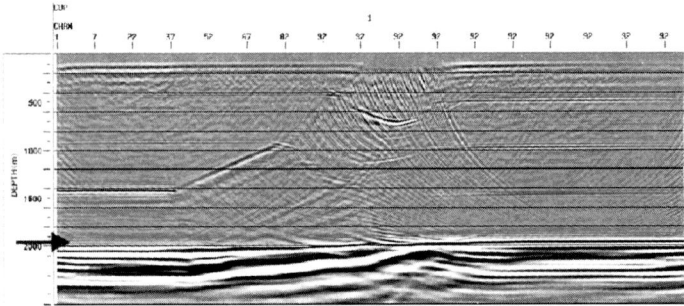

Figure 3. Resultant isotropic, pre-stack depth migration using a constant, fast velocity in the thrust sheet. The arrow indicates the true depth position of the basal reflector. Note that the basal reflector is located too deep under the thrust sheet.

nificant in the seismic data obtained from the physical model.

PRE-STACK DEPTH MIGRATION - MODEL 1

The physical model data were processed as follows: mute; pre-stack depth migration (isotropic and anisotropic); scale; and filter (bandpass Ormsby 8-12, 50-60 Hz). The data were first migrated isotropically, with a velocity model built using the constant fast velocity of the thrust sheet and its actual spatial location, for the migration. The results of this migration are shown in Figure 3 and the associated depth gathers in Figure 4 (near offsets on the right and far offsets on the left). The continuity of the basal reflector is

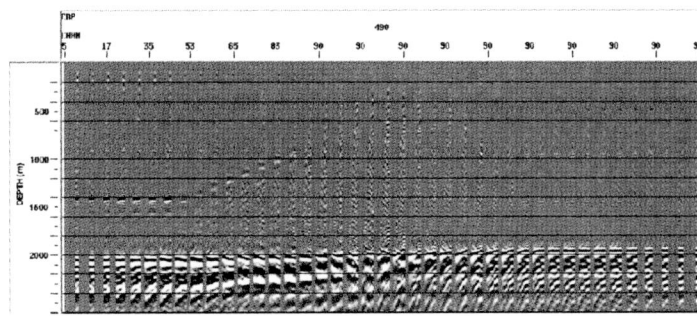

Figure 4. A sampling of depth gathers resulting from the migration in Figure 3. Note that the near offsets are at the right side of each gather and the far offsets at the left.

good; however, it is located too far in depth under the thrust sheet. In addition, the gathers in this location indicate that the migration velocity is too low and should be increased. In doing so, the reflec-

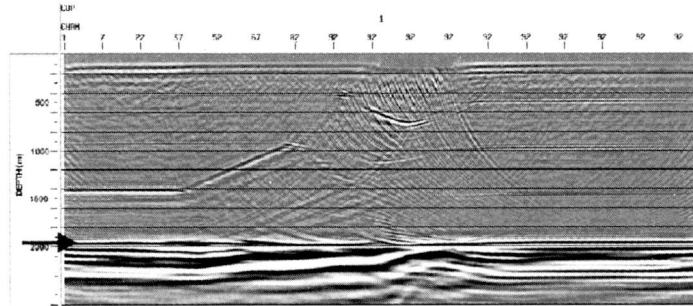

Figure 5. Resultant isotropic, pre-stack depth migration using a horizontal velocity gradient in the thrust sheet. Note that the basal reflector is located in the correct depth location under the thrust sheet; however, some residual structure is present in this reflector.

tor would be pushed even farther in depth, which is incorrect.

The second isotropic velocity model used a horizontal velocity gradient in the thrust sheet, grading from the fast velocity of the Phenolic in the top right corner of the thrust to the slow velocity in the flat lying Phenolic in the bottom left, in attempt to account for

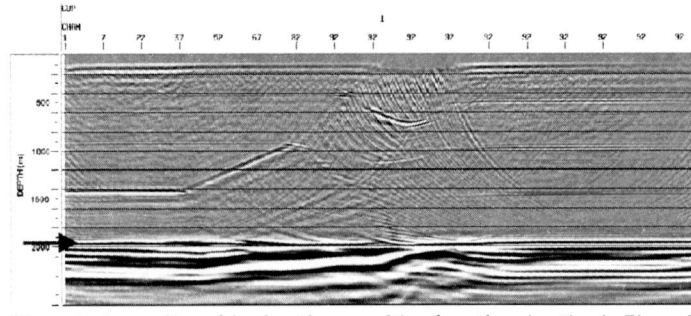

Figure 6. A sampling of depth gathers resulting from the migration in Figure 5. Note the distorted nature of the gathers, especially under the thrust sheet, which indicate that the migration velocity used was too low.

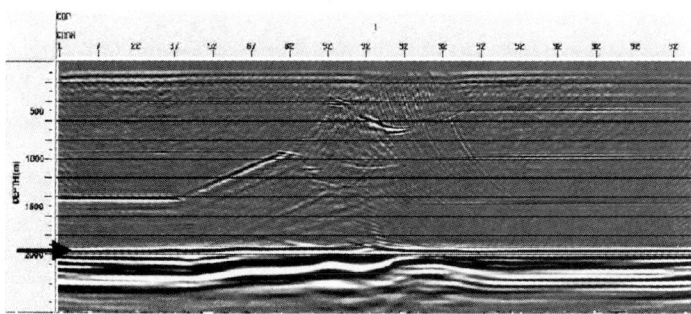

Figure 7. Resultant anisotropic, pre-stack depth migration using the correct parameters of the physical model. Note that the basal reflector is located in the correct depth location under the thrust sheet and the residual structure in the reflector of interest has been eliminated.

the anisotropy of the Phenolic material. The result is a correctly placed reflector, under the thrust sheet, with the continuity of the reflector somewhat compromised (Figure 5). The depth gather results indicate that, although the reflector is correctly located in depth, a faster migration velocity should be used to correctly migrate the data (Figure 6). This would also be incorrect, according

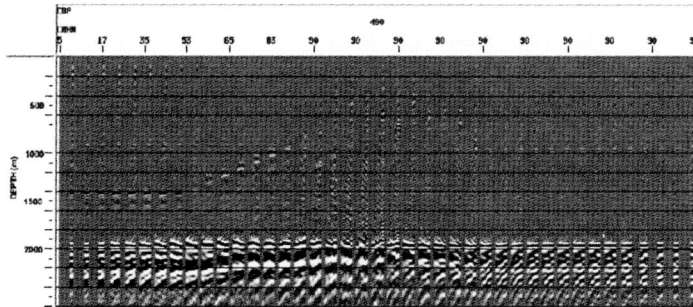

Figure 8. A sampling of depth gathers resulting from the migration in Figure 7. The minor residual moveout present in the gathers is attributed to the velocity model building process.

to the known physical model.

The only way to eliminate the discrepancy between correct depth and residual moveout on image gathers is to use anisotropic pre-stack depth migration. By using the correct velocity model, and all the correct parameters (δ, ϵ and dip) from the physical model, one obtains a correct depth image of the original model (Figure 7) which is supported by its associated depth gathers (Figure 8). Reflector continuity is maintained, it is correctly located in depth

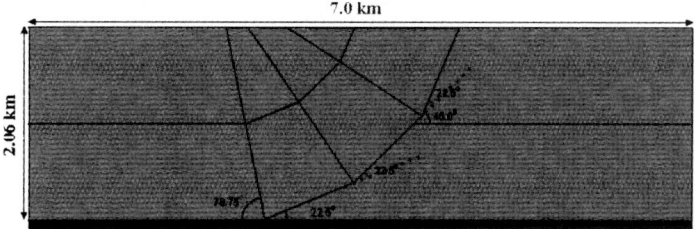

Figure 9. Diagram of second physical model built for the FRP. The reflector of interest is the base of the model at 2050 m (scaled). The top of the model was planed flat for the acquisition of the ultrasonic survey.

and events within the depth gathers are flattened. Hence, all the discrepancies from the isotropic depth migration process have been eliminated. It should be noted that the small amount of residual

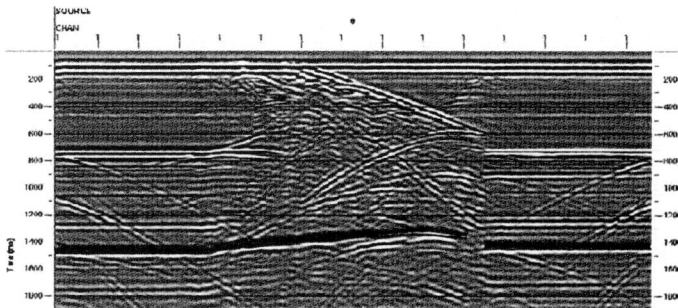

Figure 10. Resultant data from the ultrasonic survey performed on the second physical model. Note the pull-up (~50 ms) in the basal reflector located at approximately 1450 ms.

moveout in the depth gathers is associated with the velocity model building process. The migration velocity model is built in time. However, the physical data, measured from the model, are measured in depth, and the conversion from depth to time in the velocity model building process introduces small irregularities into the velocity model and, subsequently, in the migration. In conventional migration, the depth model is not known, thus the velocity

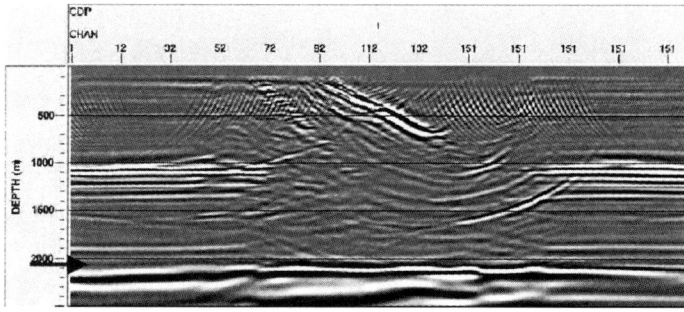

Figure 11. Resultant isotropic, pre-stack depth migration using a horizontal velocity gradient in the thrust sheet. Arrow indicates the true depth location of the basal reflector. Note that the basal reflector (trough) is located in the correct depth location under the thrust sheet; however, some residual structure is present in this reflector.

model must be built deterministically.

PHYSICAL MODEL 2

Another 1:10 000 scale model of an anisotropic thrust sheet was constructed in the physical modeling laboratory (Figure 9). The model is comprised completely of Phenolic laminate with slow and fast velocities of V_2 = 2925 m/s and V_3 = 3365 m/s, respectively, and δ = 0.081 and ϵ = 0.150 [Cheadle et al., 1991; Thomsen, 1986]. The resulting degree of anisotropy is 13%. The surrounding flat layers are also made of Phenolic. The marker of interest was again the reflection from the horizontal base of the model, at approximately 2 km depth (scaled), below the thrust sheet. This model did not have an aluminum base to it.

The 2-D, zero-offset, ultrasonic survey data are shown in Figure 10. The amount of pull-up is less than 100 ms since the surrounding isotropic material had been replaced with the higher velocity

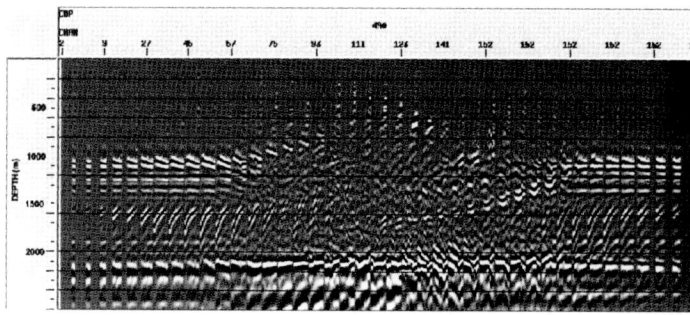

Figure 14. *A sampling of depth gathers resulting from the migration in Figure 13. Note the distorted nature of the gathers, which indicate that the dips were not assigned accurately.*

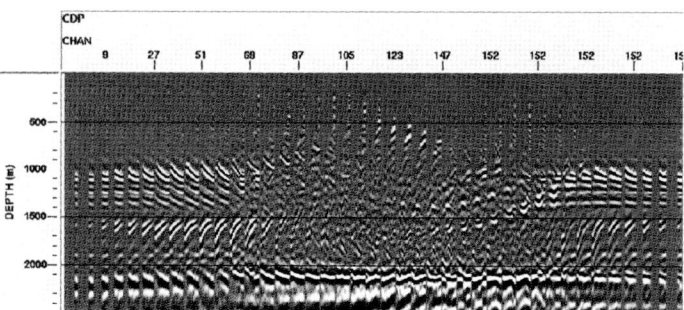

Figure 12. *A sampling of depth gathers resulting from the migration in Figure 11. Note the distorted nature of the gathers, especially under the thrust sheet, which indicate that the migration velocity used was too low.*

gathers exhibit considerable residual moveout which indicates that a higher velocity should be used to properly migrate the data (Figure 12).

It is worthy to note that the continuity of the basal reflector is compromised because the model does not have an aluminum base.

Phenolic laminate, compared with lower velocity Plexiglass in Model 1. Hence the velocity contrast is less compared to the first model, resulting in a smaller velocity anomaly, but one which is caused entirely by velocity anisotropy.

PRE-STACK DEPTH MIGRATION - MODEL 2

The physical model data were processed using a mute, pre-stack depth migration (isotropic and anisotropic), scale and filter (bandpass Ormsby 8-12, 50-60 Hz). For comparison purposes this model was first processed isotropically, using a gradient velocity model. The base of the thrust was defined and a horizontal velocity gradient was applied to the dipping part of the thrust: grading from the fast velocity of the Phenolic in the top right corner to the slow veloc-

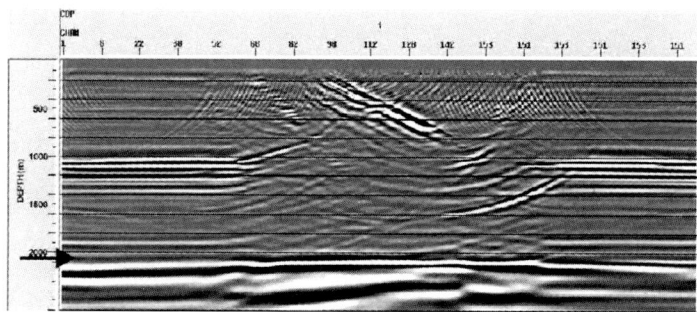

Figure 15. *Resultant anisotropic, pre-stack depth migration using distinctly dipping sections in the thrust sheet. Note that the basal reflector is located in the correct depth location under the thrust sheet and the residual structure has been eliminated.*

Since the model is sitting on a table top, thus going from a high velocity medium to a slow velocity medium, there is a negative impedance contrast at the base of the model. Also, there is a multiple (positive impedance) which constructively interferes with the

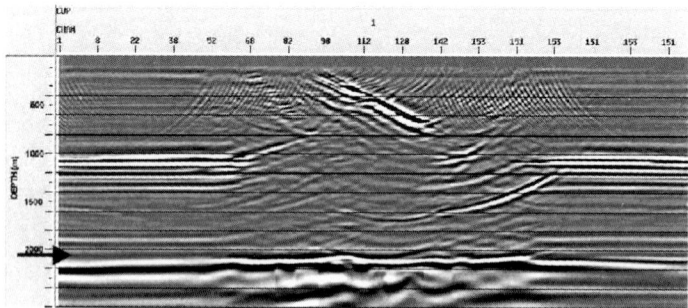

Figure 13. *Resultant anisotropic, pre-stack depth migration using a continuous dip definition (no distinctly dipping sections) in the thrust sheet. Note that the basal reflector is located in the correct depth location under the thrust sheet; however, there is an excessive amount of residual structure present in the basal reflector.*

ity of the Phenolic in the flat lying layers, in the bottom left. The resultant isotropic depth migration is comparable to the results of the first model (Figure 11). The basal reflector is correctly located in depth, with some minor structure present; however, the depth

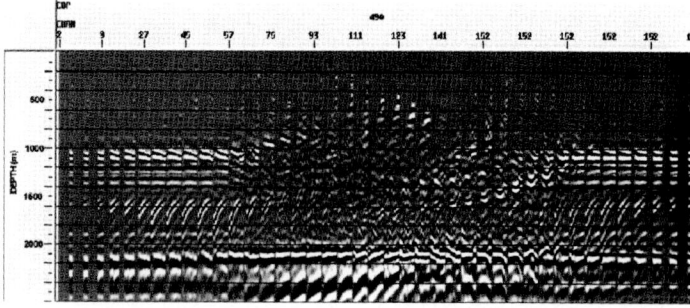

Figure 16. *A sampling of depth gathers resulting from the migration in Figure 15. The residual structure present in the gathers is attributed to the velocity model building process.*

basal reflector, resulting from the horizontal seam, halfway down the model, at approximately 1 km scaled depth. Thus the reflector is very strong in the middle of the section and weak at the edges (Figure 11).

Data from the second model were then migrated anisotropically. This velocity model consisted of the definition of the thrust base, the actual parameters of the model (slow P-wave velocity, δ and ε) and a continuous definition of dip through the thrust sheet and across the model. No distinctions were made between the different blocks of dipping Phenolic as the thrust sheet was modelled as having continuous, smooth curvature. The basal reflector is correctly located in depth, although it is severely distorted (Figure 13). The gathers are comparably distorted as well (Figure 14). Given that the input model linearly interpolates the dips between defined values, whereas the actual model has distinctly dipping sections, the results are not surprising. Thus the input model is sensitive to the dips defined within it.

In the final anisotropic velocity model, each distinct, dipping section of Phenolic was correctly located, with appropriate dip as well as all the correct parameters of the physical model. The result is a correct depth section with good reflector continuity, correct reflector location and flattened events on the depth gathers (Figures 15 and 16). Again the gathers are subject to small depth-time conversion errors, as was noted in the results of the first physical model (Figures 7 and 8). Thus, the same sensitivity to the model building process is also present in the second physical model data.

DISCUSSION AND CONCLUSIONS

Physical models are very useful for the study of seismic anisotropy. Since all the necessary parameters of the model can be determined, the data provided by the model are ideal for the testing of processing software, in particular, migration routines. In this study, it has been demonstrated that isotropic, pre-stack depth migration is limited in its ability to correctly migrate data from the anisotropic physical models, as it leads to discrepancies between the correct reflector depth and the residual moveout in image gathers. The testing completed in this study indicates that conventional pre-stack depth migration is not able to properly compensate for the effects of TI, and the results verify that anisotropy cannot be accounted for using isotropic velocities. Therefore, we conclude that anisotropic, pre-stack depth migration is required to properly process seismic data from these models and, in general, data from foothills environments.

ACKNOWLEDGMENTS

We would especially like to thank Kelman Technologies Inc. for the use of their pre-stack depth migration algorithm used in this study. Also to thank Ron Schmid and Rob Vestrum for their assistance and insight into this work.

We also gratefully acknowledge the financial support for this work provided through the sponsors of the Foothills Research Project (FRP) and the Natural Sciences and Engineering Research Council (NSERC) of Canada. Support for J. Leslie from NSERC, SEG and CSEG scholarships is also acknowledged.

REFERENCES

Banik, N.C., 1984, Velocity anisotropy of shales and depth estimation in the North Sea basin: Geophysics, **49**, p.1411-1419.

Backus, G.E., 1962, Long-wave elastic anisotropy produced by horizontal layering: J. Geophys. Res., **67**, No. 11, p.4427-4440.

Berryman, J.G., 1979, Long wavelength anisotropy in TI media: Geophysics, **44**, p.896-917.

Cheadle, S.P., Brown, R.J., and Lawton, D.C., 1991, Orthorhombic anisotropy: a physical seismic modeling study: Geophysics, **56**, p.1603-1613.

Crampin, S., 1977, A review of the effects of anisotropic layering on the propagation of seismic waves: Geophys. J. R. astr. Soc., **49**, p.9-27.

Gaiser, J.E., 1990, Transversely isotropic phase velocity analysis from slowness estimates: J. Geophys. Res., **95**, No. B7, p.11,241-11,254.

Helbig, K.,1994, Foundations of anisotropy for exploration seismics: Pergamon

Hornby, B.E., Schwartz, L.M., and Hudson, J.A., 1994, Anisotropic effective-medium modelling of the elastic properties of shales: Geophysics, **59**, p.1570-1583.

Johnston, J.E., and Christensen, N.I., 1995, Seismic anisotropy of shales: Journal of Geophysical Research, **100**, p.5991-6003.

A refraction-seismic field study to determine the anisotropic parameters of shales

Jennifer M. Leslie* and Don C. Lawton*

ABSTRACT

Shales are known to be transversely isotropic (TI), which can lead to incorrect subsurface imaging if isotropic behavior is assumed. By knowing the anisotropic parameters of the shales, it is possible to correct for the anisotropic effects in seismic data, thereby obtaining a more correct subsurface image. Two P-wave refraction-seismic field studies were undertaken in areas of steeply dipping marine shales of the Wapiabi Formation to determine the anisotropic parameters of these shales in situ. The first survey was located in Jumpingpound Creek, Alberta, Canada, and the other location was west of Longview, Alberta, Canada. Seismic lines were laid out parallel, perpendicular, and 45° to the local strike directions. At both locations, the compressional headwave velocities along strike were determined to be faster than the velocities perpendicular to strike. Analysis of the Jumpingpound Creek data location yielded anisotropic parameters of $\varepsilon = 0.14 \pm 0.05$ and $\delta = 0.00 \pm 0.08$, whereas at Longview, the anisotropic parameters were determined to be $\varepsilon = 0.25 \pm 0.06$ and $\delta = 0.00 \pm 0.06$. The difference in ε is attributed to variations in the shale facies between the two sites.

INTRODUCTION

Seismic anisotropy is of increasing interest since it can result in velocity anomalies and image mispositioning in reflection seismic data, particularly in areas of complex geological structure. Time-to-depth conversion of P-wave seismic data in structured areas has not yet taken into account traveltime and velocity distortions caused by velocity anisotropy (Leslie and Lawton, 1996). In order to properly account for anisotropy in the depth-conversion process, the anisotropic parameters of the rocks involved must be known. Shales are of particular interest because they are known to be transversely isotropic (TI) due to the periodic thin layering of their platelike compositional minerals (Backus, 1962; Postma, 1955; Levin, 1979; Schoenberg, 1994; Johnston and Christensen, 1995). The difference between the bedding-parallel and bedding-perpendicular velocities can be more than 30% (Backus, 1962; Berryman, 1979; Levin, 1979; Jones and Wang, 1981; Banik, 1984; Gaiser, 1990; Sayers, 1994; Hornby et al., 1994; Kebaili and Schmitt, 1996; Vernik and Liu, 1997). Laboratory studies can be performed on core and outcrop samples; but, it is difficult to mimic the in-situ conditions in the laboratory. Furthermore, it is very difficult to perform laboratory velocity measurements on shales since they are friable and tend to break apart, especially under the saturated conditions necessary to obtain realistic anisotropic parameters.

Clearly, there is a need to determine the anisotropic parameters of shales in a field experiment, where the bulk response of the rocks can be assessed. Namely, the anisotropic parameters can be determined on the scale of several seismic wavelengths for direct application to seismic data. The approach taken in this study was to undertake a multiazimuth refraction-seismic experiment in an area where the rocks of interest crop out at the surface and the strata have a uniform steep dip, preferably vertical. This structural geometry is relatively common in fold-thrust belts. By laying out seismic lines parallel, perpendicular, and at 45° to the local strike directions, the Thomsen (1986) anisotropic parameters, ε and δ, can be determined. For instance, for vertically dipping strata, the measured velocities along the local strike and dip directions are the bedding-normal and bedding-parallel velocities, equivalent to the vertical and horizontal velocities described by Thomsen (1986). These parameters are obtained by measurement of headwave velocities along the seismic lines, which need to be of sufficient length to ensure that the refractor velocities measured are from rocks that are below the near-surface weathered layer. A similar, successful study was undertaken by Gendzwill (1993), except that he evaluated anisotropic effects caused by vertical fractures rather than vertical bedding.

Presented at the 67th Annual International Meeting, Society of Exploration Geophysicists. Manuscript received by the Editor January 26, 1998; revised manuscript received January 8, 1999.
*University of Calgary, Dept. of Geology and Geophysics, 2500 University Drive, N.W., Calgary, Alberta T2N 1N4, Canada. E-mail: leslie@geo.ucalgary.ca; donl@geo.ucalgary.ca.
© 1999 Society of Exploration Geophysicists. All rights reserved.

DATA ACQUISITION

Multiazimuth refraction seismic surveys for anisotropic parameter determination were undertaken at two locations in the Rocky Mountain Foothills in Alberta, Canada, where uniform panels of shales crop out at the surface. These rocks belong to the Cretaceous Wapiabi Formation (Alberta Group) and are fine-grained, black, marine shales up to 300 m thick, with millimeter-scale laminations. The first site was at Jumpingpound Creek, approximately 30 km west of Calgary (Figure 1). At this location, shales of the Wapiabi Formation dip to the east at a constant angle of about 70° and form part of the eastern flank of the triangle zone in this region (Slotboom et al., 1996). There, the regional strike direction is 157°. Figure 2 is a photograph of an outcrop of Wapiabi shales immediately along strike from the field survey site. A diagram showing the layout of seismic lines occupied during the survey is given in Figure 3. Line lengths were dictated essentially by the available aperture at the site, which was located on a flat floodplain, and generally were longer in the dip direction than in the strike direction (Table 1). Each line was recorded by a fixed spread of 96 single 28-Hz geophones.

The second site chosen was located immediately west of Highway 541, approximately 20 km west of Longview (Figure 1). The terrain at this location was variable with one minor creek that cut across the dip line. At this site, the shales of the Wapiabi Formation dip west at a relatively constant angle generally greater than 80°. The regional strike of the shales at this location is 161°. Figure 4 is a photograph of the subvertical shales of this location as seen in outcrop in Flat Creek, immediately adjacent to the survey site. Each line was recorded by a fixed spread of 48 single 28-Hz geophones with a group interval of 4 m (Table 1). Figure 5 is a diagram of the seismic line layout for this site.

The source used for both locations was a Bison Elastic Wave Generator (EWG Model 3) owned and operated by the University of Calgary. This is an accelerated weight-drop source using a 270-kg mass, which is accelerated vertically downwards onto the ground by gravity, assisted by pretensioned elastic bands. Instruments used to record the data were a 24-bit, 96-channel Bison system operating at 0.25-ms sample interval and a record length of 0.5 s. A transducer mounted on the EWG source triggered the instruments, and data were recorded directly onto the hard disk in SEG-2 format and later copied to tape. Typically, two or three vertical impacts were summed at each source location to provide optimum signal-to-noise ratio for the P-wave first arrivals. Survey parameters for each line are summarized in Table 1.

In order to provide data redundancy and allow for detailed static corrections for the surficial weathering layer at the Jumpingpound Creek location, shotpoints were occupied at every second receiver location along the fixed receiver spread for each line. At each shotpoint, the source hammer was skidded 1.5 m perpendicular to the line direction. The program resulted in 49 records being obtained for each line.

At the Longview location, the weathering layer was determined to be considerably thinner than at Jumpingpound Creek, and 5 records/line were adequate for reliable headwave velocity analysis. At each shotpoint, the source hammer was skidded 2 m perpendicular to the line direction.

RESULTS

Preprocessing of the data was accomplished using ProMax software. Initially, bad traces were removed from each shot record. First breaks were then picked for the first shot record, after which the automatic first break picker was initialized. To assess the accuracy of the automatic picker, the first shot record was repicked and the new picks appeared over top of the original picks. The picks were considered to be accurate to ±2 ms, and the neural network picker was allowed to pick the rest of the first breaks. Any bad picks were then edited manually.

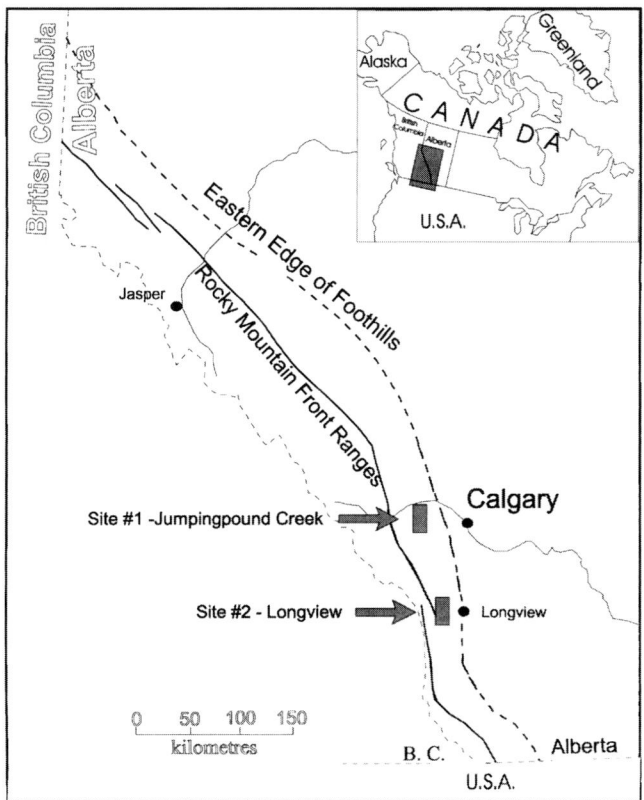

FIG. 1. Location map of the refraction survey experiment.

Table 1. **Survey parameters of refraction seismic lines.**

Site Location	Line	Azimuth (°)	Length (m)	Number of receivers	Receiver interval (m)	Shot interval (m)
Jumpingpound Creek	Strike	157	285	96	3	6
	Dip	62	380	96	4	8
	45°	114	380	96	4	8
Longview	Strike	161	188	48	4	12
	Dip	071	188	48	4	~12
	45°	116	188	48	4	12

Jumpingpound Creek

Data quality was good with clear first arrivals pickable over the full recording aperture. A plot of the first break picks from Jumpingpound Creek is shown in Figure 6. Short-wavelength distortions in the curve are interpreted to be due to variations in thickness and velocity of the surface layer. This layer is interpreted to be approximately 2 m thick and is comprised of reworked glacial sediments and alluvial sands and gravels deposited by Jumpingpound Creek. Traveltimes to the receivers closest to the shot indicate the velocity of this layer is approximately 120 m/s. By applying shot and receiver statics to the data, the surface layer was effectively reduced from the first break data. Figure 7 is a representative first break plot with the static corrections applied. It can be seen from this graph that there were two segments to each record. The segment closest to the shot location is curved, indicating that turning rays were present which needed to be incorporated into the velocity analysis. By curve-fitting the data and assuming a linear velocity function, $v(z) = v_o + kz$, the velocity gradient, k, was found to be largest in the strike direction with a value of 35 s^{-1}, whereas in the dip direction, the gradient is lower with a value of 20 s^{-1}. The velocity gradient is attributed to weathering of the shales, which extends to a depth of approximately 100 m. At far offsets, the traveltime-distance data became linear with offset, indicating that the base of the weathering layer had been reached. Velocity analysis for anisotropic parameter determination used the data from these deeper, constant values. There was insufficient overlap between the records to allow the deeper refractor velocities to be determined from reciprocal analysis, as used in the Longview study, so the inverse slope method was used.

As expected, the dip line exhibits a slower velocity than both the strike (fastest) and 45° lines (intermediate). The velocities are summarized in Table 2. The errors in the velocities

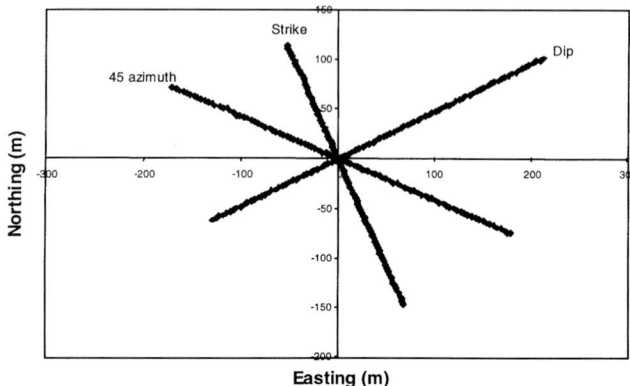

FIG. 3. Diagram indicating the layout of the refraction-seismic survey lines at the Jumpingpound Creek location.

Table 2. Summary of velocities calculated from first break data for the two locations.

Location	Velocity (m/s) ±100 m/s			Anisotropic Parameters	
	Dip	Strike	45°	ε	δ
Jumpingpound Creek	2800	3200	2900	0.14 ± 0.05	0.00 ± 0.08
Longview	3100	3800	3200	0.23 ± 0.05	−0.05 ± 0.07

FIG. 2. Photograph of outcropping shales of the Wapiabi Formation at the Jumpingpound Creek location. Note the 60°–70° dip of the shales, as indicated by the white horizons in the picture. The refraction survey was performed on the flat floodplain in the foreground.

were determined using the procedure of maximum/minimum slopes. Subsequently, the errors in the anisotropic parameters were statistically calculated from the error in the velocities. The anisotropic parameters for these deeper shales were calculated using the following Thomsen (1986) equations and are summarized in Table 2:

$$\delta = 4[(v_{45}/v_0) - 1] - [(v_{90}/v_0) - 1], \quad (1)$$

$$\varepsilon = \frac{v_{90} - v_0}{v_0}, \quad (2)$$

where v_0, v_{45} and v_{90} are the bedding-normal, 45° azimuth, and bedding-parallel velocities, respectively. In Thomsen (1986) notation, v_0 and v_{90} are the vertical and horizontal velocities, respectively. We have generalized these terms to be bedding-normal and bedding-parallel velocities in order to account for the arbitrary dip of the strata. Note also that the phase and group velocities are equal in both the bedding-normal and bedding-parallel directions. Furthermore, the small error in the 45° azimuth velocity, due to the difference between the phase and group angles, is well within the calculated error in δ (Table 2).

Longview

At the Longview location, the weathering layer was very thin (<1 m). At offsets greater than the crossover distance, the first break data plotted on straight lines (Figure 8), indicating that there is no vertical velocity gradient in these shales. The approach taken here for velocity analysis was a reciprocal method using "minus-times" as described by Hagedoorn (1959). In this method, the traveltime data from forward and reverse seismic records are used, hence requiring a fixed receiver spread with a shot at each end. The minus-time is defined as the time recorded at a receiver from the forward record, minus the time recorded at the same receiver on the reverse record, minus the total shot-to-shot traveltime of the record. Velocity analysis performed using minus-times is independent of any static effects caused by both surface or refractor topography and enables a reliable headwave velocity of the refractor to be determined.

Minus-time versus distance graphs for all three lines are given in Figure 9. Except for a few noisy data points, the data lay on straight lines, indicating that a velocity gradient was not present at this location. The velocities were calculated as twice the inverse slope of each line and are tabulated in Table 2. The anisotropic parameters were calculated using equations (1) and (2) and are also contained in Table 2. The errors in the velocities were determined using the procedure of maximum/minimum slopes. Subsequently, the errors in the anisotropic parameters were statistically calculated from the error in the velocities. The lack of velocity gradient, combined with the higher degree of anisotropy measured, suggests that the shales at this location are less weathered and less deformed than those shales at the Jumpingpound Creek location.

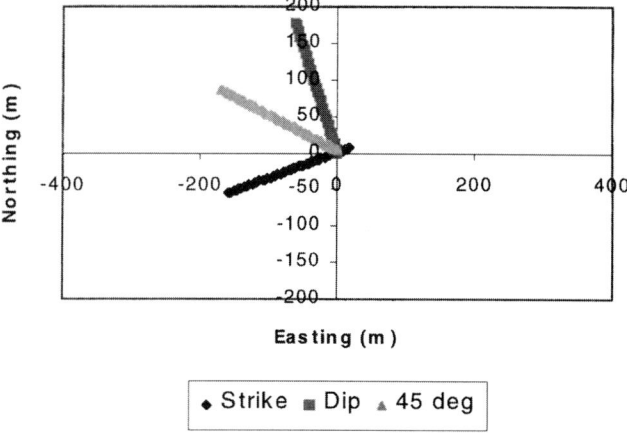

FIG. 5. Diagram indicating the layout of the refraction seismic survey at the Longview location.

FIG. 4. Photograph of the shale outcrop in Flat Creek, immediately south of the Longview refraction survey site.

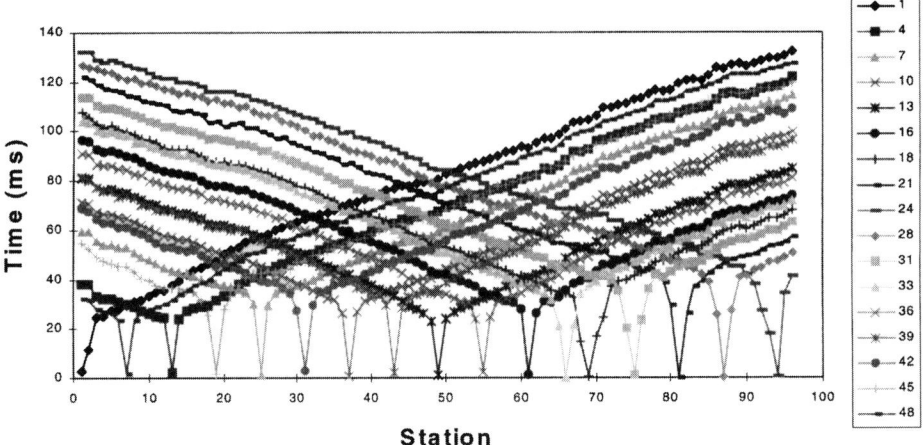

FIG. 6. Raw first break data for the strike line at Jumpingpound Creek. For clarity, only every third record was plotted.

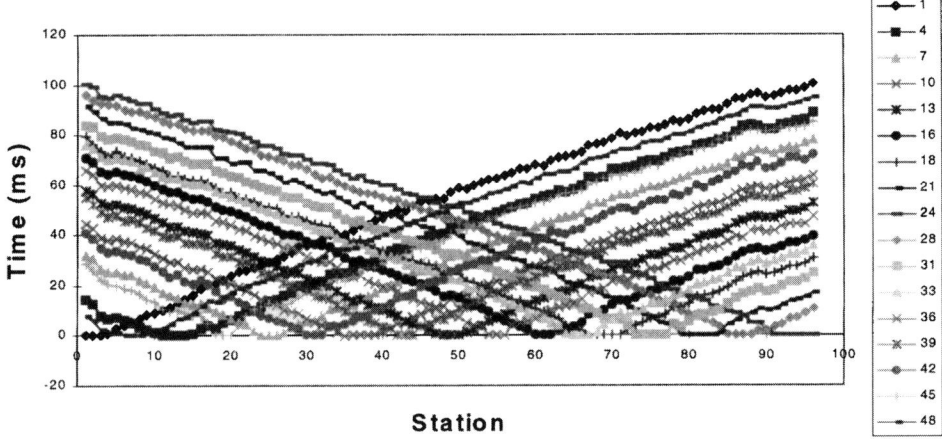

FIG. 7. First break data for the strike line at Jumpingpound Creek with the source and receiver statics applied. For clarity, only every third record was plotted.

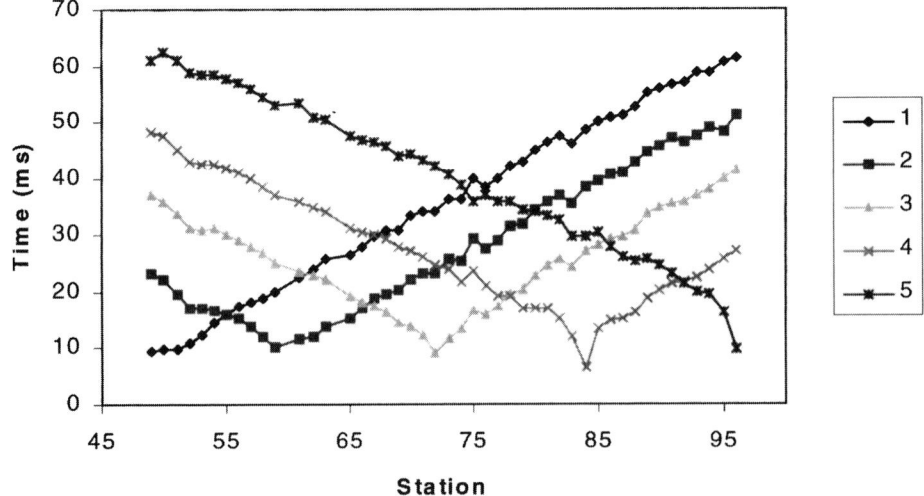

FIG. 8. Raw first break data for the strike line at the Longview location.

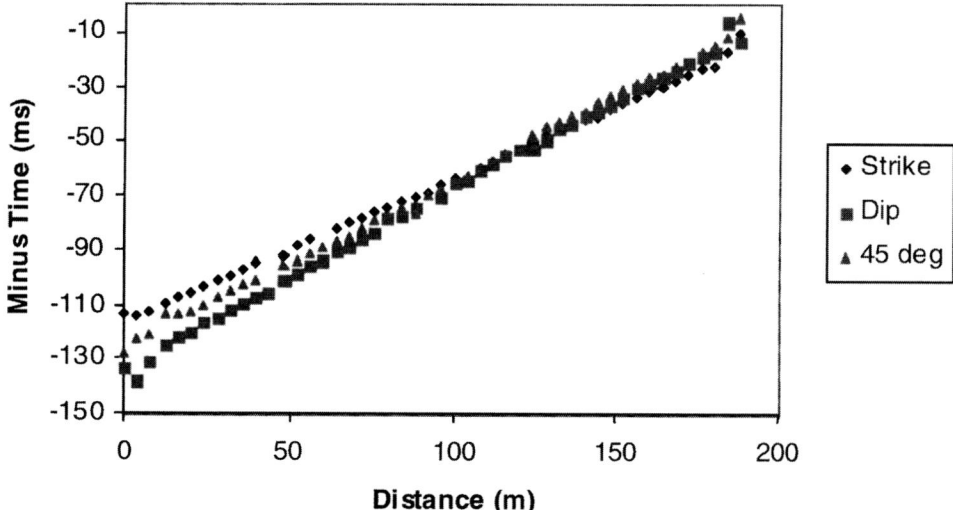

FIG. 9. Minus-time analysis for the Longview location. Note that the data for each azimuth lie on well-defined straight lines (excluding the endpoints), which have different slopes, allowing for accurate determination of the refractor velocities.

DISCUSSION

The results of this survey indicate conclusively that refraction seismic methods can be used to successfully measure velocity anisotropy in situ at locations where uniform blocks of steeply dipping strata outcrop. The Longview site shows generally higher velocities and a greater degree of anisotropy than that found in the Jumpingpound Creek survey. The difference is attributed to a change in the composition of the shales and a greater degree of penetrative strain at the Jumpingpound Creek location compared with the Longview location. The calculated anisotropic parameters for the Wapiabi Formation at the two locations lie within the range of values for shales presented by Thomsen (1986) and Vernik and Liu (1997). An alternate method of obtaining these parameters is through the analysis of nonhyperbolic moveout (Tsvankin, 1997) from reflection seismic data. However, this approach is not considered feasible in complex structural environments.

These results are significant, as anisotropic parameters are required for the correct processing of seismic data, particularly in seismic velocity analysis and migration. By incorporating these anisotropic parameters into processing routines, velocity anomalies due to anisotropy can be eliminated from the seismic data. Consequently, the mispositioning of exploration targets, both laterally as well as in depth, can be accounted for during depth conversion, particularly in structurally complex areas. As a result, a more accurate depth image of the subsurface exploration target can be obtained.

ACKNOWLEDGMENTS

We thank the staff and students of the University of Calgary 1996 Geophysics Field School and the staff of both the FRP and CREWES consortia for their assistance in acquiring the refraction data. We thank also the landowners who graciously allowed us to use their property for the acquisition of this data. This work was funded by sponsors of the Foothills Research Project and the Natural Sciences and Engineering Research Council (NSERC) of Canada. A version of this paper was presented at the 59th Conference of the European Association of Geoscientists and Engineers and is published here with EAGE's permission.

REFERENCES

Banik, N. C., 1984, Velocity anisotropy of shales and depth estimation in the North Sea basin: Geophysics, **49**, 1411–1419.

Backus, G. E., 1962, Long-wave elastic anisotropy produced by horizontal layering: J. Geophys. Res., **67**, No. 11, 4427–4440.

Berryman, J. G., 1979, Long wavelength anisotropy in TI media: Geophysics, **44**, 896–917.

Gaiser, J. E., 1990, Transversely isotropic phase velocity analysis from slowness estimates: J. Geophys. Res., **95**, No. B7, 11,241–11,254.

Gendzwill, D. J., 1993, Seismic velocity, fracture density and anisotropy of some Manitoba limestones: Can. J. Expl. Geophys., **29**, 153–162.

Hagedoorn, J. G., 1959, The plus-minus method of interpreting seismic refraction sections: Geophys. Prosp., **7**, 158–182.

Hornby, B. E., Schwartz, L. M., and Hudson, J. A., 1994, Anisotropic effective-medium modelling of the elastic properties of shales: Geophysics, **59**, 1570–1583.

Johnston, J. E., and Christensen, N. I., 1995, Seismic anisotropy of shales: J. Geophys. Res., **100**, 5991–6003.

Jones, L. E. A., and Wang, H. F., 1981, Ultrasonic velocities in Cretaceous shales from the Williston Basin: Geophysics, **46**, 288–297.

Kebaili, A., and Schmitt, D. R., 1996, Velocity anisotropy observed in wellbore seismic arrivals: Combined effects of intrinsic properties and layering: Geophysics, **61**, 12–20.

Leslie, J. M., and Lawton, D. C., 1996, P-wave traveltime anomalies below a dipping anisotropic thrust sheet: Presented at the 7th Internat. Workshop on Seismic Anisotropy.

Levin, F. K., 1979, Seismic velocities in transversely isotropic media: Geophysics, **44**, 918–936.

Postma, G. W., 1955, Wave propagation in a stratified medium: Geophysics, **20**, 294–392.

Sayers, C. M., 1994, The elastic anisotropy of shales: J. Geophys. Res., **99**, No. B1, 767–774.

Schoenberg, M., 1994, Transversely isotropic media equivalent to thin isotropic layers: Geophys. Prosp., **42**, 885–915.

Slotboom, R. T., Lawton, D. C., and Spratt, D. A., 1996, Seismic interpretation of the triangle zone at Jumping Pound, Alberta: Bull. Can. Petr. Geol., **44**, 233–243.

Thomsen, L., 1986, Weak elastic anisotropy: Geophysics, **51**, 1954–1966.

Tsvankin, I., 1997, Reflection moveout and parameter estimation for horizontal transverse isotropy: Geophysics, **62**, 614–629.

Vernik, L., and Liu, X., 1997, Velocity anisotropy in shales: A petrophysical study: Geophysics, **62**, 521–532.

Anisotropic true-amplitude migration

Tong Fei*, Joe A. Dellinger, and Gary E. Murphy, Amoco EPTG, Tulsa
Jeffrey L. Hensley, University of Tulsa
Samuel H. Gray, Amoco Canada Petroleum Company

SUMMARY

True-amplitude prestack depth migration attempts to preserve the amplitude and phase information of recorded seismic data during the migration process. The resulting image contains information about angle-dependent reflectivity (AVA) at every image point, and thus, indirectly, information about elastic-parameter and lithology contrasts within the earth. Here, we use a Kirchhoff true-amplitude migration technique to migrate a synthetic dataset for an anisotropic thrust model. The resulting migrated true-amplitude image compares well with the correct model.

INTRODUCTION

True-amplitude migration or inversion has recently become a very important tool in seismic imaging. Various authors (Berkhout 1985; Bleistein et al. 1987; Hanitzsch, 1995; Tarantola, 1984) have developed true-amplitude migration techniques for acoustic/elastic isotropic media. Gray (1997) gave a comprehensive review of the various methods.

In this work, true-amplitude migration techniques have been applied to 2D synthetic data for an elastic anisotropic media. In addition to positioning the reflectors, true-amplitude migration also provides angle-dependent reflection coefficients and AVO information.

NUMERICAL MODEL

A 2D numerical model, resembling the physical model of an anisotropic thrust sheet constructed by the Foothills Research Project at the University of Calgary (Leslie and Lawton 1996), is shown in Figure 1. The lightly shaded part of the model is an isotropic medium with a P-wave velocity of 2.74 km/s and an S-wave velocity of 1.38 km/s. The darkly shaded part of the model is anisotropic, with fast and slow p-wave velocities of 3.365 km/s and 2.925 km/s respectively, and v_s/v_p ratio of 0.5214. The associated weak anisotropy parameters (Thomsen 1986) of the anisotropic sheet are $\delta=0.081$, $\epsilon=0.15$ and $\gamma=0.035$. The symmetry axis of the anisotropic sheet is perpendicular to the long axis of the sheet.

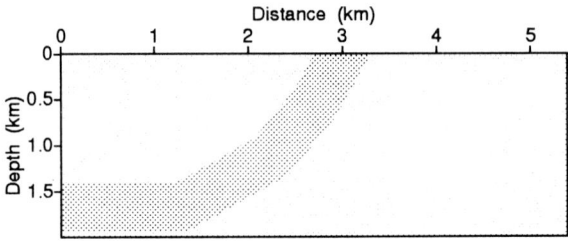

FIG. 1. Anisotropic thrust model. Light shading indicates an isotropic medium, dark shading indicates an anisotropic sheet.

The synthetic data for this model were generated by a finite-difference modeling technique. Figure 2 shows an example of a shot record with the source located at the surface, 1 km to the right of the origin.

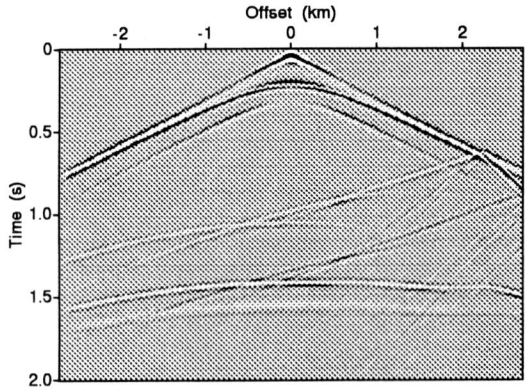

FIG. 2. Synthetic shot record. The source is located at the surface, 1 km to the right of the origin.

MIGRATION RESULT

Since the traveltime and amplitude play a central role in Kirchhoff true-amplitude migration, they have to be computed with accuracy. Here, we use ray tracing through a triangulated anisotropic model (Ruger 1997) to compute the first-arrival traveltime and amplitude fields. Figures 3 and 4 show examples of traveltime and amplitude maps for the source at the surface, 1 km to the right of the origin. From the figures, we can see that there are holes in the traveltime and amplitude maps, which are caused by ray tracing.

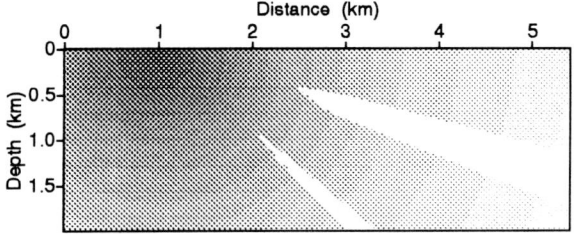

FIG. 3. Example of a first-arrival traveltime map for the anisotropic thrust model.

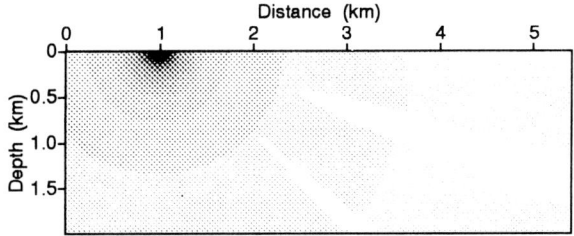

FIG. 4. Example of a first-arrival amplitude map for the anisotropic thrust model.

Using the traveltime and amplitude maps generated by ray tracing, we obtain a true-amplitude migration for the anisotropic model. Figure 5 shows the final stack of the migration image. From the figure, we see that the interfaces of the model are well imaged; moreover, the flat reflector below the dipping anisotropic layer is correctly imaged. From the prestack migration gather, we can also analyze the amplitude variation with offset (AVO), as seen in Figure 6.

CONCLUSIONS

True-amplitude migration has been applied to synthetic data generated by finite-differencing a 2D anisotropic thrust model. The migrated image shows that the reflectors are well positioned; moreover, from the amplitude of the image, one could estimate the AVO effect and obtain angle-dependent reflection coefficients for the model.

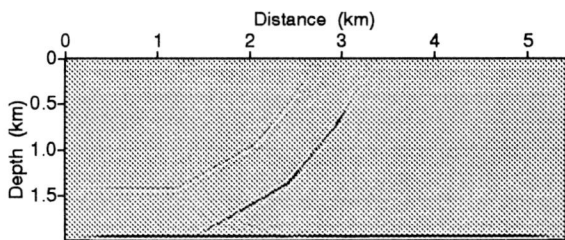

FIG. 5. Final stacked image of the prestack true-amplitude migration for the anisotropic model.

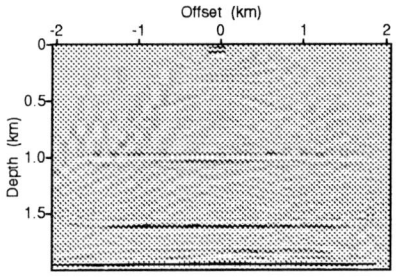

FIG. 6. Amplitude variation with offset (AVO) at location 2 km.

REFERENCES

Berkhout, A. J., 1985, Seismic migration: Imaging of acoustic energy by wavefield extrapolation A: Theoretical aspects, Elsevier Science Publ.

Bleistein, N., Cohen, J. K., and Hagin, F. G., 1987, Two and one-half dimensional born inversion with an arbitrary reference: Geophysics, **52**, no. 1, 26–36.

Gray, S. H., 1997, True-amplitude seismic migration: A comparison of three approaches: Geophysics, **62**, no. 3, 929–936.

Hanitzsch, C., 1995, Amplitude preserving prestack Kirchhoff depth migration/inversion in laterally inhomogeneous media: Ph.D. dissertation, University of Karlsruhe.

Leslie, J. M., and Lawton, D. C., 1996, Structural imaging below dipping anisotropic layers: predictions from seismic modeling: 66th Annual Internat. Mtg., Soc. Expl. Geophys., Expanded Abstracts, pages 719–722.

Ruger, A., 1997, Reflection coefficients and azimuthal AVO analysis in anisotropic media: Ph.D. dissertation, Colorado School of Mines.

Tarantola, A., 1984, Linearized inversion of seismic reflection data: Geophys. Prosp., **32**, 998–1015.

Thomsen, L., 1986, Weak elastic anisotropy: Geophysics, **51**, no. 10, 1954–1966.

Prestack depth migration in TI media: examples with numerical and physical modeling data
Nanxun Dai*, Scott Cheadle, Veritas DGC Inc. and J. Helen Isaac, The University of Calgary

Theoretical studies as well as laboratory and field experiments have shown that the physical properties of many rocks are anisotropic. In particular, the sedimentary rocks formed through rapid cyclical sedimentation patterns often exhibit transverse isotropy (TI). Velocities from laboratory studies indicate values measured parallel to the bedding of sedimentary rocks are often higher than values perpendicular to the bedding. The observation has also been supported by field studies (Banik, 1984, Leslie and Lawton, 1997). The difference can be as high as 30%. In areas with complex subsurface geology, such as the Rocky Mountain Foothills, the strata are distorted, forming folds and thrusts, so that the orientation of the TI axes of symmetry are spatially variable.

Conventional depth migration assumes an isotropic earth. When anisotropy is encountered the velocity model is modified in an empirical fashion in an attempt to obtain the best image. Since prestack depth migration depends on an accurate earth model, velocity distribution along with its directivity will have an important effect on the positioning of geological boundaries in migrated images. Ignoring anisotropy in prestack depth migration may cause defocussing and mispositioning of events.

For prestack depth migration in TI media, we have developed a finite difference algorithm which is capable of correctly handling spatially variable axes of symmetry. A first order hyperbolic differential system is formulated to describe the down-going seismic waves in TI media. The system is rotated locally according to the variation of the symmetry axis direction. The algorithm is applied in processing several sets of seismic data collected over physical models of TI media. The physical modeling studies were conducted as part of the Foothill Research Project at the University of Calgary for the purpose of investigating imaging errors incurred by the use of isotropic processing algorithms where there is seismic velocity anisotropy present in the subsurface (Isaac and Lawton, 1997). Numerical forward modeling of anisotropic wavefield propagation to complement the physical modelling was also conducted to generate shot gathers used for testing the migration algorithm.

Two types of model were used in the experiments to be described here. The first model featured a structural step below a scaled 1500 m thick layer of anisotropic material with the symmetry axis dipping at 45o. Using an isotropic algorithm, the imaged position of the step was shifted laterally about 300 m (Isaac and Lawton, 1997). Our prestack depth migration process, using the known transversely isotropic velocity model with a proper description of the axes of symmetry, properly locates the feature both vertically and laterally. The second model simulated a dipping anisotropic thrust sheet where the symmetry axis is spatially variable, as shown in Figure 1a. One of the challenges in imaging such a structure is properly flattening the basal reflector beneath the steeply dipping thrust sheet, which can not be achieved with time domain processing or isotropic depth migration algorithms (Isaac and Lawton, 1997). Also, the underside of the thrust is difficult to properly focus without taking anisotropy into account. Figure 1b shows the prestack depth migrated results from the numerical modeling data. The underside of the thrust is clearly imaged and the basal reflector is properly flattened beneath the steepest dipping portions of the sheet. Similar results from the other models will be shown in the presentation. Ongoing work is primarily geared to improving our ability to determine the anisotropic velocities and symmetry axis orientation as an extension to conventional velocity analysis.

References

Banik, N.C., 1984, Velocity anisotropy of shales and depth estimation in North Sea basin, Geophysics, 49.

Isaac, J. H. and Lawton, D. C., 1997, Anisotropic physical modeling, in Foothills Research Project Research Report, vol. 3.

Leslie, J.M. and Lawton, D.C., 1997, In situ anisotropic parameter determination of the Belly River and Wapiabi Formations, west of Longview, Alberta, in Foothills Research Project Research Report, vol. 3.

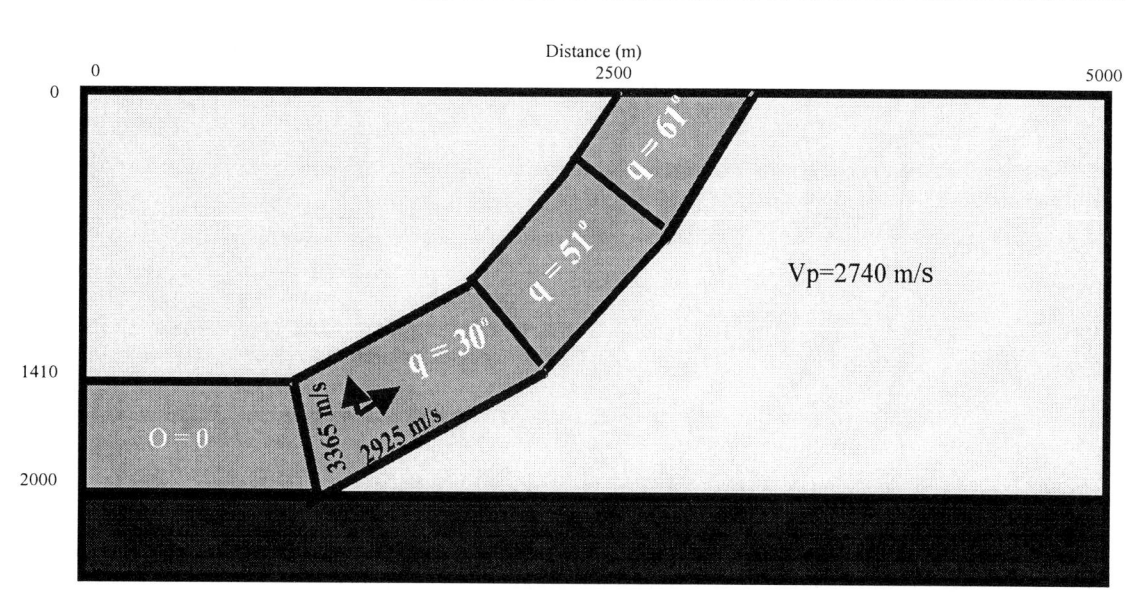

(a) Earth model with a tranversely isotropic thrust

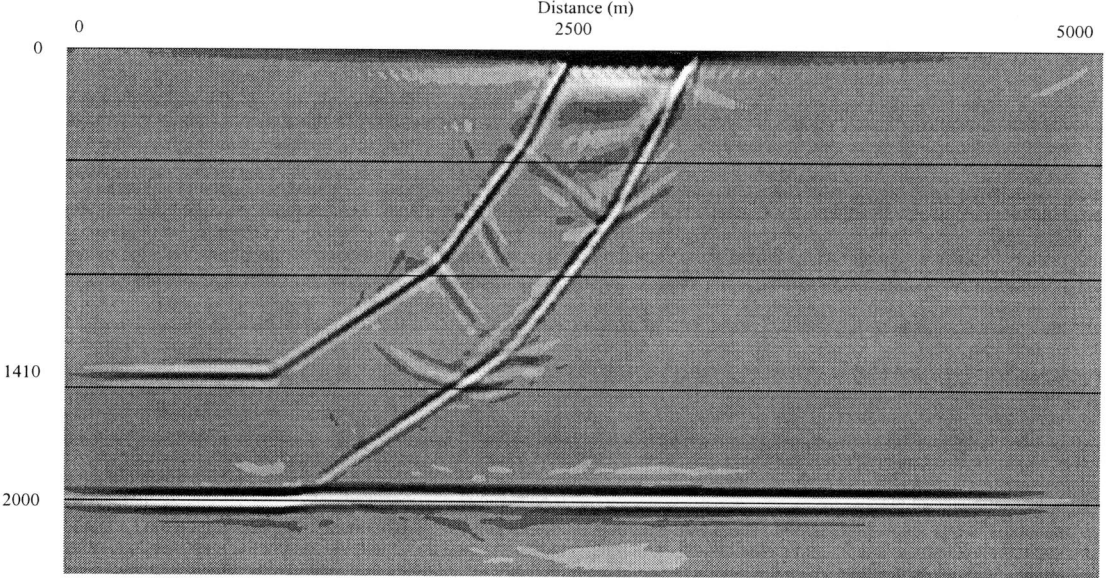

(b) Depth migration section

Figure 1   (a) A transversely isotropic thrust model used in numerical simulation and physical modeling.
(b) The anisotropic prestack depth migration of the numerical modeling data.

COMPARISON OF STRUCTURAL IMAGING IN ANISOTROPIC MEDIA USING *P*-WAVE AND *S*-WAVE DATA.

M.Graziella Kirtland Grech, J. Helen Isaac and Don C. Lawton, Department of Geology & Geophysics, University of Calgary

Various laboratory and in situ measurements have shown that certain rocks, including sandstone, shale and limestone exhibit seismic anisotropy (Thomsen, 1984; Bamford et al., 1979; Leslie et al., 1998). If not accounted for, anisotropy can cause several problems in the processing and imaging of seismic data. Typical errors are incorrect lateral position of events and errors in depth estimates.

Physical seismic modeling experiments using *P*-wave data have shown that when isotropic processing code is used to image the location of a target underneath dipping anisotropic beds, the imaged location is shifted laterally from its true location (Isaac et al., 1999). This lateral shift is attributed to two effects: velocity effects and sideslip effects. The former is the effect that, as a result of anisotropy, the velocity in one direction is faster than in other directions. The sideslip effect is brought about by the difference between phase and group velocities.

In this work, four seismic modeling experiments involving *P*-wave and *S*-wave data were carried out over a physical model to determine whether *S*-wave data can provide a better image of a target underneath dipping anisotropic beds with high *P*-wave anisotropy. Normal isotropic processing was used.

The physical model

A scaled physical model was built to simulate an environment typical of fold and thrust belts, where dipping anisotropic shale beds often overlie potential reservoir targets. The model used in the experiments is shown in Figure 1. The upper part was made up of a phenolic block with its axis of symmetry dipping at 45°. The phenolic block exhibits orthorhombic symmetry and Thomsen's anisotropic parameters ε, γ and δ were measured to be 0.24, 0.1 and 0.1 respectively. The slab underneath is made of isotropic plexiglass, with a step which represents a fault or a reef edge, cut in it at an angle of 45°. When scaled by a factor of 10,000, the anisotropic layer is 1500 m thick.

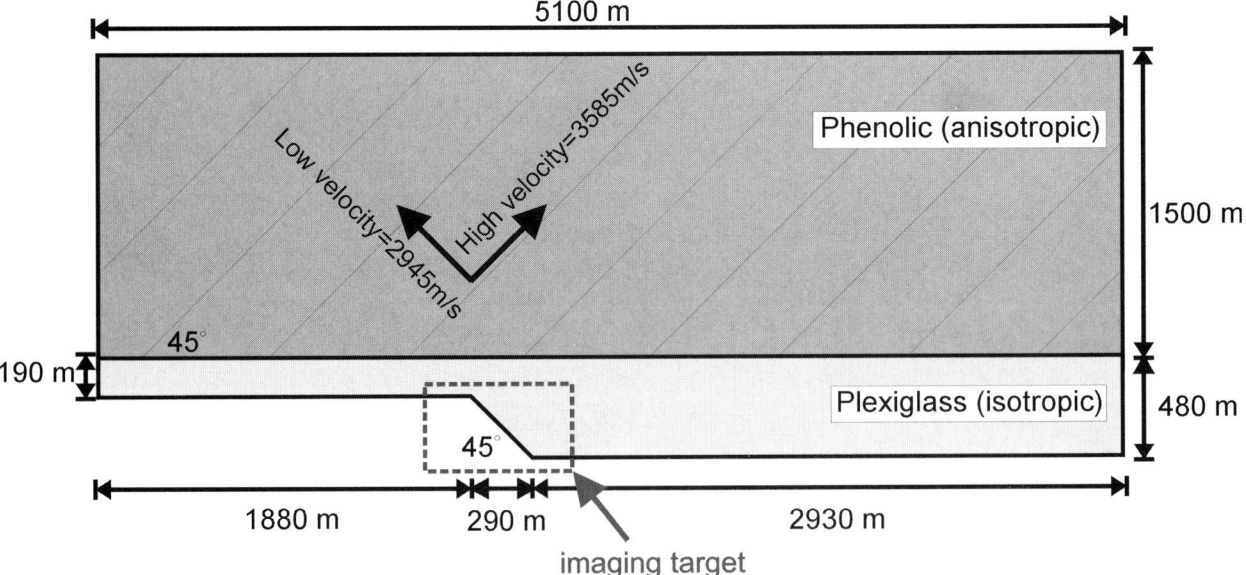

Figure 1. The physical model. The phenolic represents dipping anisotropic shale beds, and the step represents a fault or reef edge. Annotated on this figure are the dimensions scaled by a factor of 10,000. The dotted box highlights the imaging target.

The physical modeling experiments

The seismic data were acquired in the Department of Geology and Geophysics at the University of Calgary. Four datasets were acquired over the model shown in Figure 1, using (1) a vertical component source and receiver, (2) a transverse component source and receiver, (3) a radial component source and receiver and (4) a vertical component source and a radial component receiver. Radial component means polarization in the line of survey, whereas transverse component means polarization perpendicular to the line of survey. The data were collected along a symmetry plane of the material.

The same processing flow was applied to the four datasets, with some additional processing on the radial component and converted-wave data to reduce the *P-P* and *P-SV* reflections caused by *P*-wave energy leakage on the radial transducer. Isotropic processing code was used.

Figures 2 to 6 show different steps from the processing sequence. Figure 2 shows a typical shot gather from each of the four datasets. The gathers are dominated by direct arrivals and surface waves and the radial and transverse component records have poor S/N. Pure *P*-wave reflections are evident on the converted-wave gather and the radial component gather at about 1.2 s. The negative offsets on the converted-wave gather have been reversed in polarity for further processing.

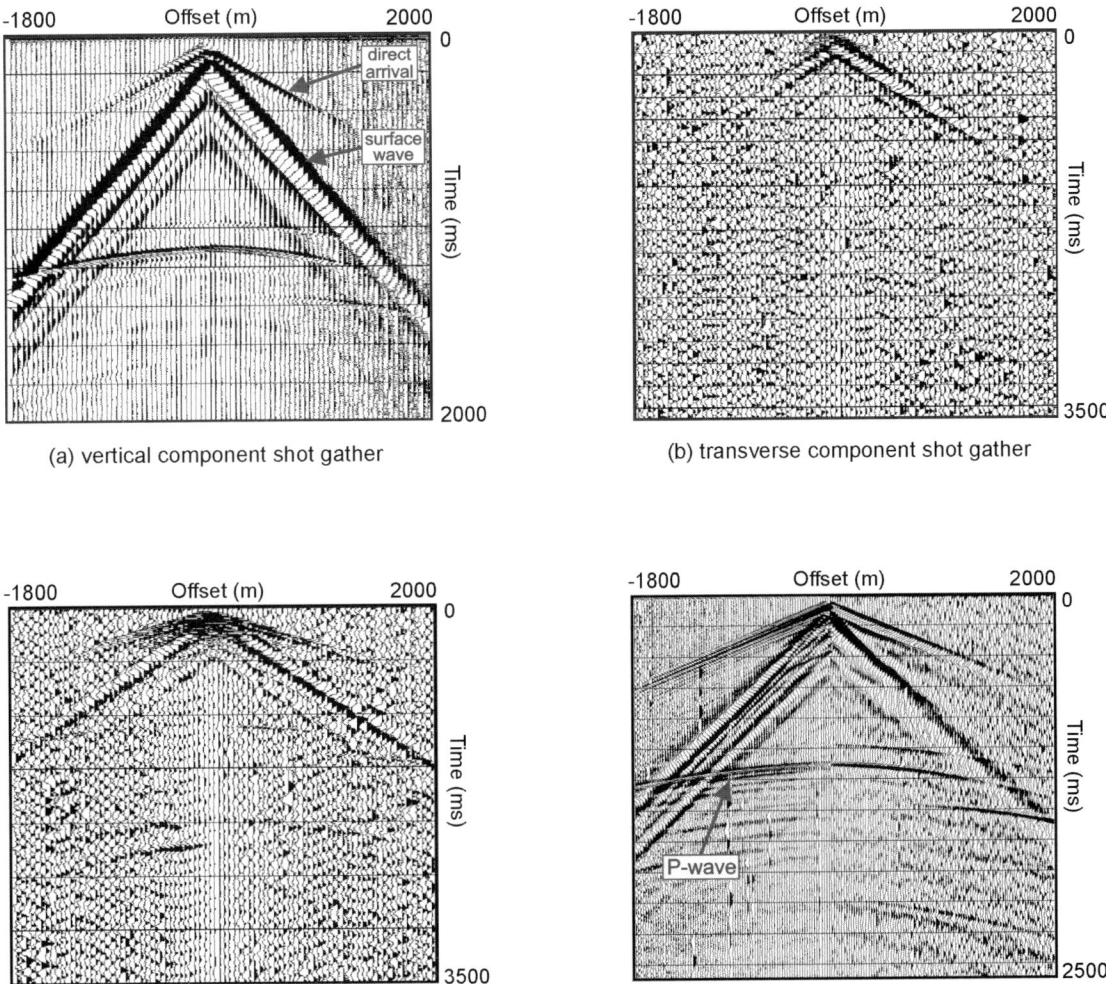

Figure 2. Typical shot gathers from each data set.

Figure 3 shows a series of bandpass filter panels from the vertical component dataset (Figure 3a) and from the transverse component dataset (Figure 3b). The radial component and converted-wave datasets had frequency content similar to the transverse component and are therefore not shown. These panels were used to determine the pass band for bandpass filtering. The vertical component dataset had a broad frequency band ranging between 5-75 Hz, whereas the other three datasets had approximately half that band, with 5-35Hz. An *f-k* filter was used to remove low velocity noise.

Filtering was followed with conventional velocity analysis, NMO correction and stacking. The converted-wave data was stacked using Common Conversion Point binning and an rms velocity function was derived from the vertical rms velocities, assuming Vp/Vs=2. Figure 4 shows the stacked sections of the target step. The top and base of the step can be easily identified on the vertical component and transverse component stacks. The pure *SV* events on the radial component stack are masked by *P-P* and *P-SV* reflections, brought about by the *P*-wave energy leakage mentioned earlier. Figure 4b shows the radial component stack with the different events identified. These events were verified by using average velocities obtained from transmission tests and the corresponding thicknesses

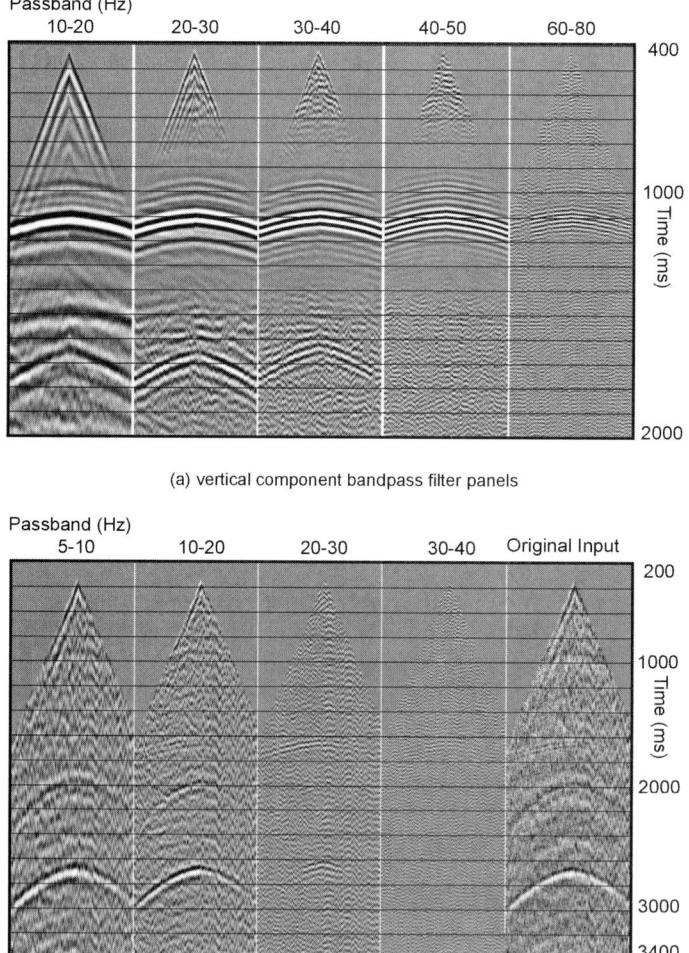

(a) vertical component bandpass filter panels

(b) transverse component bandpass filter panels

Figure 3. Bandpass filter panels from the vertical component and transverse component datasets. The frequency content of the radial component and converted-wave dataset had similar frequency content to the transverse component shown above.

from the model in order to determine the arrival times. Some additional processing was therefore carried out to eliminate these events and enhance the pure *SV* reflections. The converted-wave stack (Figure 4d) also had pure *P*-wave reflections from the top and base of the step, but these were less of a concern as they did not interfere with the converted-wave events from the top and base of the step.

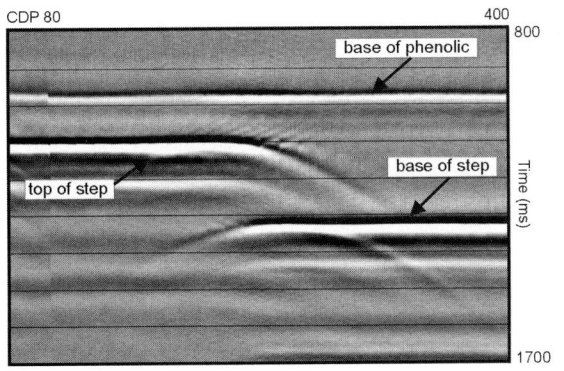

(a) Vertical component stack

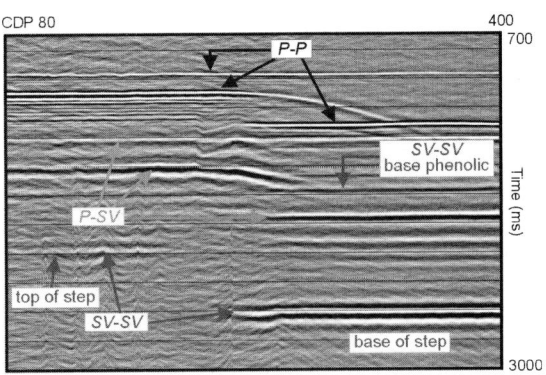

(b) Radial component stack

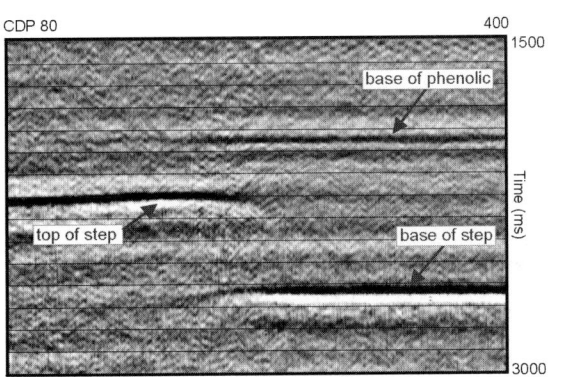

(c) Transverse component stack

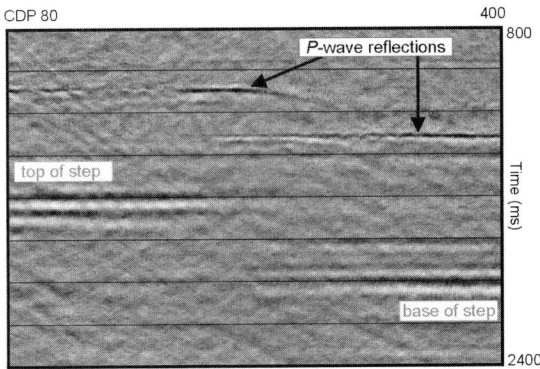

(d) Converted-wave (vertical-radial) stack

Figure 4. The stacked section from each of the four surveys.

Figures 5a and 5b show the radial-component stack after the additional processing to eliminate the unwanted reflections. Figure 5a was obtained after "biased velocity analysis" - this means that the velocity function after NMO was biased towards the lower velocities associated with the *SV*-events in order to attenuate the pure *P*-wave and converted-wave reflections. One significant improvement was in the attenuation of the *P*-wave reflections and enhancement of the pure *SV* reflection from the top of the step. Another attempt involved using a Radon filter after NMO (Figure 5b), which also resulted in a significant improvement of the pure *SV* reflection from the top of the step.

Again, the pure *P*-wave reflections were attenuated while the converted-wave reflections were still prominent. As the target could now be easily identified the "biased velocity analysis" stack was migrated.

Figure 6 shows the migrated section of the step target from each of the four surveys. Each section was migrated using an isotropic steep-dip finite-difference algorithm and a constant velocity for both the anisotropic and isotropic sections. A shift in the up-dip direction of the beds is observed on the vertical (Figure 6a), transverse (Figure 6c) and converted-wave (Figure 6d) sections, whereas no shift is observed on the radial section (Figure 6b).

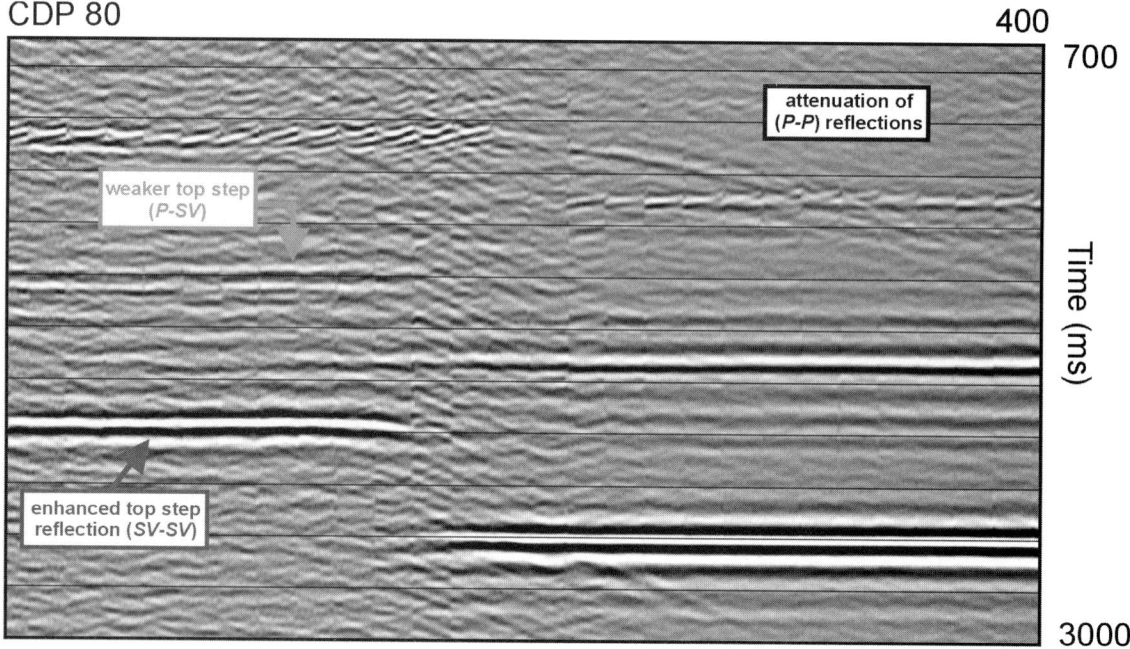

(a) radial component stack with biased velocity analysis.

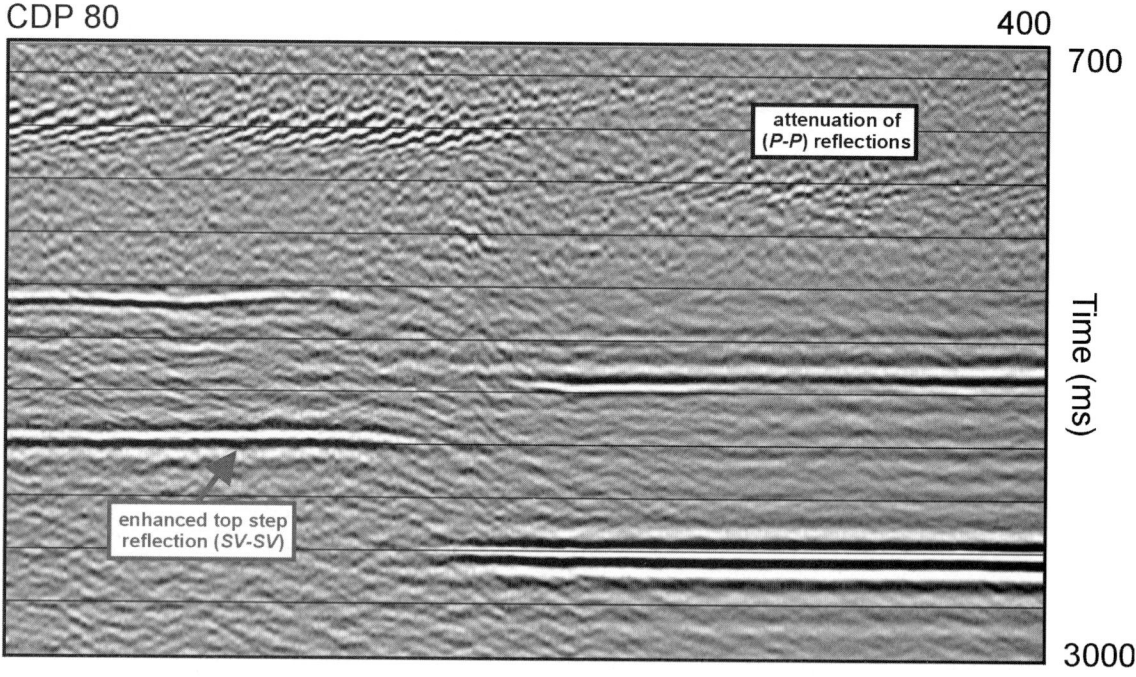

(b) radial component stack with radon filter after NMO.

Figure 5. The radial component stack section with extra processing to attenuate the unwanted reflections.

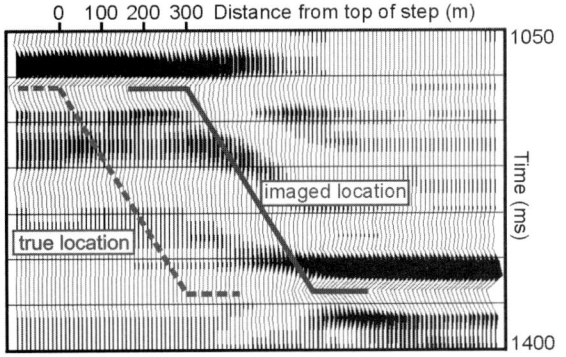

(a) The vertical component section migrated with a constant velocity of 3200 m/s. The top of the step is located about 300 m in the updip direction from the true location.

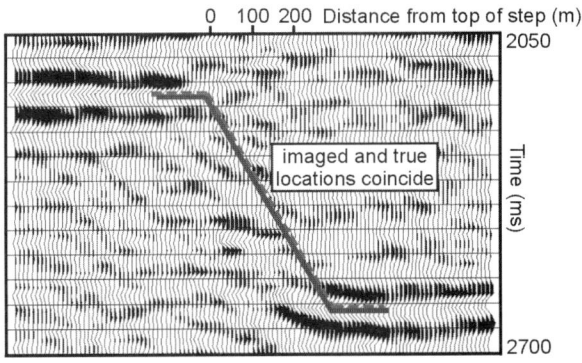

(b) The radial component section migrated with a constant velocity of 1500 m/s. The top of the step is located at the true location.

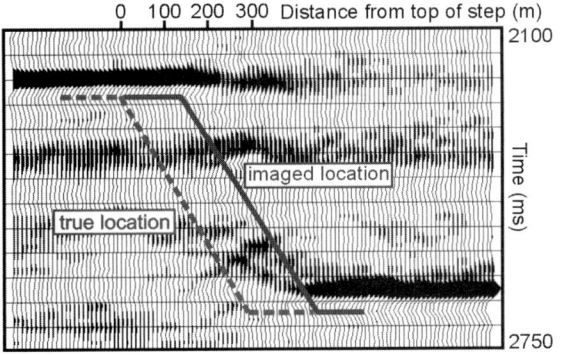

(c) The transverse component section migrated with a constant velocity of 1600 m/s. The top of the step is located about 140 m in the updip direction from the true location.

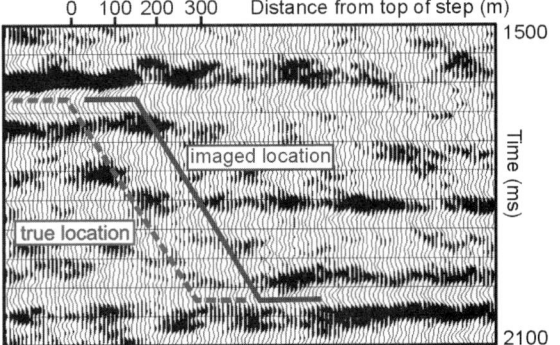

(d) The converted-wave section migrated with a constant velocity of 2224 m/s. The top of the step is located about 150 m in the updip direction from the true location.

Figure 6. The time migrated section from each of the four surveys.

Conclusions

The velocity anisotropy calculated using velocities obtained from transmission tests is 18% for the vertical component case, 9% for the transverse component case and 2% for the radial case. These results show that the greatest lateral shift (of about 300 m) is observed in the vertical component case, where the velocity anisotropy is highest, a smaller shift of about 140 m is observed in the transverse case where the velocity anisotropy is less (9%), and the step is correctly imaged in the radial case where the velocity anisotropy is least. On the converted-wave data, the shift is about 150 m. Hence in this case, radial-component data successfully image the true location of the step using isotropic processing code, while the *P*-wave section exhibits a severe lateral positioning error of 300 m.

This work implies that multicomponent datasets may help with lateral positioning issues in complex thrust belts, where potential hydrocarbon targets often lie under dipping anisotropic shales and are not imaged correctly by *P*-wave data.

Acknowledgements

We gratefully acknowledge the financial support for this work by Sponsors of the Foothills Research Project and the Natural Sciences and Engineering Research Council of Canada. We would like to thank Eric Gallant of the CREWES project for his assistance with model building and data collection. We also acknowledge Landmark Graphics Corporation for access to ProMAX processing software used in this study.

References

Bamford D. and Nunn, K.R., 1979, In situ seismic measurements of crack anisotropy in the carboniferous limestones of Northwest England: Geophysical Prospecting, **27**, 322-338.

Isaac, J.H. and Lawton, D.C., 1997, Image mispositioning due to dipping TI media: A physical seismic modeling study: Geophysics **64,** 1230-1238.

Leslie, J.M. and Lawton, D.C., 1998, A refraction seismic field study to determine the anisotropic parameters of shales: The Leading Edge, **17**, No. 8, 1127-1129.

Thomsen, L., 1986, Weak elastic anisotropy: Geophysics **51**, 1954-1966.

Image mispositioning due to dipping TI media: A physical seismic modeling study

J. Helen Isaac* and Don C. Lawton*

ABSTRACT

A scaled physical model was constructed to investigate the magnitudes of imaging errors incurred by the use of isotropic processing code when there is seismic velocity anisotropy present in the dipping overburden. The model consists of a block of transversely isotropic (TI) phenolic material with the TI axis of symmetry dipping at an angle of 45°. Its scaled thickness is 1500 m, and it is intended to simulate the dipping clastic sequences found in many fold-thrust belts. A piece of isotropic Plexiglas, affixed to the underside of the anisotropic block, has a step function in it to simulate a target reef edge or fault. The anisotropy parameters of the material are $\delta = 0.1$ and $\varepsilon = 0.24$.

On zero-offset data the imaged position of the target is shifted laterally 320 m in the updip direction of the beds, whereas on time- and depth-migrated multichannel sections the shift is 300 m. The lateral shift is offset dependent, with the amount of shift in any common-midpoint gather decreasing from 320 m on the near offsets to 280 m on the far offsets. Prestack depth-migration velocity analysis based upon obtaining consistent depth images in the common-offset domain results in the base of the anisotropic section being imaged 50 m (about 3%) too deep.

INTRODUCTION

Many hydrocarbon resource exploration and development plays in different tectonic settings involve dipping clastic sequences, which in many cases lie above the reservoir or target zones. Thick sequences of dipping sandstones and shales often overlie the reservoir in fold-and-thrust belts, for example in the Canadian Foothills (Lebel et al., 1996), the Caucasus, Russia, (Sobornov, 1994), and Argentina (Fielding and Jordan, 1988). Movement of salt bodies, such as those encountered in the Gulf of Mexico and the North Sea, also gives rise to dipping beds, whereas prograding clastic wedges associated with continental margins may have appreciable thicknesses.

Many of these clastic sequences contain considerable amounts of shales, which often possess some degree of seismic velocity anisotropy. Laboratory experiments have measured velocity anisotropy in shales (e.g., Banik, 1984; Vernik and Liu, 1997) and also, to a lesser degree, in sandstones (Lo et al., 1986; Thomsen, 1986; Levin, 1990). This anisotropy is caused by various factors, such as the orientation of bedding planes (shales), the orientation of cracks and flat pores (sandstones), or thin layers. Laboratory experiments have measured velocity anisotropy ranging from 6% to 33% (Jones and Wang, 1981; Lo et al., 1986; Vernik and Nur, 1992; Johnston and Christensen, 1994; Vernik and Liu, 1997); however, few in-situ measurements have been made. Seismic refraction surveys acquired by the University of Calgary in the foothills of southern Alberta measured velocity anisotropy of 12% and 20% in the near-surface Upper Cretaceous Wapiabi shales [Leslie and Lawton, 1998, 1999 (this issue)]. In many areas in the Alberta foothills, dipping panels of otherwise relatively undeformed Wapiabi shales are found in abundance, overlying the deeper carbonate reservoirs.

We investigate the magnitude of positioning errors in seismically imaged structures by creating an anisotropic physical model that simulates dipping shales. We show that the presence of dipping, anisotropic shale beds above a flat target reflector causes a lateral shift from the true position of the target to the imaged position, if isotropy is assumed during data processing, as is common practice. The magnitude of this shift is not well appreciated by many interpreters working in areas where the phenomenon occurs. Mispositioning of dipping reflectors in vertical transversely isotropic (VTI) media has been demonstrated (Larner and Cohen, 1993; Jaramillo and Larner, 1994, 1995) while Uren et al. (1991) discussed the lateral shift on zero-offset physical modeling data and the offset-dependent lateral shift on multichannel numerical data for a model similar to ours.

Manuscript received by the Editor February 2, 1998; revised manuscript received January 15, 1999.
*The University of Calgary, Dept. of Geology and Geophysics, 2500 University Dr. N.W., Calgary, Alberta T2N 1N4, Canada. E-mail: isaac@geo.ucalgary.ca; donl@geo.ucalgary.ca.
© 1999 Society of Exploration Geophysicists. All rights reserved.

THE PHYSICAL MODEL

The material we used to simulate a shale formation is composed of layers of linen saturated and bonded with a phenolic resin. These thin layers may be regarding as being analogous to fine-scale stratigraphic laminations. The model is composed of a rectangular slab of this material, with the transversely isotropic (TI) axis of symmetry dipping at an angle of 45°. The actual thickness of the slab is 0.15 m and its length is 0.51 m, which, when scaled by 10 000, correspond to a thickness of 1500 m and a length of 5.1 km. After conducting initial calibration surveys, we modified the block model by affixing a piece of isotropic Plexiglas material to its underside with a very thin layer of leg wax. This Plexiglas has a step function in it to simulate a reef edge or fault block, and was introduced to observe image mispositioning beneath a dipping anisotropic overburden.

The seismic surveys were acquired in the physical modeling laboratory of the Department of Geology and Geophysics at the University of Calgary. The equipment consists of a modified flatbed plotter that can accommodate models up to 1 m long × 0.6 m × 0.6 m. One plotter arm controls the source transducer; an added second arm controls the receiver transducer. The transducers are flat-faced cylindrical piezoelectric contact transducers, which were vertically polarized for these seismic surveys. The active element is 12.6 mm in diameter, and the transducers have an absolute positioning accuracy of 0.03 mm, which scales to 0.3 m (Gallant and Bertram, 1992). A thin layer of treacle (molasses) achieves coupling between the transducers and the surface of the model. Emplacement of the transducers for the first shot is set manually; thereafter, the motion of the transducers is preprogrammed. Multichannel data are recorded for each source station by moving the receiver transducer station by station while keeping the source location fixed. In the following discussion all units of time and distance have been scaled by 10 000 from the model size to real world values. The scaled seismic data have a bandwidth of 4 to 70 Hz, with a central frequency of 20 Hz.

CALIBRATION EXPERIMENT

Calibration experiments were conducted initially to measure a subset of the compressional- and shear-wave velocities of the phenolic material. P- and S-waves were transmitted through a cube of the material at angles of 0° and 90° to the axis of symmetry, which was normal to the plane of the face of the transducers. P-waves were also transmitted through a block of the material which had its axis of symmetry rotated to angle of 45°. The compressional-wave velocities were measured as $V_p(0) = 2945 \pm 5$ m/s, $V_p(45) = 3145 \pm 5$ m/s, and $V_p(90) = 3585 \pm 5$ m/s, and the transverse shear-wave velocity as $V_{sh}(0) = 1540 \pm 5$ m/s.

A transmission test was conducted through the 45°-dipping anisotropic block before the isotropic step was added beneath it (Figure 1). A single source was located underneath, and a total of 510 receivers, spaced at 10-m intervals along the surface, recorded the energy propagating one way through the anisotropic material. Figure 2a shows the raw data with a hyperbolic curve having a moveout velocity of 3250 m/s superimposed in white. This curve was positioned to overlie the first arrivals on the near offsets. These first arrivals are close to hyperbolic on the near offsets but deviate slightly from hyperbolic at longer offsets. The position of the minimum P-wave traveltime, which occurs at about 0.48 s, is shifted laterally 290 m in the updip direction of the dipping anisotropic material from the position vertically above the source location. The first-break arrival time of the true zero-offset trace was used to calculate the 45° compressional-wave velocity, $V_p(45)$, as 3145 ± 5 m/s. There is some controversy over which velocity, group or phase, is measured in laboratory experiments employing transducers of a finite size (Dellinger and Vernik, 1994; Johnston and Christensen, 1994; Vestrum, 1994) but, in our experiments, the transducer is very small (12.6 mm) compared to the thickness of the model (150 mm). Thus, we assume a point source and measurement of group velocities.

The data were migrated using an isotropic Kirchhoff time migration with a constant velocity of 3145 m/s for the anisotropic section, yielding the image in Figure 2b. The source image is quite clear, but it is again shifted 290 m in the updip direction from the true location. The deeper event, with a minimum traveltime of 0.95 s in Figure 2a, is an S-wave arrival. Its

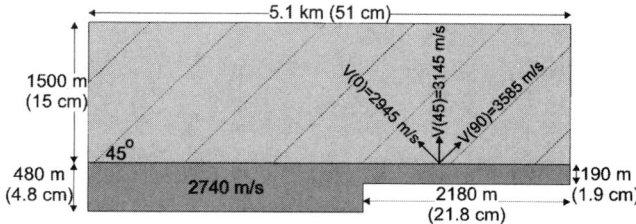

FIG. 1. The scaled physical model after the addition of a piece of isotropic Plexiglas underneath, with a step function intended to simulate a reef edge or a fault. The step was added to investigate imaging errors of targets beneath the anisotropic section. The actual dimensions of the model are given in brackets after the scaled dimensions.

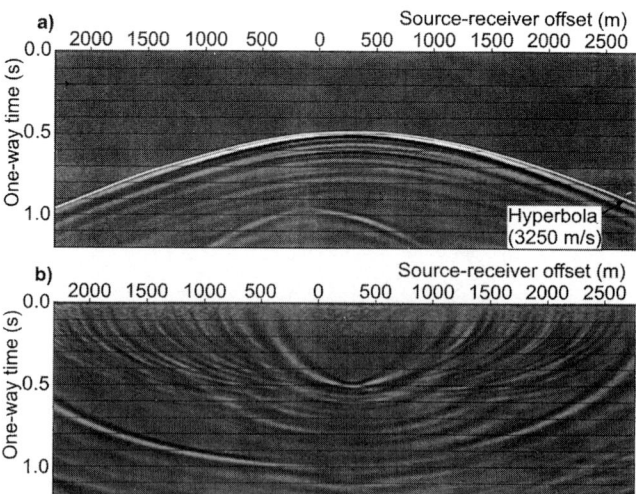

FIG. 2. Raw (a) and migrated (b) data recorded by the transmission survey, in which the source is placed beneath the anisotropic block. Note that the time scale is one-way time. The first arrivals in (a) gradually deviate from hyperbolic (the curve annotated in white). The location of the apex of the first direct P-wave arrival curve is shifted laterally 290 m from the true location of the source. The deeper event at 0.95 s in (a) is the S-wave arrival, and the apex of this curve is shifted only slightly in the opposite direction.

apex is much closer to the true source position than that of the *P*-wave because the *S*-wave velocities in this plane exhibit little anisotropy. This *S*-wave event does not collapse to a point after migration because of the high migration velocity used.

SURFACE SEISMIC SURVEYS

To investigate the mispositioning effects on a target beneath the anisotropic section which result from using isotropic processing code, the model was modified by the addition of a piece of isotropic Plexiglas to the underside of the phenolic block (Figure 1). This Plexiglas layer, with a velocity of 2740 m/s, has a 90° step function in it to simulate a target fault or a reef edge. We conducted both zero-offset and multichannel seismic surveys over this model.

Zero-offset survey

The zero-offset survey was recorded by 510 source-receiver pairs spaced at 10-m intervals along the top of the model. Due to the physical limitations of the recording equipment, the source and receiver were actually 220 m apart in the plane orthogonal to the direction of the survey. The data were time migrated using a steep-dip finite difference algorithm and an isotropic velocity of 3145 m/s for the anisotropic overburden. This velocity was chosen because it is the actual vertical velocity through the anisotropic block as measured from the transmission experiment.

For zero-offset data, we are accustomed to regarding the reflection point as being vertically beneath the recording station, but actually it is displaced laterally, even for flat reflectors, when the overlying anisotropic rocks are dipping. Byun (1982) comments on the "side-slip" effect on a surface seismic section of zero-offset reflection events because the ray (or group) incident angle is not 0° for normal incidence. Figure 3a depicts the group velocity direction when the wavefront incident angle is 0° (i.e., when the phase angle, as measured from the axis of symmetry, is 45°). For this model, zero-offset data record reflected energy that has a wavefront incident angle of 0° at the basal interface. The reflected raypath is the reverse of the incident raypath, resulting in a reflection point that is displaced in the downdip direction from the point vertically beneath the source/receiver station. The calculated lateral shift of the true reflection point is 324 m, whereas the observed shift on our migrated zero-offset physical modeling data (Figure 3b) is 320 ± 10 m. Enlarged versions of the zone of interest on the raw and migrated images are shown in Figures 4a and 4b, respectively. On the migrated data (Figures 3b and 4b), the step exhibits a lateral shift of about 320 m from its true position. This lateral shift is in the updip direction of the dipping anisotropic material.

Multichannel survey

A multichannel survey was also acquired over the step model. The source station interval was 120 m, and the receiver

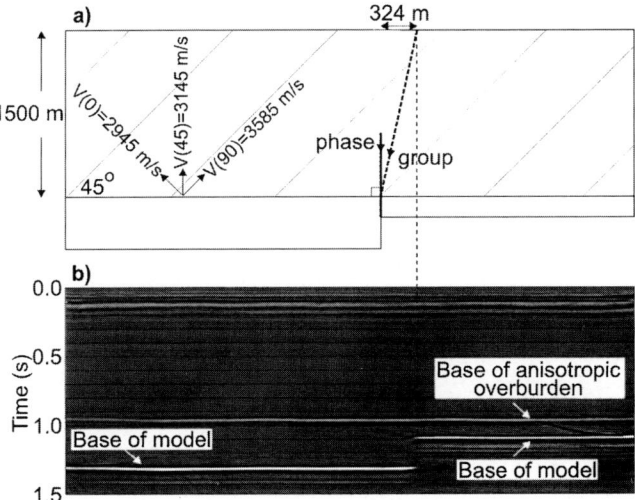

FIG. 3. (a) Sketch of the raypath for a zero-offset trace. The source and receiver station is located 324 m away from the true imaged point. The wavefront incident angle for this raypath is 0° at the boundary between the anisotropic and isotropic materials. Thus the reflected raypath is the reverse of the incident raypath, demonstrating that zero-offset data acquired through a dipping anisotropic medium do not image the subsurface directly beneath the surface location; a phenomenon we observe on the migrated zero-offset data (b).

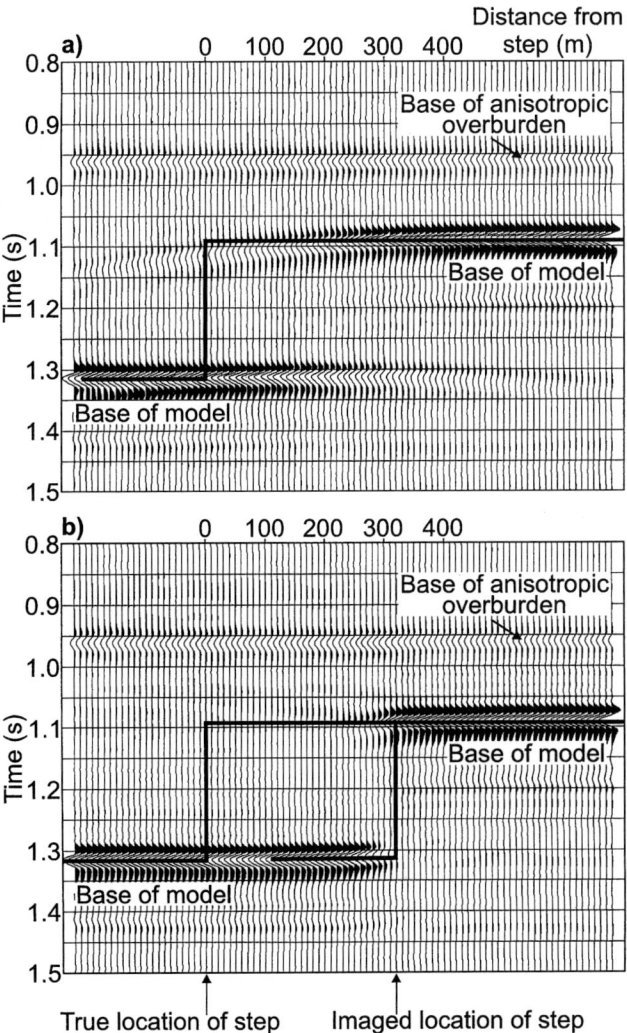

FIG. 4. Enlarged portion of the unmigrated (a) and migrated (b) seismic data from the zero-offset survey run across the model in Figure 1. The migrated data show a lateral shift of about 320 m in the updip direction.

station interval was 20 m. Source-receiver offsets ranged from 240 to 2120 m; the 240-m source-receiver near offset being caused by the physical limitations of the recording equipment. Offsets were limited because of the deterioration in data quality on offsets longer than about 2000 m.

Figure 5a shows a raw shot gather from the multichannel survey and Figure 5b the same gather after f-k filtering, which results in high-quality target reflections. Reflections from the base of the anisotropic block occur just before 1.0 s. The two levels of the base of the isotropic slab are seen from 1.1 to 1.3 s on the negative offsets and from 1.3 to 1.5 s on the positive offsets. A shot gather (Figure 6) shows that the moveout on the reflectors is very close to hyperbolic, even at the far offsets, and that the moveout velocity falls within the range of the anisotropic material's velocities. The hyperbolic moveout velocities annotated on Figure 6 were determined from velocity semblance analysis. In the multichannel data, the reflections have total traveltimes which exhibit only small departures from hyperbolic moveout. Hence, standard velocity analysis of common-midpoint (CMP) gathers gives the impression of isotropic data with hyperbolic moveout and yields interval velocities that are within the range of the true velocities of the material. Thus, there is no indication on the shot gathers or in the velocity analysis that there might be anisotropy present. Several authors (e.g., Tsvankin and Thomsen, 1994, 1995; Alkhalifah, 1997) have discussed the derivation of anisotropy parameters of TI media from the degree of nonhyperbolic moveout observed in data, but in our experimental data we are unable to detect measurable nonhyperbolic moveout over our offset range even though the material has 18% anisotropy [percentage anisotropy here is $[V_p(90)-V_p(0)]/V_p(90)]$. Di Nicola-Carena (1997) suggests that the two anisotropic parameters may be determined from seismic data if the target reflectors have a wide distribution in dip. We are unable to apply this method because in our model the target is horizontal under a dipping overburden.

Stacked sections were produced using two velocity models. The first model has interval velocities calculated from the picked stacking velocities (Figure 7a); the second model has the correct geometry with an isotropic velocity of 3145 m/s for the anisotropic section (Figure 7b). The two stacked sections (Figure 8) exhibit few differences in data character. Each

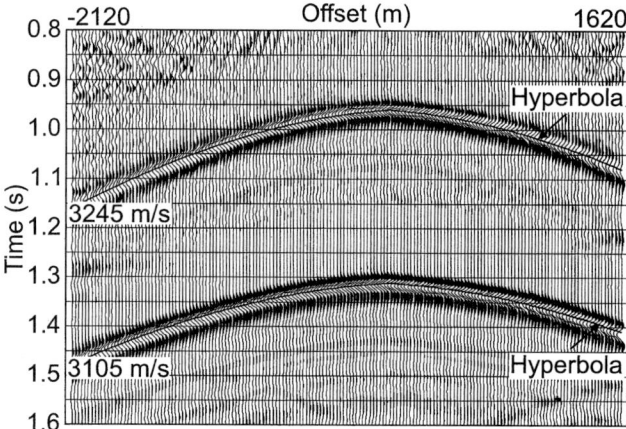

FIG. 6. A shot gather exhibiting near-hyperbolic moveout, as indicated by the curves. In this example, the base of the anisotropic section has a moveout velocity of 3245 m/s, whereas the base of the isotropic section has a moveout velocity of 3105 m/s (based on semblance analysis). Automatic gain control (AGC) has been applied to the gathers to show the shallow event better.

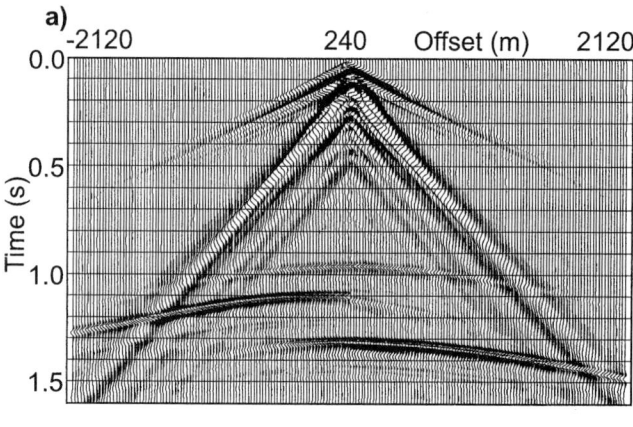

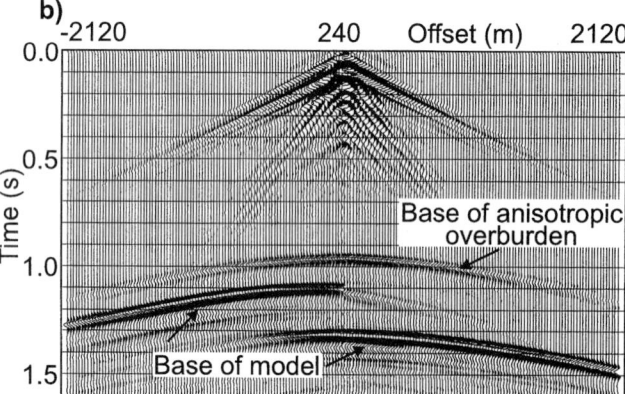

FIG. 5. Raw (a) and filtered (b) shot gathers from the multichannel survey acquired over the model in Figure 1. The gather was filtered to remove the wave trains of energy that were degrading the quality of the reflections. The base of the anisotropic section is just before 1.0 s, and the two levels of the base of the isotropic section are seen from 1.1 to 1.3 s on the negative offsets and from 1.3 to 1.5 s on the positive offsets.

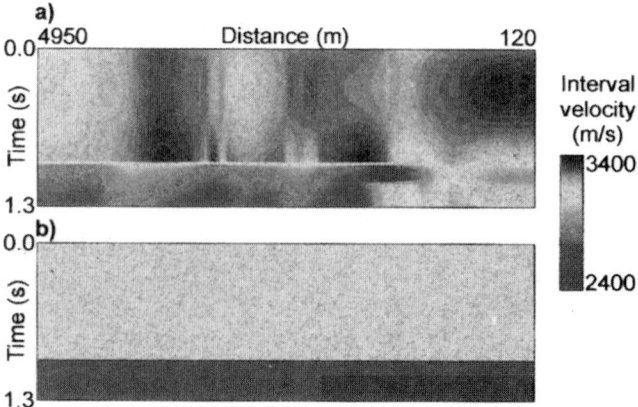

FIG. 7. Interval velocities derived from (a) stacking velocities and (b) a model with the geometry of our physical model, using the measured velocity of 3145 m/s for the anisotropic layer.

section was time migrated by a steep-dip finite-difference algorithm with the appropriate velocity model used to produce the corresponding stacked section. Figure 9a displays the data stacked and migrated with the stacking velocities; Figure 9b displays the data stacked and migrated using the model velocities. These figures both demonstrate a 300-m lateral shift of the imaged step in the updip direction and have very similar character.

The data were also prestack time migrated using an isotropic Kirchhoff algorithm incorporating the model velocities. This migrated section (Figure 10) also exhibits a 300-m lateral shift in the imaged location of the step. To investigate spatial imaging differences at various offsets, the prestack time migrated data were sorted into the common-offset domain. The near offsets ranged from 240 to 470 m, the mid offsets from 960 to 1190 m, and the far offsets from 1680 to 1910 m. The imaged locations on the near, mid, and far offset migrated gathers (Figures 11a, 11b, 11c, respectively) are shifted 320, 300, and 280 m, respectively, from the true location. Thus, CMP stacking traces with different offsets leads to a smearing of the actual reflection points, in this case over about 40 m. With longer offsets, the effect would be more severe.

Prestack isotropic depth migration in the common-offset domain, using both the stacking velocities and the model velocities, generates the sections in Figures 12a and 12b, respectively. On both sections, the position of the step is shifted 300 m updip, but the depth images are different. Use of stacking velocities results in the imaged base of the anisotropic overburden being too deep by about 50 m, the upper side of the step too deep by 50 m, and the lower side of the step too deep by about 80 m (Figure 12a). Model velocities result in more accurate depths although they are too shallow by 20–30 m (Figure 12b). This figure also shows a pull-up in depth over the part of the model where there is a correct lower velocity beneath the isotropic slab.

As with the prestack time migrated data, common-offset gathers were created from the prestack depth migrated data. The imaged locations on the near, mid, and far offset prestack

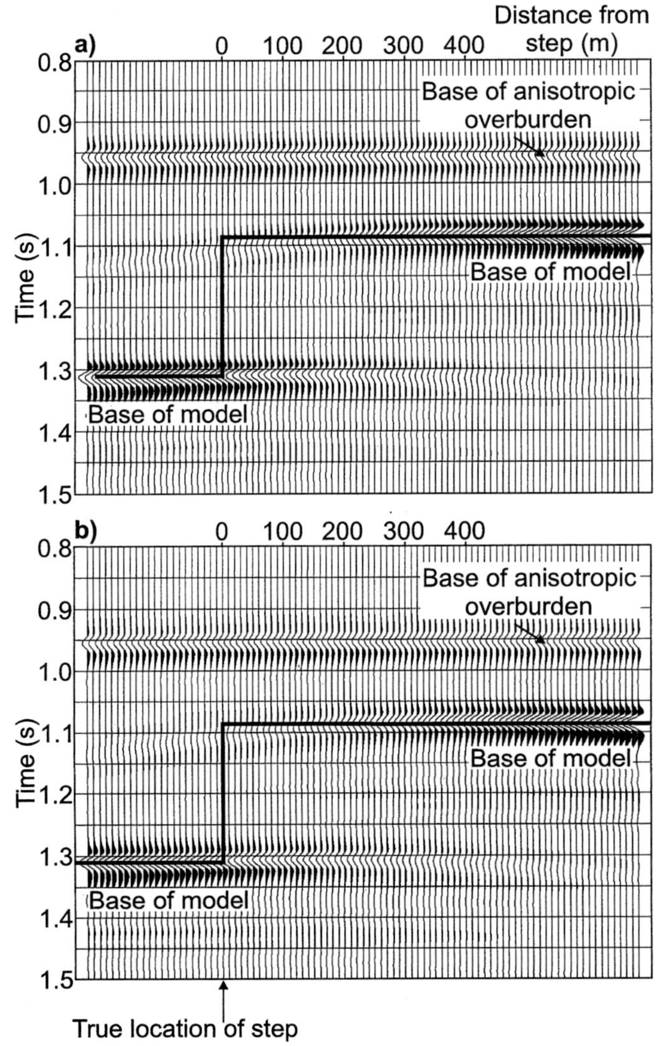

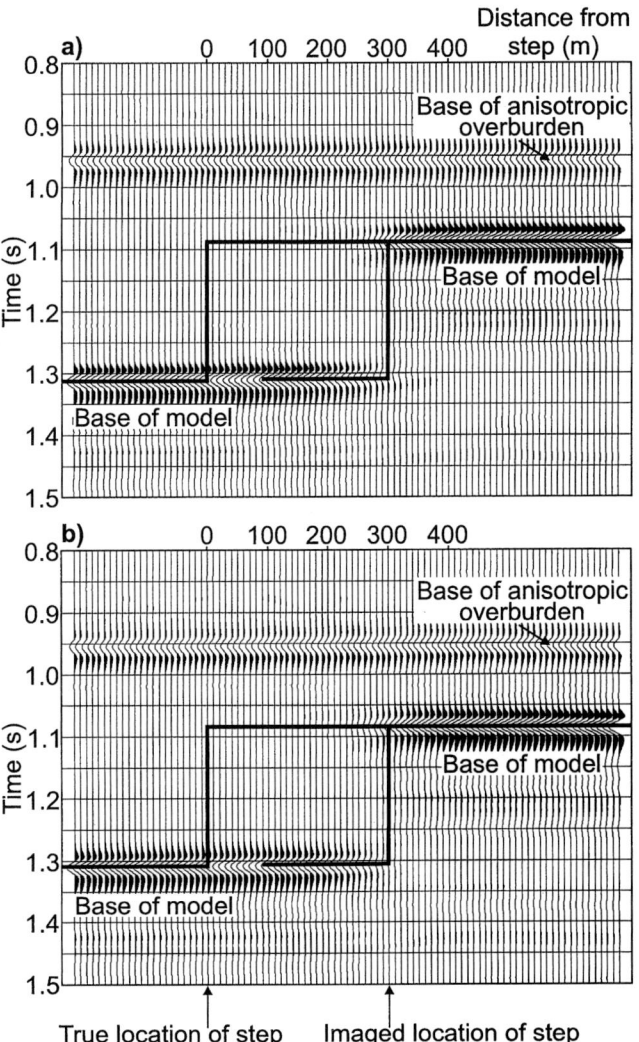

FIG. 8. Sections stacked using the (a) picked stacking velocities and (b) a velocity model having the geometry of the physical model and a velocity of 3145 m/s for the anisotropic overburden.

FIG. 9. Migrated multichannel data, stacked and migrated using (a) the picked stacking velocities and (b) a velocity model having the geometry of the physical model and a velocity of 3145 m/s for the anisotropic overburden.

depth migrated gathers (Figures 13a, 13b, 13c, respectively) are shifted 320, 300, and 280 m, respectively, from the true location; the same values that were observed on the corresponding time migrated offset stacks in Figure 11. The imaged depths also vary with offset. For example, the base of the anisotropic section is imaged correctly at 1500 m on the near offsets but is too shallow on the others, so that when these offsets are stacked, the image is slightly too shallow.

The depth-migration velocity analysis software used here works on the assumption that the correct velocity field produces consistent depth images in the common-offset domain. Residual moveout of reflectors in a common-reflection-point (CRP) gather is assumed to be indicative of an incorrect velocity field and that the correct velocity is the one which flattens the events in these gathers. Prestack depth migrated gathers obtained using a constant velocity of 3145 m/s in the anisotropic section have the correct depth on the near offsets but show residual moveout with offset (Figure 14a). A velocity of 3250 m/s in the anisotropic section flattens the CRP gathers (Figure 14b), but the depths obtained in these flattened gathers are too deep by 50 m (about 3% of the true depth).

NUMERICAL MODELING

The following equations (1) for arbitrary anisotropy (Daley and Hron, 1977; Berryman, 1979) were used to calculate the phase velocity, $v(\theta)$, group angle, ϕ, and group velocity, $V(\phi)$, for an input range of given phase angle, θ.

$$\rho v^2(\theta) = \tfrac{1}{2}(C_{33} + C_{44} + (C_{11} - C_{33})\sin^2\theta + D(\theta))$$

$$\tan\phi = \left(\tan\theta + \frac{1}{v}\frac{dv}{d\theta}\right) \bigg/ \left(1 - \frac{\tan\theta}{v}\frac{dv}{d\theta}\right) \quad (1)$$

$$V^2(\phi) = v^2(\theta) + \left(\frac{dv}{d\theta}\right)^2$$

where $D(\theta) \equiv \{(C_{33} - C_{44})^2 + 2[2(C_{13} + C_{44})^2 - (C_{33} - C_{44})(C_{11} + C_{33} - 2C_{44})]\sin^2\theta + [(C_{11} + C_{33} - 2C_{44})^2 - 4(C_{13} + C_{44})^2]\sin^4\theta\}^{\tfrac{1}{2}}$.

The elastic constants, C_{ij}, were determined from the following relationships between them and the measured group velocities (Cheadle et al., 1991): $C_{11} = \rho V_p(90)^2$, $C_{33} = \rho V_p(0)^2$, $C_{44} = \rho V_{sh}(0)^2$. C_{13} was determined by iterating its value in the anisotropy equations (1) until the calculated $V_p(45)$ matched the measured $V_p(45)$. The density of the material is 1360 kg/m^3. These values for the elastic constants were then used to determine the anisotropy parameters ε and δ (Thomsen, 1986) to be 0.24 and 0.1, respectively.

Figure 15 depicts the lateral shift between the CMP and the true reflection point as a function of the source-receiver offset/thickness. It is calculated using equations (1) with the

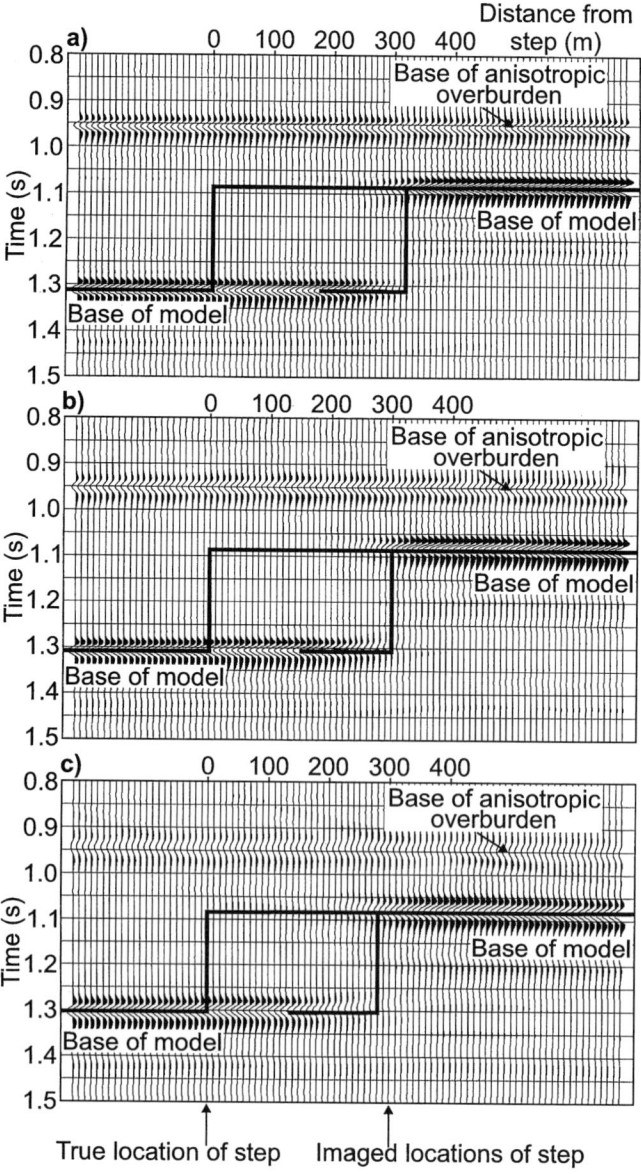

FIG. 11. Prestack time migrated offset gathers, migrated using the model velocities. Near offsets of 240–470 m (a) display a shift of 320 m, mid offsets of 960–1190 m (b) a shift of 300 m, and far offsets of 1680–1910 m (c) a shift of 280 m in the updip direction from the true location.

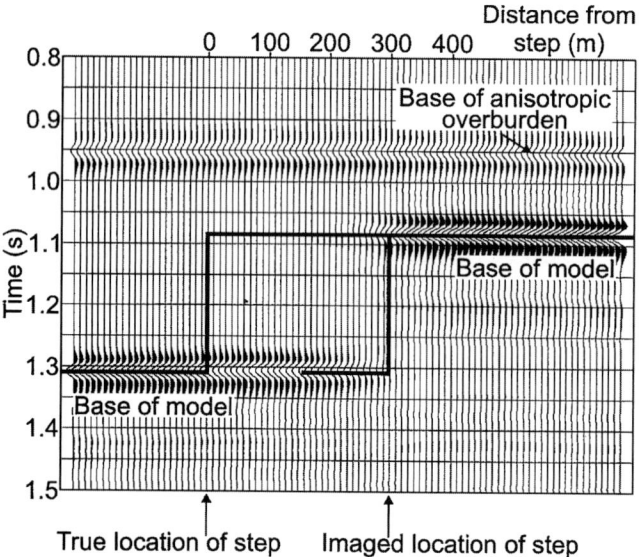

FIG. 10. Prestack time migrated section, migrated using the model velocities. The location of the step is shifted 300 m in the updip direction.

parameters of our physical model and is expressed as a percentage of the anisotropic layer thickness. The lateral shift is almost 22% at zero-offset, then decreases to less than 15% for a source-receiver offset/thickness of 3.5. For our model, which has a 1500-m-thick overburden, the lateral shift is 324 m at zero-offset and decreases to 220 m at an offset of 5000 m. Thus, in any CMP gather, the imaged subsurface location changes with offset, so that CMP binning does not correctly gather common reflection points, and it introduces a smear in the stacked section. Regarding the zero-offset reflection point as the base point, the lateral smear of reflection points in any CMP gather is calculated and is annotated on Figure 15, also expressed as a percentage of the anisotropic layer thickness. This numerical modeling predicts that for our physical model surveys, we should see a lateral shift of 320 m on the near-offset gathers and a lateral shift of 275 m on the far offset gathers. These values compare very well to the shifts of 320 m and 280 m, respectively, observed in the physical modeling data.

In order to observe the influence of δ on the amount of lateral smear, the shift was calculated for values of δ of 0.25, 0.2, 0.15, 0.1, and 0.05, keeping ε constant at 0.24. In Figure 16, we see that the greatest smear occurs when δ has a low value. When $\delta = 0.05$, the lateral shift decreases from nearly 22% at zero-offset to under 12% at the far offsets, causing a large amount of smear in a CDP gather. When δ is smaller than ε, the amount of shift decreases with offset, but when δ is larger than ε, the shift increases with offset. The amount of smear depends heavily on δ, which has been found to be highly variable in laboratory experiments, as tabulated by Vernik and Liu (1997). Both the numerical and physical modeling suggest that if such a smear of an imaged target is observed on common offset gathers of real seismic data, then anisotropy should be suspected.

The magnitude of the imaging position error is also affected by the dip angle of the beds, γ, with respect to the normal to the reflectors. This error was calculated for zero-offset data only with the parameters of our physical model, for dip angles

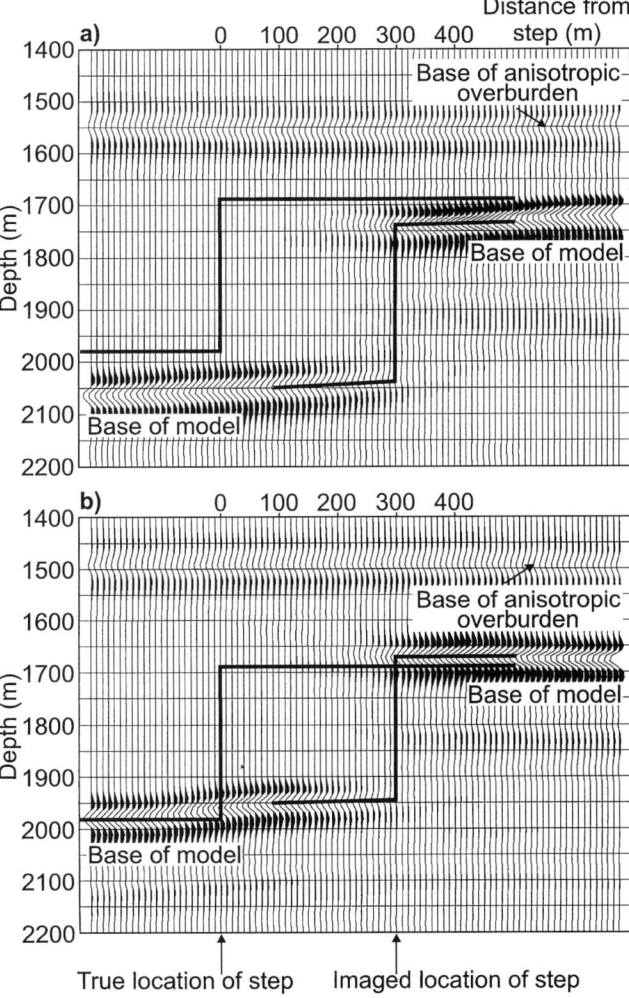

FIG. 12. Prestack depth migrated sections, migrated using the (a) stacking velocities and (b) the model velocities. The location of the step is shifted 300 m in the updip direction. Use of the stacking velocities results in an image that is too deep, whereas use of the model velocities results in an image that is too shallow.

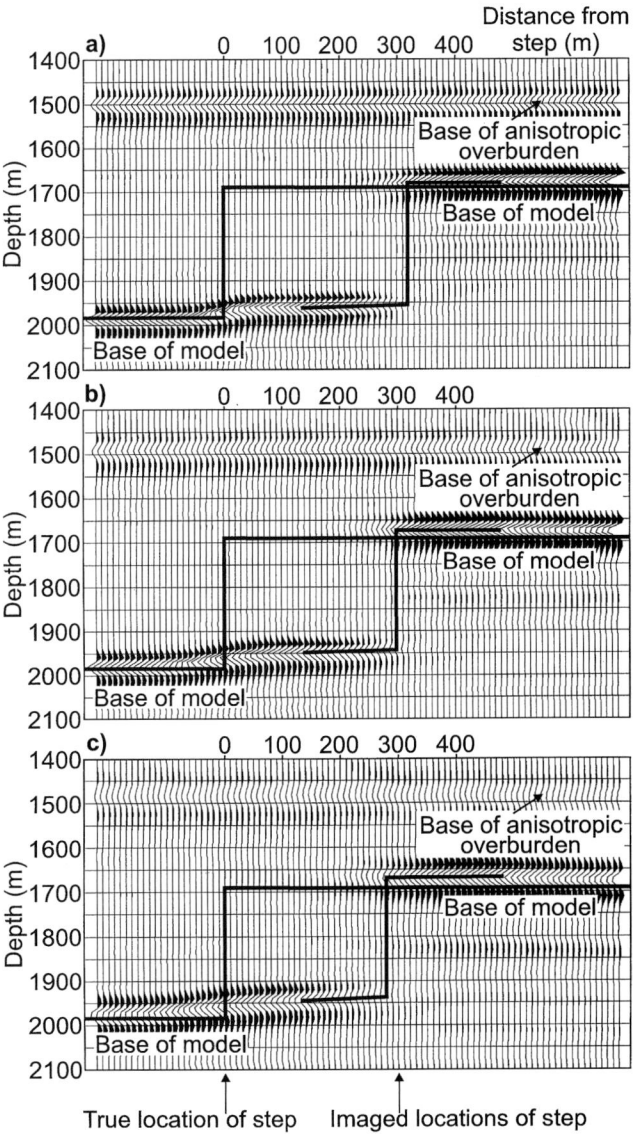

FIG. 13. Prestack depth migrated offset gathers, migrated using the model velocities. Near offsets of 240–470 m (a) display a shift of 320 m, mid offsets of 960–1190 m (b) a shift of 300 m, and far offsets of 1680–1910 m (c) a shift of 280 m in the updip direction from the true location.

ranging from 0° to 90°. It is zero when the beds are vertical or horizontal, and increases to a maximum of almost 22% of the overburden thickness at a dip of 49° (Figure 17).

DISCUSSION

These physical modeling experiments demonstrate the magnitude of the image mispositioning incurred by the use of inappropriate isotropic processing code when velocity anisotropy is present in the overburden. In a transmission experiment through a 1500-m-thick anisotropic layer with its axis of symmetry dipping at 45°, the observed lateral shift of the true source point is 290 m in the updip direction. The lateral shift of an imaged target beneath this dipping anisotropic overburden is significant. Zero-offset data exhibit a shift in the imaged location of 320 m in the updip direction of the dipping beds, whereas the shift on time and depth migrated multichannel data is 300 m. Migrated common-offset gathers demonstrate that the amount of lateral shift is offset dependent, being 320 m on the near offsets and 280 m on the far offsets of our data set. Thus, a smear of reflection points is introduced by CMP stacking the data, even after prestack migration. This case illustrates clearly the need for anisotropic depth migration, which is currently being developed.

ACKNOWLEDGMENTS

We gratefully acknowledge the financial support for this work by Sponsors of the Foothills Research Project and the Natural Sciences and Engineering Research Council of Canada. Eric Gallant of the Department of Geology and Geophysics at

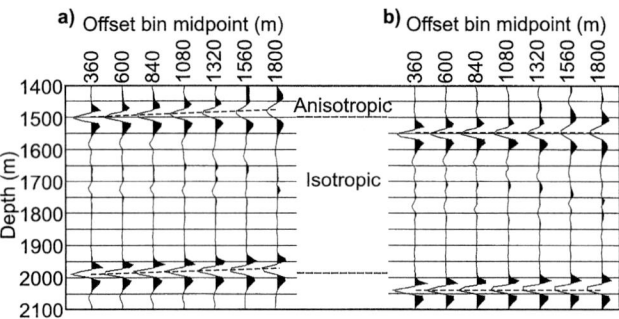

FIG. 14. Prestack depth migrated offsets from a single CRP gather. The velocity field that gives the correct depths on the nearest offsets does not flatten the gathers (a) whereas the velocity field that flattens the gathers (b) results in depths too great by 50 m. AGC has been applied to the gathers to display better the event representing the base of the anisotropic section.

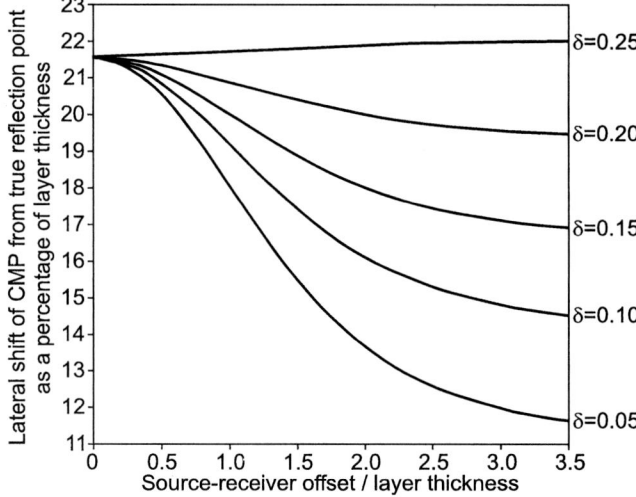

FIG. 16. The error between the CMP and the true reflection point, expressed as a percentage of the anisotropic layer thickness. The values are calculated using the general equations for anisotropy with the parameters of our physical model, keeping ε constant at 0.24 but varying δ.

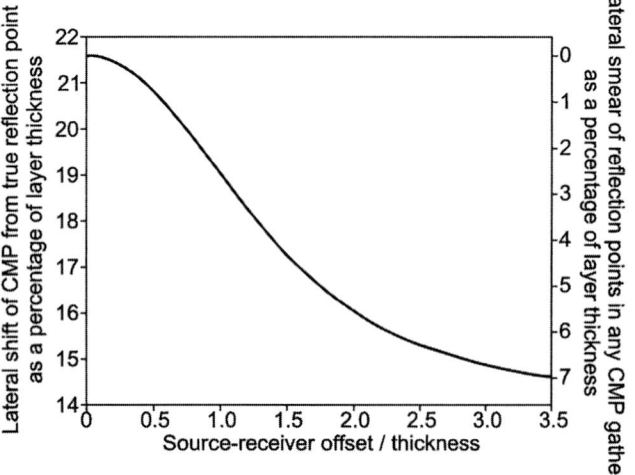

FIG. 15. The error between the CMP and the true reflection point, calculated using the general equations for anisotropy with the parameters of our physical model. The left-hand scale shows the offset-dependent lateral shift of CMPs from the true reflection point, whereas the right-hand scale shows the lateral smear in any CMP gather from the zero-offset reflection point, both expressed as a percentage of the anisotropic layer thickness.

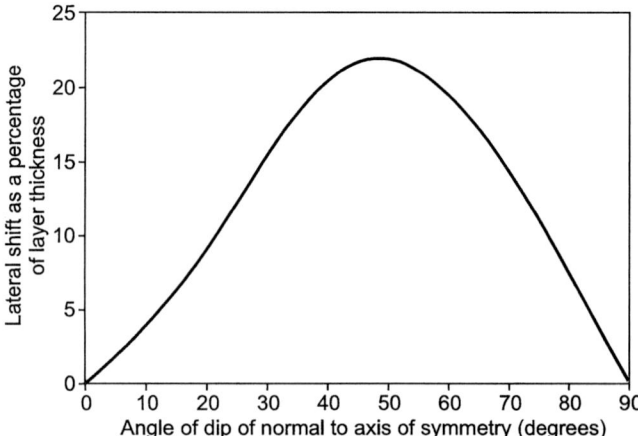

FIG. 17. Lateral shift of the zero-offset CMP location from the true reflection point as a function of the dip angle of the axis of symmetry, expressed as a percentage of the layer thickness. The shift is calculated using the general equations for anisotropy with the parameters of our physical model. The maximum deviation is 21.9% at an angle of 49°.

the University of Calgary is thanked for building the model and for his cooperation in acquiring the physical modeling data. All the seismic data processing described in this paper was done using Landmark Corporation's ProMAX data processing system.

REFERENCES

Alkhalifah, T., 1997, Velocity analysis using nonhyperbolic moveout in transversely isotropic media: Geophysics, **62**, 1839–1854.

Banik, N. C., 1984, Velocity anisotropy of shales and depth estimation in the North Sea basin: Geophysics, **49**, 1411–1419.

Berryman, J. G., 1979, Long-wave elastic anisotropy in transversely isotropic media: Geophysics, **44**, 896–917.

Byun, B. S., 1982, Seismic parameters for media with elliptical velocity dependencies: Geophysics, **47**, 1621–1626.

Cheadle, S. P., Brown, R. J., and Lawton, D. C., 1991, Orthorhombic anisotropy: A physical modeling study: Geophysics, **56**, 1603–1613.

Daley, P. F., and Hron, F., 1977, Reflection and transmission coefficients for transversely isotropic media: Bull. Seis. Soc. Am., **67**, 661–675.

Dellinger, J., and Vernik, L., 1994, Do traveltimes in pulse-transmission experiments yield anisotropic group or phase velocities?: Geophysics, **59**, 1774–1779.

Di Nicola-Carena, E., 1997, Solving lateral shift due to anisotropy: 67th Ann. Internat. Mtg., Soc. Expl. Geophys., Expanded Abstracts, 1525–1528.

Fielding, E. J., and Jordan, T. E., 1988, Active deformation at the boundary between the Precordillera and Sierras Pampeanas, Argentina, and comparison with ancient Rocky Mountain deformation, in Schmidt, C. J., and Perry, W. J., Jr., Eds., Interaction of the Rocky Mountain foreland and the Cordilleran thrust belt: Geol. Soc. Am. Memoir **171**, 143–163.

Gallant, E. V., and Bertram, M. B., 1992, Update on the physical modeling system: CREWES Research Report, **4**, 3.1–3.4.

Jaramillo, H., and Larner, K., 1994, A demonstration of pre-stack migration error in transversely isotropic media: 64th Ann. Internat. Mtg., Soc. Expl. Geophys., Expanded Abstracts, 1224–1227.

——— 1995, Prestack migration error in transversely isotropic media: 65th Ann. Internat. Mtg., Soc. Expl. Geophys., Expanded Abstracts, 1204–1207.

Johnston, J. E., and Christensen, N. I., 1994, Elastic constants and velocity surfaces of indurated anisotropic shales: Surveys in Geophysics, **15**, 481–494.

Jones, L. E. A., and Wang, H. F., 1981, Ultrasonic velocities in Cretaceous shales from the Williston Basin: Geophysics, **46**, 288–297.

Larner, K. L., and Cohen, J. K., 1993, Migration error in transversely isotropic media with linear velocity variation in depth: Geophysics, **58**, 1454–1467.

Lebel, D., Langenberg, W., and Mountjoy, E. W., 1996, Structure of the central Canadian Cordilleran thrust-and-fold belt, Athabasca-Brazeau area, Alberta: A large, complex intercutaneous wedge: Bull. Can. Petr. Geol., **44**, 282–298.

Leslie, J. M., and Lawton, D. C., 1998, A refraction seismic field study to determine the anisotropic parameters of shales: The Leading Edge, **17**, 1127–1129.

——— 1999, A refraction seismic field study to determine the anisotropic parameters of shales: Geophysics, **64**, 1247–1252, this issue.

Levin, F., 1990, Reflections from a dipping plane—Transversely anisotropic solid: Geophysics, **55**, 851–855.

Lo, T.-W., Coyner, K. B., and Toksöz, M. N., 1986, Experimental determination of elastic anisotropy of Berea sandstone, Chicopee shale, and Chelmsford granite: Geophysics, **51**, 164–171.

Sobornov, K. O., 1994, Structure and petroleum potential of the Dagestan thrust belt, northeastern Caucasus, Russia: Bull. Can. Petr. Geol., **42**, 352–364.

Thomsen, L., 1986, Weak elastic anisotropy: Geophysics, **51**, 1954–1966.

Tsvankin, I., and Thomsen, L., 1994, Nonhyperbolic reflection moveout in anisotropic media: Geophysics, **59**, 1290–1304.

——— 1995, Inversion of reflection traveltimes for transverse isotropy: Geophysics, **60**, 1095–1107.

Uren, N. F., Gardner, G. H. F., and McDonald, J. A., 1991, Anisotropic wave propagation and zero-offset migration: Expl. Geophys., **22**, 405–410.

Vernik, L., and Liu, X., 1997, Velocity anisotropy in shales: A petrophysical study: Geophysics, **62**, 521–532.

Vernik, L., and Nur, A., 1992, Ultrasonic velocity and anisotropy of hydrocarbon source rocks: Geophysics, **57**, 727–735.

Vestrum, R. W., 1994, Group- and phase-velocity inversions for the general anisotropic stiffness tensor: M.Sc. thesis, Univ. of Calgary.

Imaging structures below dipping TI media

Robert W. Vestrum*, Don C. Lawton‡, and Ron Schmid*

ABSTRACT

Seismic anisotropy in dipping shales causes imaging and positioning problems for underlying structures. We developed an anisotropic depth-migration approach for P-wave seismic data in transversely isotropic (TI) media with a tilted axis of symmetry normal to bedding. We added anisotropic and dip parameters to the depth-imaging velocity model and used prestack depth-migrated image gathers in a diagnostic manner to refine the anisotropic velocity model.

The apparent position of structures below dipping anisotropic overburden changes considerably between isotropic and anisotropic migrations. The ray-tracing algorithm used in a 2-D prestack Kirchhoff depth migration was modified to calculate traveltimes in the presence of TI media with a tilted symmetry axis. The resulting anisotropic depth-migration algorithm was applied to physical-model seismic data and field seismic data from the Canadian Rocky Mountain Thrust and Fold Belt. The anisotropic depth migrations offer significant improvements in positioning and reflector continuity over those obtained using isotropic algorithms.

INTRODUCTION

Dipping anisotropic strata overlying a target of interest can be characterized as a lens for propagating seismic energy. Below this lens, dipping as well as horizontal reflectors at boundaries between isotropic strata will be incorrectly positioned if isotropic models are assumed during data processing—particularly depth migration. Whereas isotropic depth migration corrects imaging problems and positioning errors associated with lateral, but isotropic, velocity heterogeneity, anisotropic depth migration is required to correctly locate images when transversely isotropic (TI) strata with a dipping axis of symmetry are present.

Larner and Cohen (1993) and Alkhalifah and Larner (1994) document migration errors in TI media. Uzcategui (1995) and Alkhalifah (1995) address the problem of depth imaging in the presence of vertical transverse isotropy (VTI). Kitchenside (1992), Ball (1995), and Vestrum and Muenzer (1997) address seismic imaging in the presence of tilted transversely isotropic (TTI) media. Isaac and Lawton (1999) show dramatic positioning errors of horizontal reflectors below TI media with a tilted symmetry axis. In this paper, we illustrate an instance in which anisotropic depth migration is necessary to create an accurate depth image of physical seismic modeling data. We also apply anisotropic depth migration to a seismic data set from the Canadian Rocky Mountain Foothills.

In this approach, we do not attempt to derive physical properties of individual stratigraphic layers; rather, we find anisotropic parameters for large intervals in our velocity model that offer the best seismic depth image. The approach to the TTI problem presented here is based on traditional prestack depth migration analysis. Inspection of prestack image gathers from the resulting anisotropic depth migration directs modification of the anisotropic velocity model. Image-gather analysis for anisotropic depth migration is similar to the analysis used commonly in isotropic depth migration, except that additional anisotropic parameters in the velocity model are estimated.

ANALYTIC BACKGROUND

Clastic sediments, particularly shales, typically exhibit transverse isotropy, where P-wave seismic velocity is constant in all directions parallel to bedding and typically slower in all other directions. Dipping anisotropic strata in the overburden cause mispositioning errors on seismic reflectors below (Di Nicola-Carena, 1997; Leslie et al., 1997; Vestrum and Muenzer, 1997).

Thomsen (1986) defines a simple and useful method for parameterizing a TI medium. The weak anisotropy approximation using Thomsen's parameters [equation (1)] gives the phase velocity, v, as a function of phase angle, θ, from the symmetry axis or bedding-plane normal:

Presented at the 67th Annual Meeting, Society of Exploration Geophysicists. Manuscript received by the Editor January 27, 1998; revised manuscript received January 13, 1999.
*Kelman Seismic Processing, 600-540 5th Ave. SW, Calgary, Alberta T2P 0M2, Canada; e-mail: robv@kelman.com; ron@kelman.com.
‡Dept. of Geology and Geophysics, Univ. of Calgary, 2500 University Drive NW, Calgary, Alberta T2N 1N4, Canada; e-mail: donl@geo.ucalgary.ca.
© 1999 Society of Exploration Geophysicists. All rights reserved.

$$v(\theta) = v_0(1 + \delta \sin^2\theta \cos^2\theta + \varepsilon \sin^4\theta), \quad (1)$$

where v_0 is the velocity normal to bedding, δ determines how fast or slow the P-wave velocity propagates at small angles oblique to the symmetry axis, and ε is approximately the fractional difference between the velocities parallel and perpendicular to bedding. While Thomsen's approximation is for weak anisotropy, the parameters v_0, ε, and δ may be used to describe P-wave velocities and other kinematic signatures for TI media with any strength of anisotropy (Tsvankin, 1996).

Figure 1 illustrates a propagating wavefront in a TI medium with a tilted axis of symmetry. The sideslip velocity, $\vec{s}$, is the rate at which energy moves transverse to the wavefront (Dellinger, 1991). Adding the sideslip-velocity and phase-velocity vectors yields the group velocity, $\vec{g}$, or velocity of energy transport (in a lossless medium) away from the source. The equation for the magnitude of $\vec{s}$, which is obtained by differentiating equation (1) with respect to θ, is

$$s \equiv \frac{\partial v}{\partial \theta} = v_0[2\delta(\cos^3\theta \sin\theta - \cos\theta \sin^3\theta) + 4\varepsilon \cos\theta \sin^3\theta]. \quad (2)$$

This relationship between phase and group velocity is shown in equations (3) and (4). Equation (3) gives the relationship between the magnitudes of the phase and group velocity vectors. Equation (4) gives the relationship between the angles of the two vectors relative to the symmetry axis (i.e., normal to bedding):

$$g = \sqrt{v^2 + \left(\frac{\partial v}{\partial \theta}\right)^2} \quad (3)$$

and

$$\phi = \theta + \arctan\left[\frac{(\partial v/\partial \theta)}{v}\right], \quad (4)$$

where ϕ is the angle between the group-velocity vector and the symmetry axis and θ is the angle between the phase-velocity vector and the symmetry axis.

It is convenient to define sideslip distance, S, as a function of the vertical thickness of the dipping anisotropic strata, T, and the magnitudes of the sideslip velocity, s, and the phase velocity v:

$$S \equiv \frac{s}{v}T. \quad (5)$$

The sideslip distance is the displacement of a reflection point from the common midpoint for a horizontal reflector below a layer of vertical thickness T with a tilted axis of symmetry. It is the lateral mispositioning error shown in Figure 2.

Figure 3 is a plot of S normalized by T as a function of both dip of the anisotropic strata and δ for $\varepsilon = 0.1$ in the zero-offset case. The phase angle for a zero-offset reflection from a horizontal reflector is $0°$ from vertical; therefore, the phase angle

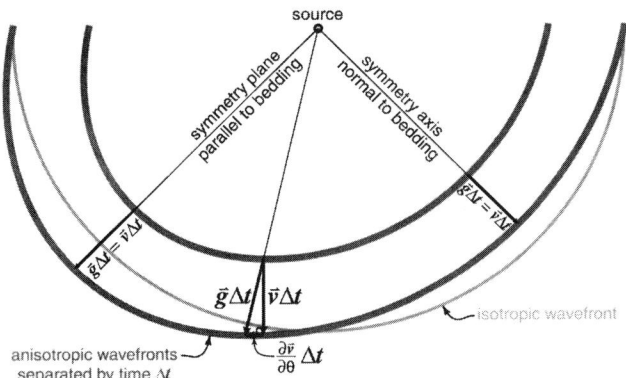

FIG. 1. Wavefronts separated by time Δt. The group velocity $\vec{g}$ is the velocity of energy propagation, and the phase velocity $\vec{v}$ is the velocity normal to the wavefront. The distance between wavefronts normal to the wavefront is $\vec{v}\Delta t$, and the distance between wavefronts in the direction away from the source is $\vec{g}\Delta t$.

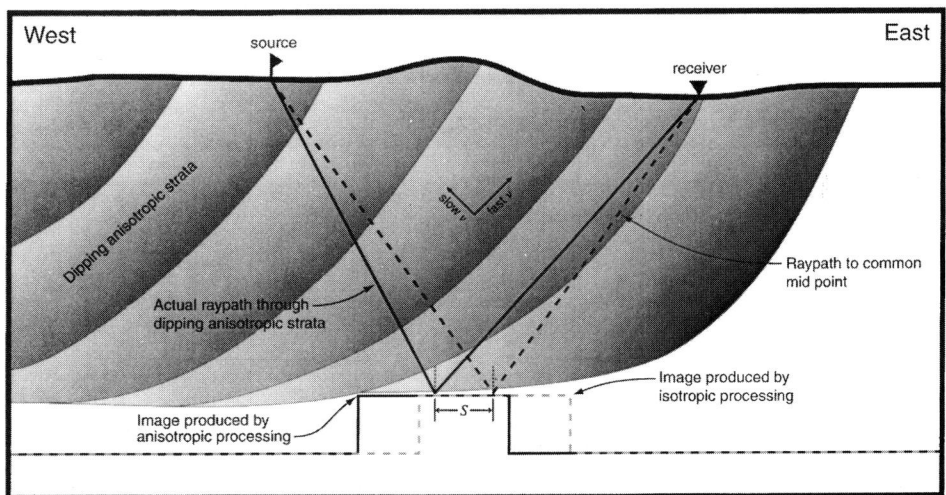

FIG. 2. Cross-section showing lateral movement predicted from isotropic processing of data where transverse isotropy with a tilted axis of symmetry is present. The dashed lines represent the result of isotropic processing. S is the distance between the actual subsurface reflection point and the reflection point when isotropy is assumed.

from the axis of symmetry, θ, is equal to the dip for the zero-offset reflection from a horizontal reflector. From this graph we can see, for example, that for a reflector below 1000 m of anisotropic strata dipping at 45° with $\varepsilon = 0.1$, the zero-offset reflection point will be displaced laterally by about 100 m for any value of δ between 0 and 0.1. Figure 3 also shows that δ has a greater impact than ε on reflection-point displacement for dips less than about 20°.

Figure 2 shows the shift one can expect on a subsurface structure below dipping shales. Because the velocity parallel to the shale bedding is higher than the velocity perpendicular to bedding, rays from the eastern or updip side of the structure will travel in the fast direction and rays from the western or cross-dip side of the structure will travel in the slow direction. The least-time raypath will have a longer travelpath in the faster direction, as shown in this figure and predicted by the anisotropic Snell's law (Byun, 1982). If we process these data with an isotropic assumption, the subsurface structure will image in the updip direction of the anisotropic strata from where it is in the subsurface.

Anisotropic depth migration

The only difference between the isotropic and anisotropic Kirchhoff depth migrations implemented in this study is in the traveltime computations. Both algorithms use a traveltime generator that propagates wavefronts from each shot and receiver location through a gridded velocity model. Wavefronts separated by constant time increments populate the velocity field. A migration traveltime field is then calculated by adding the shot and receiver traveltimes at each point in the model. In the isotropic case, each point along the wavefront curve moves forward in the direction normal to the wavefront for each time step in accordance with the model velocity local to that point. We modified the existing traveltime calculator such that anisotropic group velocities, calculated using equations (3) and (4), are used to propagate the wavefront forward each step. In an anisotropic medium, the point on the wavefront moves forward along the group velocity vector, which is generally oblique to the wavefront normal. Once source and receiver traveltimes are calculated, migration operators for a Kirchhoff depth migration are generated from the traveltimes.

DEPTH IMAGING OF PHYSICAL-MODEL SEISMIC DATA

Seismic data from an anisotropic physical model described by Isaac and Lawton (1999) were used to test the migration. A cross-section of this model is shown in Figure 4a; it consists of a TI overburden layer, 1500 m thick, with the axis of symmetry dipping at 45°. The layer is made of phenolic resin and has parameters $v_0 = 2945$ m/s, $\varepsilon = 0.241$, and $\delta = 0.100$. An isotropic layer of plexiglas containing a fault (Figure 4a) underlies this anisotropic overburden. Isotropic depth migration of the seismic data, shown in Figure 4b, exhibits errors in both depth of horizontal reflectors and position of the fault.

The position error arises because isotropic migration places the reflection point at the midpoint for reflections from horizontal interfaces. The depth error arises because the isotropic imaging velocities (i.e., those velocities which best flatten image gathers) are greater than the vertical velocities in the model. As a result, the reflectors are imaged over 100 m too deep.

The isotropic depth image (Figure 4b) was generated using the optimum imaging velocities, as shown by the flat events in the image gathers displayed on the section. The image gather on the far left shows some residual moveout. The difference in moveout on this gather is likely from the different azimuth distribution on this gather: it is at the downdip end of the line. There are more data migrating into this gather from the fast direction, and the image gather indicates a faster model velocity is required to flatten the gather.

Anisotropic depth migration produced consistently flat prestack image gathers with correct depths, as shown in Figure 4c. In this case, the input to the migration consisted of a grid containing values of v_0, ε, δ, and tilt of the symmetry axis at each node. The grid spacing was 30 m by 50 m in depth. The basement fault is also imaged close to its true position. Although the reflectors have zero dip, the fault (interpreted by the reflection terminations) was moved laterally by 300 m in the anisotropic depth migration (Figure 4c) as compared with the isotropic depth migration (Figure 4b) to be placed in the correct position. Using ε for this material of 0.241, we predicted a sideslip error of 334 m for the zero-offset reflection. There is a difference of 34 m between the lateral-positioning error observed in the multioffset data and the error predicted by calculating the sideslip error for a zero-offset reflection. Some lateral-positioning discrepancies may also be from a misinterpretation of the reflection terminations. There is also a migration artifact on the upper reflection termination that makes it difficult to precisely pick the top edge of the structure.

FIELD DATA EXAMPLE

Anisotropic depth migration was also tested on the Husky/Talisman data set from the southern Alberta Foothills, described by Stork et al. (1995). This line runs over rough topography with dipping reflectors coming to surface (Lines et al., 1996). The upper 2.5 km of section is shale-dominated clastics, and the lower part of the section is mostly carbonates.

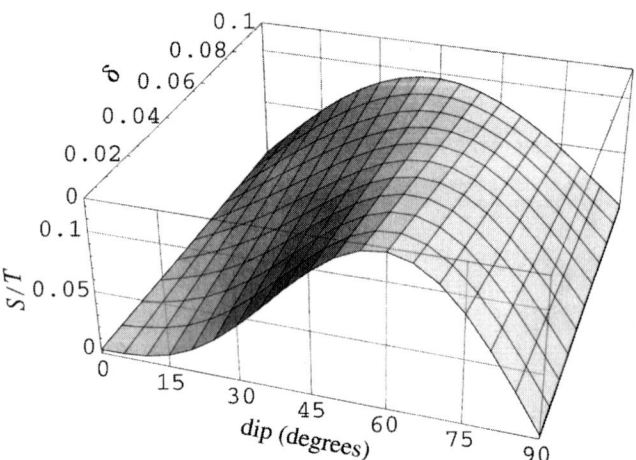

FIG. 3. S/T, the sideslip distance S normalized by vertical thickness of the anisotropic strata T, versus δ and dip of anisotropic strata; $\varepsilon = 0.1$, and the reflector is horizontal. This graph shows the lateral-positioning error we can expect below dipping anisotropic strata. For example, 1 km of shale dipping at 45° will create a lateral-positioning error of approximately 100 m for $\varepsilon = 0.1$ and any value of δ. We can also observe the subtle effect δ has on the lateral-positioning error.

Although anisotropic velocities are suspected in the upper clastic section, we are concerned particularly about imaging below the dipping strata. The result of an isotropic depth migration of this section is displayed, in time, in Figure 5.

During isotropic depth-migration velocity analysis, we first observed that the image gathers were flat below flat-lying strata in the overburden but showed significant residual moveout beneath steeply dipping strata. Identical stratigraphic units required a much higher model velocity when these units were dipping steeply. We felt that using velocity gradients such that steeply dipping rocks had a higher model velocity than the same rocks with zero dip was not geologically justified. Figure 6a

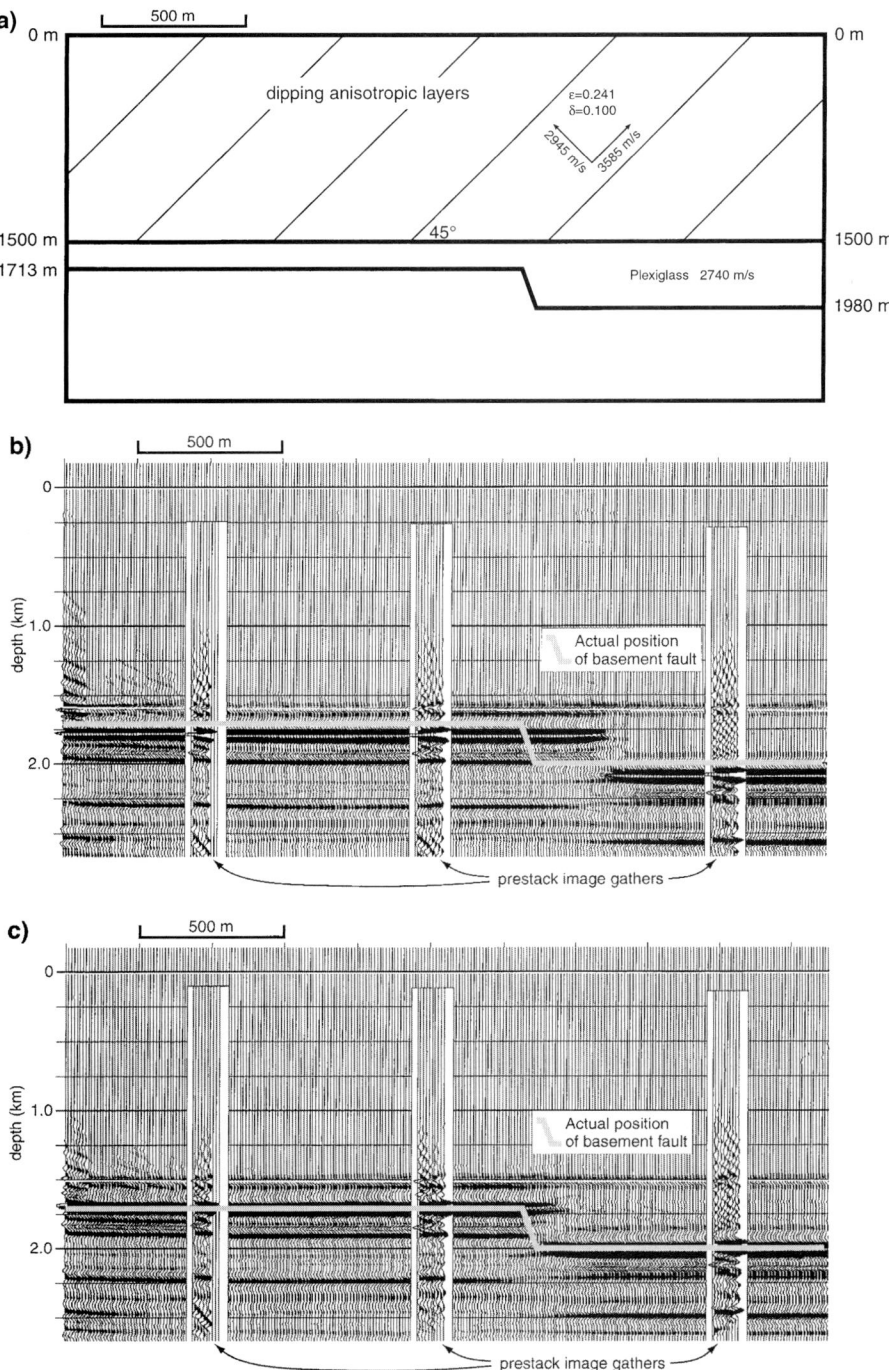

FIG. 4. (a) Anisotropic physical model from the Foothills Research Project. Note the 45° dipping anisotropic overburden above horizontal reflectors. The plexiglas layer is isotropic. (b) This isotropic depth migration yields flat image gathers but incorrect reflector depths and lateral position. (c) The anisotropic image shows flat image gathers with correct depths and improved lateral position.

shows the result of isotropic depth migration of the data set, including image gathers and some selected raypaths of events that contribute to the gathers. The gather on the right of Figure 6a shows residual moveout on the basement reflector. This indicates the dipping zone in the upper part of the section needs a slower velocity to better flatten the gather. The gather in the center, below the steeply dipping shales, shows that the moveout on the basement reflector is significantly overcorrected, indicating the same dipping zone on the model needs a higher velocity to improve the imaging. This conflict in imaging-velocity requirements of the two gathers was reconciled using anisotropic velocities that are laterally consistent, with lower velocities perpendicular to bedding and higher velocities parallel to bedding.

The velocity model was then parameterized for an anisotropic depth migration with the required dip information obtained from the interpretation in Figure 5. Only the shale-dominated clastic layers in the upper part of the section were considered anisotropic in our model. We used parameters determined from refraction-survey measurements on near-vertical shales by Leslie and Lawton (1998) on dipping shales in the Alberta Foothills to help us constrain the anisotropic velocity model. In two separate field locations, δ was found to be near zero for an ε of 0.14 and 0.26. Since δ is small and changes in δ have a subtle influence on s, we parameterized our model with ε. We set δ to be one-quarter of ε and iterated to find the ε that produced the flattest prestack image gathers. If possible, a complete anisotropic velocity model should contain estimates for both δ and ε.

Initial values of anisotropic parameters were $\varepsilon = 0.15$ and $\delta = 0.038$, based on field measurements by Leslie and Lawton (1998) for similar rocks, and the initial v_0 was based on an isotropic depth-migration velocity analysis beneath the horizontal anisotropic layers. After a few revisions to the model,

guided by the inspection of image gathers, optimum imaging was achieved for $\varepsilon = 0.10$ and $\delta = 0.025$. Once the image-gather conflict shown in Figure 6a was resolved, we inspected the image gathers again for improvements in flatness of gathers with a change in v_0.

With the anisotropic migration (Figure 6b), imaging was improved below the dipping strata without compromising image quality below horizontal strata. The laterally varying dip was advantageous in refining the anisotropic model parameters. When confronted with residual moveout on an image gather, we had difficulty deciding whether to change v_0 or ε and δ to flatten the gather. By assuming that the properties of the strata do not change with dip, however, only a limited range in v_0, ε, and δ could result in a good image below both flat-lying and steeply dipping strata. Although the solution is not unique, the resulting image is better than that obtained from isotropic migration.

A comparison between the final anisotropic depth migration and the final isotropic depth migration is shown in Figure 7. Differences observed between the sections are attributed only to differences between the isotropic and anisotropic migrations.

The highlighted rectangles on both sections in Figure 7 show some of the key improvements of anisotropic depth migration when compared to isotropic depth migration. Boxes A and A' show differences beneath the dipping strata; there is a lateral movement and improvement in reflector continuity below the fault contact in the upper left corner of A' compared with that of A. The reflector that spans the bottom of A' also shows improved continuity. The anticline interpreted in box B (Figure 7a) is not present in B' (Figure 7b). The anticline in B is considered to be a false depth structure caused by a significant change in near-surface dips. Basement reflector continuity is also improved after anisotropic depth migration, as evident by comparing boxes C and C'.

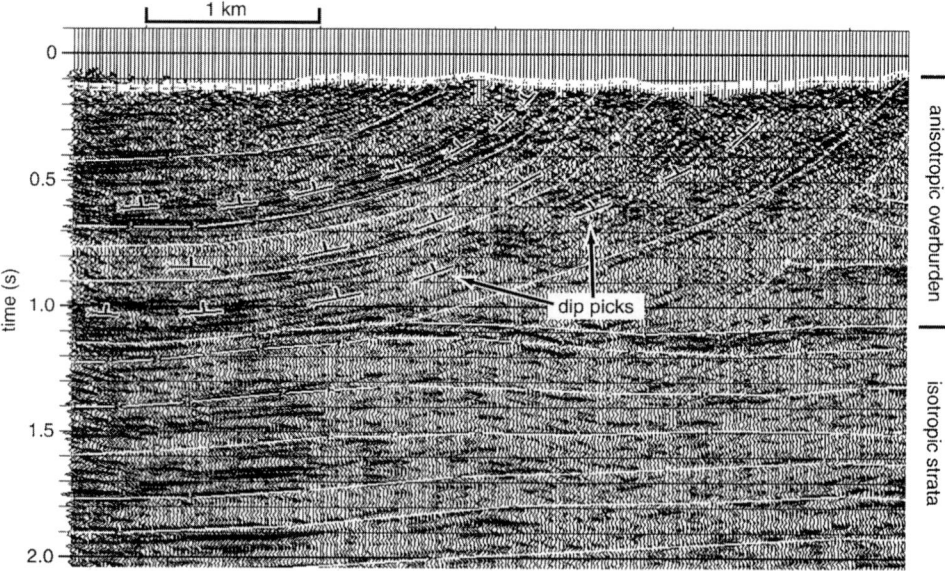

FIG. 5. Interpreting dip from the seismic section. The white lines on the section are velocity-model zone boundaries. The dip is allowed to vary laterally within a zone, as shown here. The upper section contains the shale-dominated clastic sequence; the lower section contains carbonates, assumed to be isotropic in our velocity model.

Thus, the depth-migrated sections in Figure 7 show a significant improvement in image quality when seismic anisotropy is taken into account, and demonstrate the need for anisotropic depth migration in areas of complex geological structure.

CONCLUSIONS

We draw the following conclusions from this study:

1) an isotropic assumption causes mispositioning of imaged structures below dipping anisotropic strata because sideslip is ignored;

2) we carried out anisotropic traveltime calculations to account for sideslip in anisotropic depth migration;

3) for anisotropic strata dipping at angles near 45°, ε has a greater influence on sideslip than does δ;

4) tests using the anisotropic migration on physical-model seismic data resulted in flat image gathers and more accurate target depths and positions than did isotropic migration; and

5) application of the anisotropic depth migration to field data using focusing criteria to interpret the anisotropic model produced a depth image with image quality superior to that of the isotropic migration.

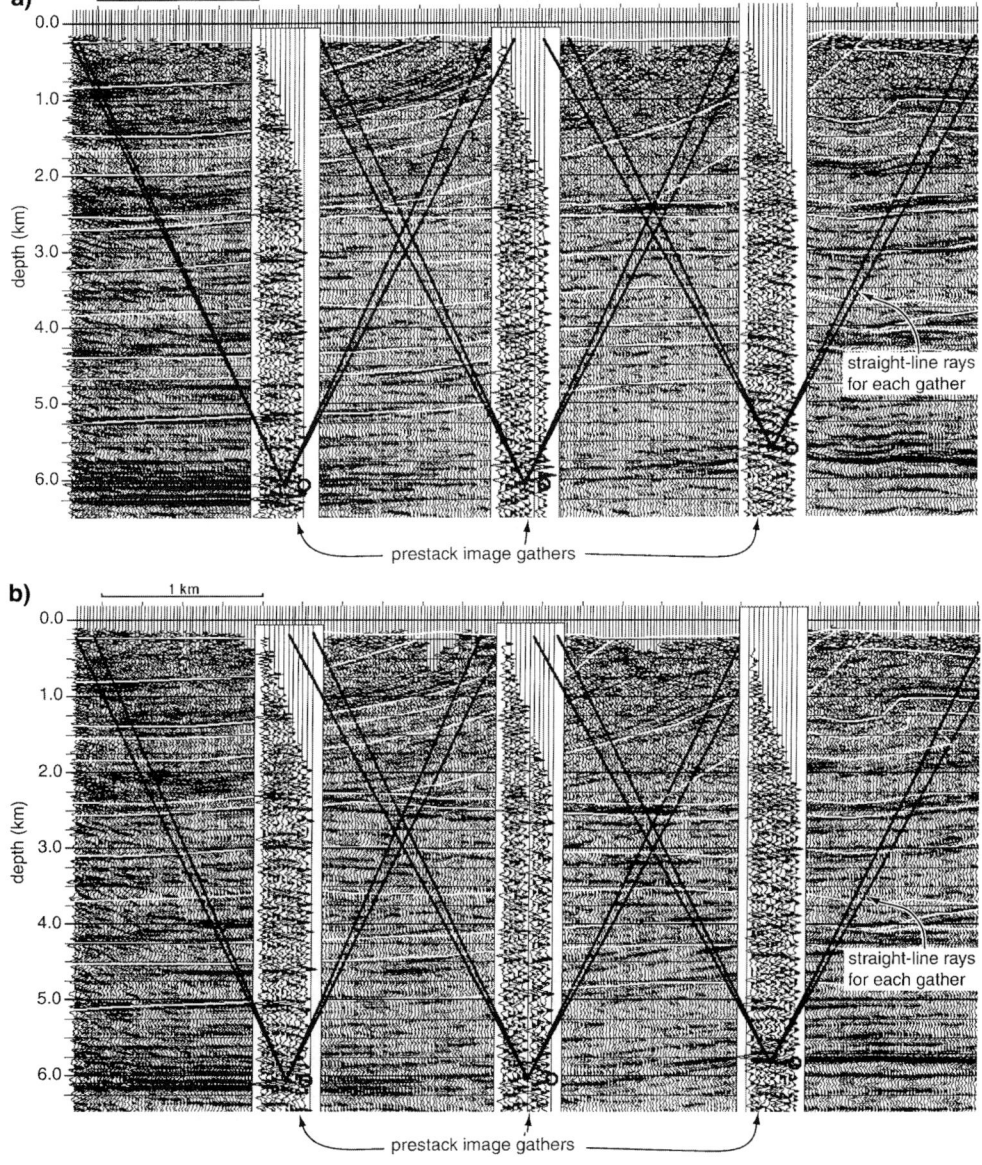

FIG. 6. (a) Image-gather displays for the isotropic model. The gather in the center is overcorrected; therefore, a faster model velocity is needed to flatten it. The gather on the right is undercorrected and needs a slower velocity. The black lines represent a straight-line approximation of ray tracing to the basement reflector at 3 km. Note that both the center gather and the right gather contain seismic energy that traveled through the same clastic layers between the two gathers and above 2.4 km depth. The gather on the right contains energy that traveled normal to bedding—the slow-velocity direction. The gather in the center contains energy that traveled parallel to bedding—the fast-velocity direction. (b) Image-gather displays for the anisotropic velocity model. Model parameters $\varepsilon = 0.1$ and $\delta = 0.025$ are used for the clastic layers in the upper 2.5 km of the section. The gather in the center has improved relative to the same gather in (a), and the reflector on the stack at 3 km shows improved continuity. The gather on the right is flat—again, an improvement over the same gather in (a).

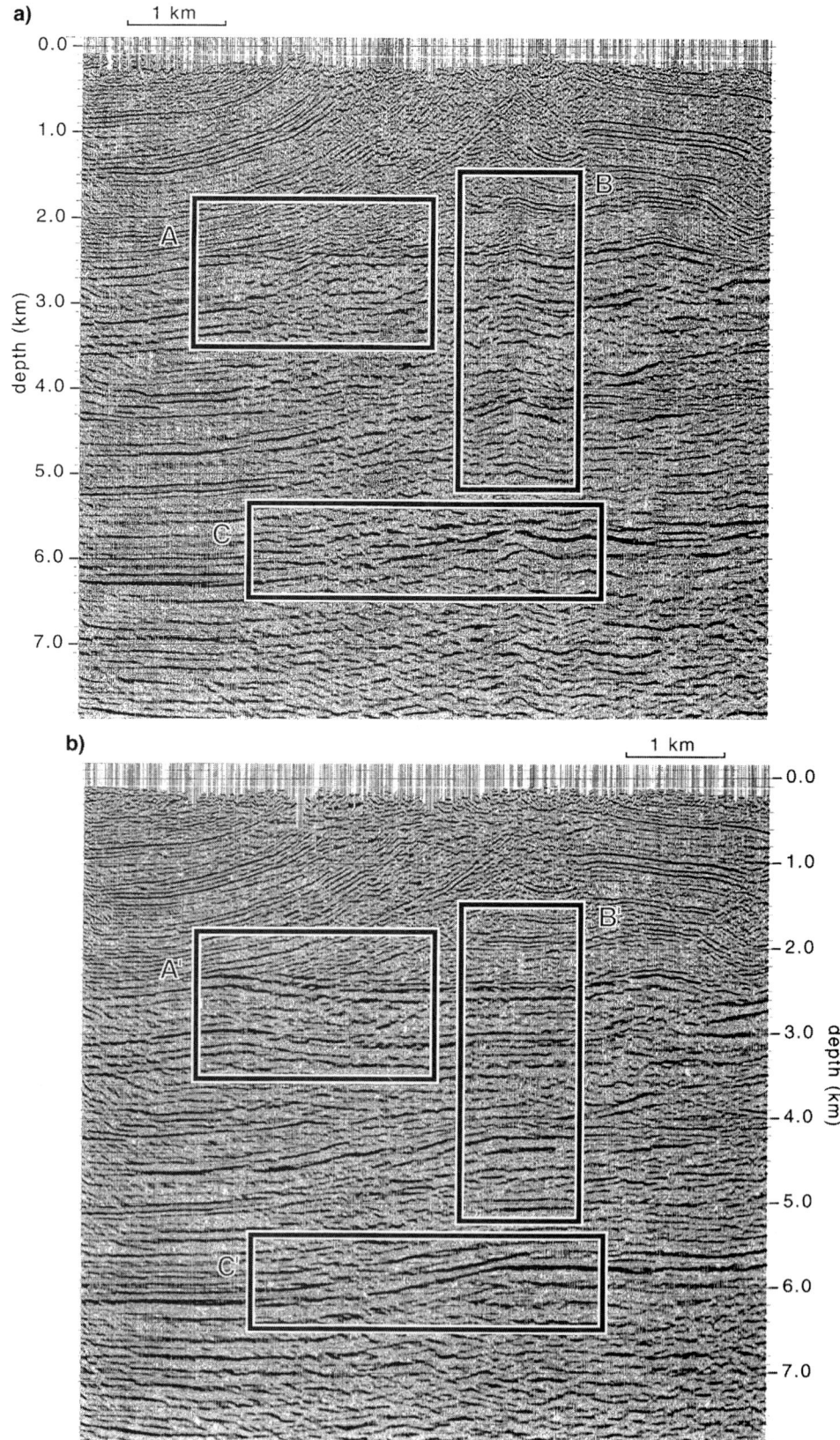

FIG. 7. (a) Final depth-migrated image using the isotropic model. Note the lack of continuity in boxes A and C and the structures on several reflectors in B for comparison with the boxes in (b), the final depth-migrated image using the anisotropic model with $\varepsilon = 0.1$ and $\delta = 0.025$. The boxes highlight some of the more significant improvements over the standard depth migration in (a). Note the improvement in continuity of reflectors in A' and C' over that for A and C, and that the imaged structures in B are flattened in B'.

ACKNOWLEDGMENTS

The authors thank the sponsors of the Foothills Research Project and the Natural Science and Engineering Research Council for their financial support of this research and Kelman Seismic Processing for data processing services. We also thank Ken Larner, Ilya Tsvankin, and an anonymous reviewer for their constructive reviews of this paper.

REFERENCES

Alkhalifah, T., 1995, Gaussian beam depth migration for anisotropic media: Geophysics, **60**, 1474–1484.

Alkhalifah, T., and Larner, K., 1994, Migration error in transversely isotropic media: Geophysics, **59**, 1405–1418.

Ball, G., 1995, Estimation of anisotropy and anisotropic 3-D prestack depth migration, offshore Zaire: Geophysics, **60**, 1495–1513.

Byun, B. S., 1982, Seismic parameters for media with elliptical velocity dependencies: Geophysics, **47**, 1621–1626.

Dellinger, J., 1991, Anisotropic seismic wave propagation: Ph.D. thesis, Stanford Univ.

Di Nicola-Carena, E., 1997, Solving lateral shift due to anisotropy: 67th Ann. Internat. Mtg., Soc. Expl. Geophys., Expanded Abstracts, 1525–1528.

Isaac, J. H., and Lawton, D. C., 1999, Image mispositioning due to dipping TI media: A physical seismic modeling study: Geophysics, **64**, 1230–1238, this issue.

Kitchenside, P. W., 1992, An implementation of anisotropic migration: Some issues and examples: 54th Mtg., Eur. Assoc. Expl. Geophys., Abstracts, **92**, 234–235.

Larner, K., and Cohen, J. K., 1993, Migration error in transversely isotropic media with linear velocity variation in depth: Geophysics, **58**, 1454–1467.

Leslie, J., and Lawton, D. C., 1998, A refraction seismic field study to determine the anisotropic parameters of shales: The Leading Edge, **17**, 1127–1129.

Leslie, J., Lawton, D. C., and Cunningham, J. D., 1997, A refraction seismic field method to determine the anisotropic parameters of Wapiabi shales: Proc. 1997 National Meeting, Can. Soc. Expl. Geophys., 61–63.

Lines, L., Wu, W., Lu, H., Burton, A., and Zhu, J., 1996, Migration from topography: Experience with an Alberta Foothills data set: Can. J. Expl. Geophys., **32**, 24–30.

Stork, C., Welsh, C., and Skuce, A., 1995, Demonstration of processing and model building methods on a real complex structure data set: Ann. Conv., Workshop 6, Soc. Expl. Geophys., Proceedings.

Thomsen, L., 1986, Weak elastic anisotropy: Geophysics **51**, 1954–1966.

Tsvankin, I., 1996, P-wave signatures and notation for transversely isotropic media: An overview: Geophysics, **61**, 467–483.

Uzcategui, Omar, 1995, 2-D depth migration in transversely isotropic media using explicit operators: Geophysics, **60**, 1819–1829.

Vestrum, R. W., and Muenzer, K., 1997, Imaging below dipping anisotropic shales: Proc. 1997 National Meeting, Can. Soc. Expl. Geophys., 64–65.

Nonstationary phase shift (NSPS) for TI media

Robert J. Ferguson and Gary F. Margrave

ABSTRACT

Nonstationary phase shift (NSPS) can be modified, using vertical and horizontal slowness, to extrapolate wavefields through media that are transversely isotropic (TI), and whose elastic parameters vary laterally. No restriction is placed on the weakness of the anisotropy, or the angle of TI symmetry. The significance of this method lies in the fact that velocity variation is allowed simultaneously in space and spatial frequency, and that wavefield extrapolation proceeds using only phase velocity.

We present impulse responses for P-waves for a number of different TI media. The first, weathered gypsum, was chosen due to its extreme P-wave anisotropy (82%). The second, Mesaverde shale, was chosen due to its more typical anisotropy (13%). Impulse responses are computed for the isotropic, vertical TI and TI with different axes of symmetry (-45 degrees for weathered gypsum, 60 degrees for Mesaverde shale). The responses for the weathered gypsum show significant anisotropic effects, particularly for dipping angle of symmetry. The Mesaverde shale responses are less significant due to the lower level of anisotropy.

To demonstrate NSPS for materials with lateral variation in anisotropy, we present a scenario where two spatially separate impulses are extrapolated simultaneously through 300-meters of Gypsum and Mesaverde shale. The weathered Gypsum grades abruptly into Mesaverde shale, both having different axis of TI symmetry. The resulting responses match those from the constant anisotropy cases.

INTRODUCTION

The notion that seismic anisotropy is intrinsic to wavefield propagation in the subsurface is swiftly gaining recognition as a factor in imaging. Though anisotropy is reported throughout the earth's' crust and upper mantle (Crampin, 1984) the special case of transverse isotropy (TI) is sufficiently common to warrant consideration (Byun, 1984). One of the most mathematically tractable forms of anisotropy, TI describes a medium in which a large number of homogeneous, isotropic layers are arranged in alternating planes, such that five elastic parameters provide a complete description of the mediums' elastic properties (see, for example, Postma, 1955). In such a medium, for wavelengths much longer than the layer thicknesses, we find that all directions relative to the same axis of symmetry are equivalent (again, Postma, 1955).

Thomsen (1986) points out the fundamental inconsistency of trying to image a potentially anisotropic subsurface using the assumption of isotropy. Isaac and Lawton (1997) demonstrate the peril of imaging TI media with standard isotropic processing. In a physical modeling experiment they show that a medium with ~20% P-wave anisotropy and a TI symmetry of 45 degrees causes large errors in the lateral position of a simulated reef edge.

Authors such as Uzcategui (1995), Kitchenside (1993), and Kitchenside (1991) have proposed algorithms by which TI media might be imaged. However, their algorithms are based on ω - x extrapolation techniques which Margrave and Ferguson (1997) and Black et al. (1984) point out are space domain approximations to a nonstationary Fourier operator. Authors such as Margrave and Ferguson (1997), Black et al. (1984) and Wapenaar (1994) show that a wavefield extrapolation scheme can be formulated in Fourier space, that is $(\omega - k_x)$, which requires no approximation to cope with laterally varying velocities. Margrave and Ferguson (1997) refer to this scheme as NSPS (nonstationary phase shift).

In this paper we present an extension of NSPS to handle TI media. This, we will show, is done without approximation of the extrapolator, without restriction of the TI axis of symmetry, and without restricting the weakness/strongness of the anisotropy. We will use Thomsen parameters to describe the anisotropy.

Examples of impulse responses are presented to demonstrate the ability of TI NSPS to handle laterally variant anisotropic parameters.

NONSTATIONARY PHASE SHIFT

Wavefield extrapolation by nonstationary phase shift is an extrapolation method suitable for depth migration or modeling. It proposes that, based on the scalar wave equation, the Fourier domain $(\omega - k_x)$ provides a complete description of the subsurface. (For a detailed description of NSPS we refer the reader to Margrave and Ferguson, 1997.)

Though Margrave's filter theory (1997) provides prescriptions for space $(\omega - x)$ (ω - x) and dual $(\omega - (x, k_x))$ as well as the Fourier domain, we presently prefer to use the Fourier domain description. This is due mainly to the fact that there are significant runtime advantages to using this domain (Ferguson and Margrave, 1997, Margrave and Ferguson, 1997, and Wapenaar, 1994).

The basic equation for Fourier domain NSPS is:

$$\varphi(k_x, \Delta z, \omega) = \int_{-\infty}^{\infty} \varphi(k'_x, 0, \omega) A(k_x, k_x - k'_x, \omega) dk'_x \qquad (1)$$

where $\varphi(k_x, \Delta z, \omega)$ is the Fourier transform of the wavefield (output) which has been extrapolated one depth step (Δz) from a reference depth $z = 0$; $\varphi(k'_x, \Delta z, \omega)$ is the Fourier transform of the wavefield (input) recorded at the reference depth $z = 0$ (the primes in equation (1) are used to distinguish input from output wavenumbers - see Margrave and Ferguson, 1997). The operator $A(k_x, k - k'_x, \omega)$ is essentially a shifted Fourier transform of the nonstationary wavefield extrapolator and is computed as:

$$A(k_x, k_x - k_x', \omega) = \int_{-\infty}^{\infty} \alpha(k_x, x', \omega) e^{-ix'(k_x - k_x')} dx' \qquad (2)$$

where,

$$\alpha(k_x, x', \omega) = \begin{cases} e^{ik_z(k_x, x', \omega)\Delta z}, & \dfrac{\omega^2}{v^2(x')} - k_x^2 \geq 0 \\ e^{i|\mathrm{Im}(k_z(k_x, x', \omega))|\Delta z}, & \dfrac{\omega^2}{v^2(x')} - k_x^2 < 0 \end{cases}, \quad k_z(k_x, x', \omega) = \sqrt{\dfrac{\omega^2}{v^2(x')} - k_x^2} \qquad (3)$$

is the nonstationary extrapolator (Margrave and Ferguson, 1997).

Equation (3) insures exponential decay of energy in the evanescent region (k_x is imaginary). The operator in equation (3) is an isotropic function of laterally variable velocity. We propose that $v(x')$ be extended to vary with angle of incidence thus, $v(x') \to v(k_x, x')$, and Equation (3) becomes the TI nonstationary extrapolator:

$$\alpha(k_x, x', \omega) = \begin{cases} e^{ik_z(k_x, x', \omega)\Delta z}, & \dfrac{\omega^2}{v^2(x', k_x)} - k_x^2 \geq 0 \\ e^{i|\mathrm{Im}(k_z(k_x, x', \omega))|\Delta z}, & \dfrac{\omega^2}{v^2(x', k_x)} - k_x^2 < 0 \end{cases}, \quad k_z(k_x, x', \omega) = \sqrt{\dfrac{\omega^2}{v^2(x', k_x)} - k_x^2} \qquad (4)$$

A major strength of this approach, compared to time domain Kirchoff methods for example, is that we need only consider the phase velocities. The group velocities, which are required for raytracing, are not needed because we treat each spectral component explicitly.

TRANSVERSE ISOTROPY

The angle dependant velocities of a TI media can be completely described by five elastic constants whose mathematical forms are well established. However, these constants are difficult to interpret physically. As a remedy, Thomsen (1986) derives the five elastic constants such that two of them are the physically realizable values of P-wave velocity (α_o) and S-wave velocity (β_o), measured at zero offset, plus three additional constants. Due to this simplicity we adopt the Thomsen derivations as input parameters for wavefield extrapolation.

In this paper we concentrate on the extrapolation of P-waves. It is our intention to provide an S-wave solution in the near future. By concentrating on P-waves the required number of parameters reduces to four. The P-wave velocity as a function of angle of incidence θ is:

$$v_p(\theta) = \alpha_o^2 \left[1 + \varepsilon \sin^2 \theta + D^*(\theta)\right] \qquad (5)$$

with

$$D^*(\theta) = \frac{1}{2}\left(1 - \frac{\beta_o^2}{\alpha_o^2}\right)\left\{\left[1 + \frac{4\delta^*}{\left(1 - \beta_o^2/\alpha_o^2\right)}\sin^2\theta\cos^2\theta + \frac{4\left(1 - \beta_o^2/\alpha_o^2 + \varepsilon\right)\varepsilon}{\left(1 - \beta_o^2/\alpha_o^2\right)^2}\sin^4\theta\right]^{\frac{1}{2}} - 1\right\} \qquad (6)$$

We remind the reader that in the above TI equations each of the anisotropic parameters is a function of input position x'. For simplicity we have suppressed the x' dependence.

TI NSPS

Wavefield extrapolation by NSPS is an operation that loops over frequency; one monochromatic wavefield is extrapolated per loop. For efficiency we want to compute required TI calculations in such a way that as much computation as possible is done outside of the loop. Following Kitchenside (1991), we compute a set of horizontal slownesses (p) for a range of evenly spaced angles (θ). (For a non-zero symmetry axis we simply relate the angle of incidence of waves to the rotated axis of anisotropy). The set of p values are used to find a set of vertical slownesses (q) for a range of evenly spaced wavenumbers (k_x). The slowness values are computed using:

$$p(x',\theta) = \frac{\sin(\theta)}{v(x',\theta)} \qquad (7)$$

$$q(x',\theta) = \frac{\cos(\theta)}{v(x',\theta)} \qquad (8)$$

where $v(x',\theta)$ is computed using equations (5) and (6). Because p and q are independent of temporal frequency (ω) they both are computed outside of the main recursion. Then, for each ω, a new set of p values are computed using:

$$p(k_x,\omega) = \frac{k_x}{\omega} \qquad (9)$$

which are then compared to $p(x',\theta)$ from equation (7). For the non-evanescent region a set of θ values are produced from this comparison and are used to compute a set of q values by equation (8), thereby relating k_x to θ. In the evanescent region the q values are computed using:

$$q(k_x,x',\omega) = -ip(x',\theta)_{max}\sqrt{k_x^2 - w^2 p^2(x',\theta)_{max}} \qquad (10)$$

where we use the maximum horizontal slowness $(p(x',\theta)_{max})$ of the media to damp evanescent energy. The resulting set of q values are then used to calculate the values of the vertical wavenumbers $k_x = \omega q$ required for extrapolation:

$$\alpha(k_x, x', \omega) = e^{i\omega q(k_x, x', \omega)\Delta z}. \tag{11}$$

It is useful to combine equations (8), (10) and (11) into a single statement for comparison with equation (4):

$$\alpha(k_x, x', \omega) = e^{i\omega q(k_x, x')\Delta z}, \begin{cases} q(k_x, x') = Interp\left(q(x',\theta), p(x',\theta), \dfrac{k_x}{\omega}\right), \dfrac{k_x}{\omega} \leq p(x',\theta)_{max} \\ q(k_x, x') = -ip(x',\theta)_{max}\sqrt{k_x^2 - \omega^2 p^2(x',\theta)_{max}}, \\ \dfrac{k_x}{\omega} > p(x',\theta)_{max} \end{cases} \tag{13}$$

Equation (13) shows that, in the non-evanescent region, $\alpha(k_x, x', \omega)$ is computed with the *interp* function. This function represents the interpolation scheme that finds $q(k_x, x')$ from $q(x',\theta)$ using $p(x',\theta)$ as an intermediary. In the evanescent region, $q(k_x, x')$ is computed using the familiar square root equation parameterized with the maximum value of p (the isotropic velocity).

EXAMPLES

To demonstrate TI NSPS we present a set of impulse responses for a number of different geologic settings. Each setting has unique anisotropic parameters and angle of symmetry. The resulting impulse responses are compared to those of the material with a vertical axis of TI symmetry, and the case of isotropy. From these comparisons we find some notable differences. These differences which are discussed in the context of vertical slowness.

Weathered gypsum

The anisotropic parameters for weathered gypsum are in Thomsen (1986) notation:

$$\alpha_o = 1991, \beta_o = 795, \varepsilon = 1.161, \delta^* = -1.075$$

If we think of the anisotropy of a material as the percent difference between the fast and slow directions we find that weathered gypsum has a maximum anisotropy of 80% at +/-90 incidence (Figure 1). Figures (2) through (4) show 3D plots of respectively, the velocity (Vp), horizontal slowness (p) and vertical slowness (q) of the material as a function of angle of incidence (θ) and angle of TI symmetry (φ). We can see that the p and q surfaces are relatively simpler than the velocity surface.

This is more clearly indicated by Figure (5) which is a plot of q anisotropy, achieving a maximum of 20% at +/- 52 degrees.

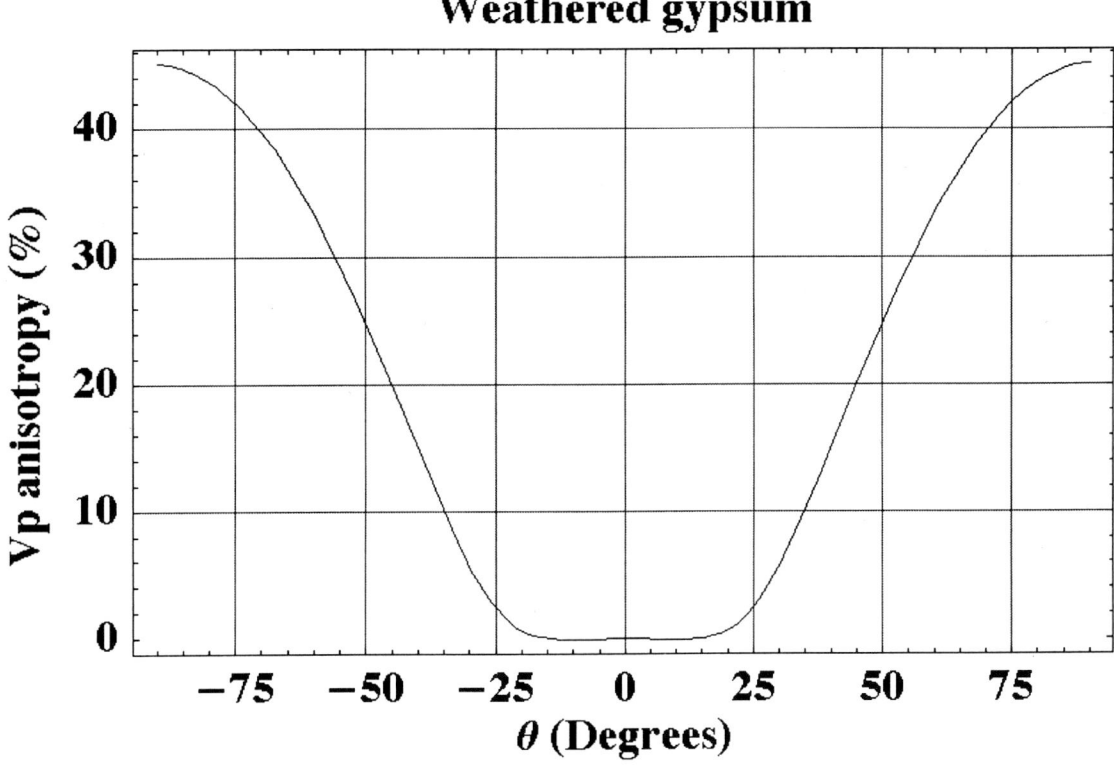

Fig. 1. P-wave anisotropy for weathered gypsum. The maximum anisotropy is 82% at +/- 90 degrees (assuming vertical TI symmetry).

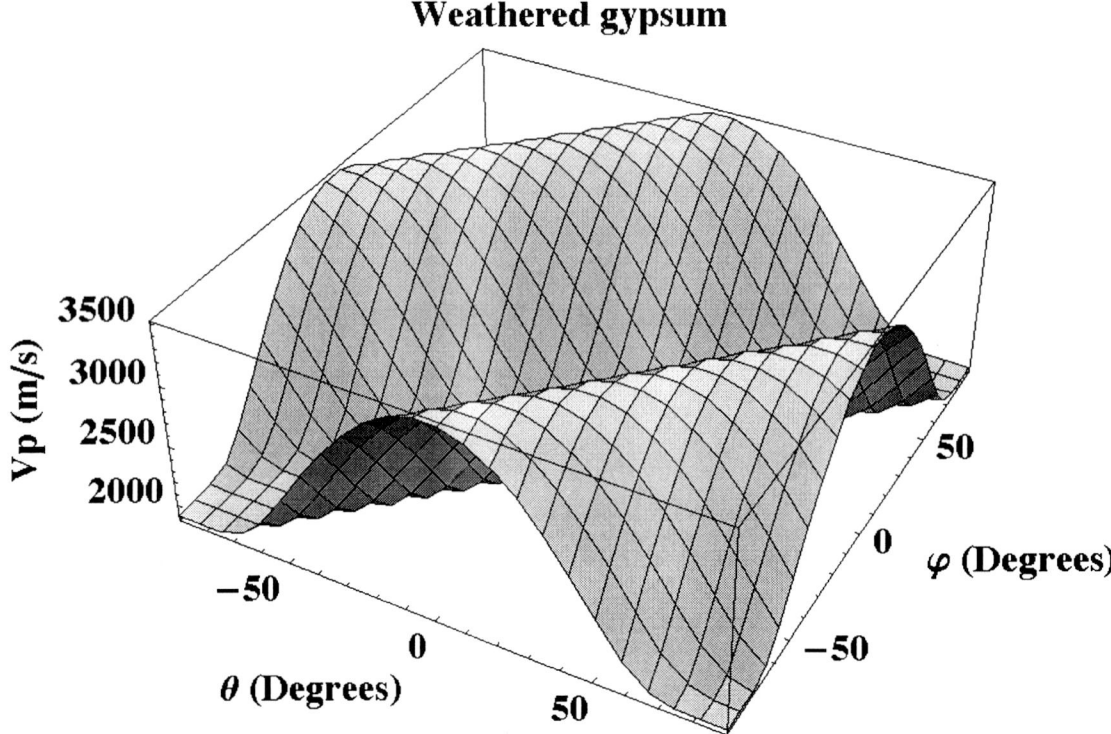

Fig. 2. P-wave velocity as a function of angle of incidence (θ) and TI symmetry axis (φ) for weathered gypsum.

Weathered gypsum

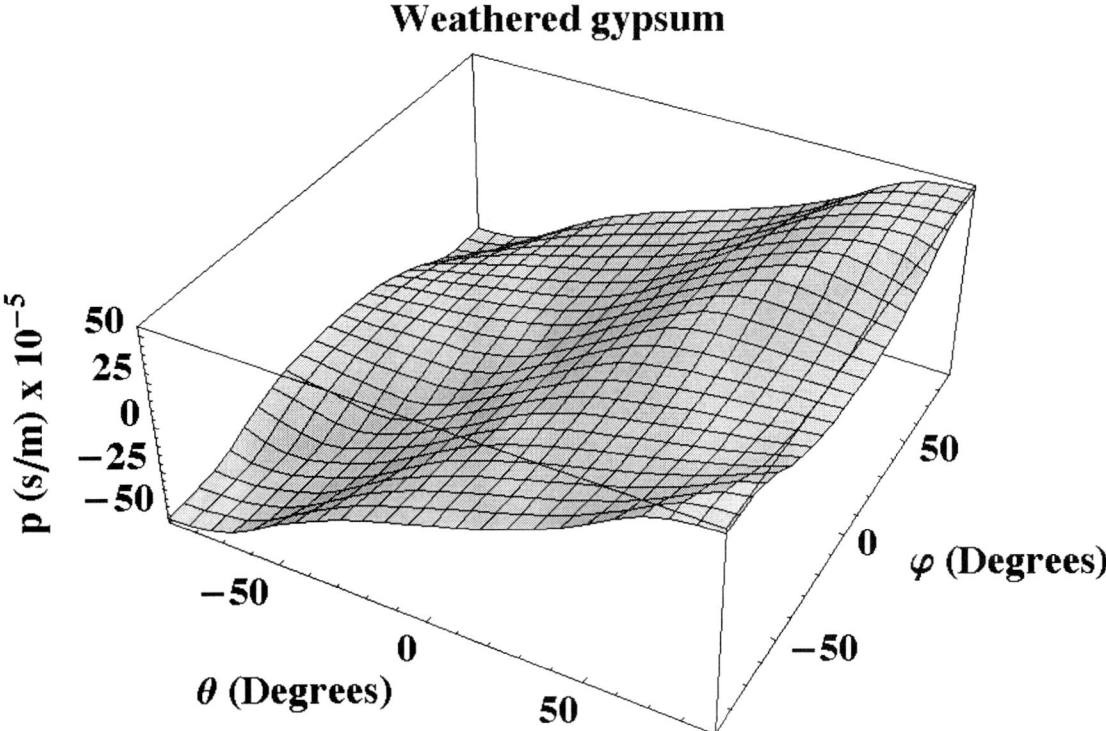

Fig. 3. Horizontal slowness (p) as a function of angle of incidence (θ) and TI symmetry axis (φ) for weathered gypsum.

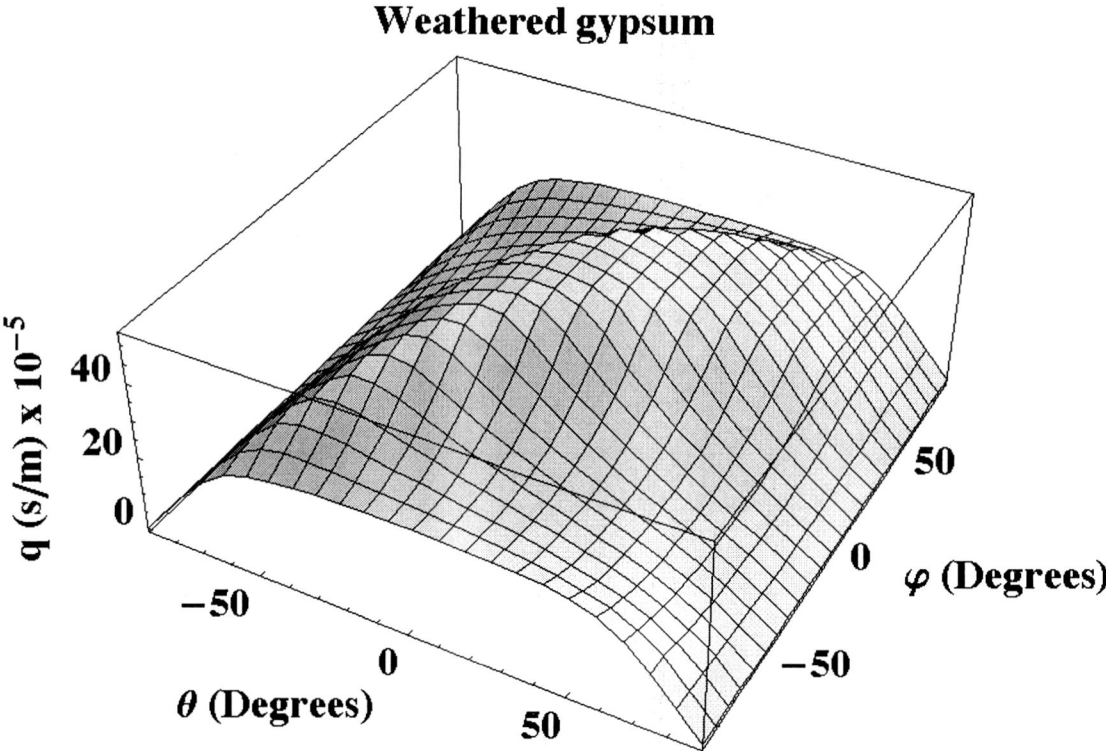

Fig. 4. Vertical slowness (q) as a function of angle of incidence (θ) and TI symmetry axis (φ) for weathered gypsum.

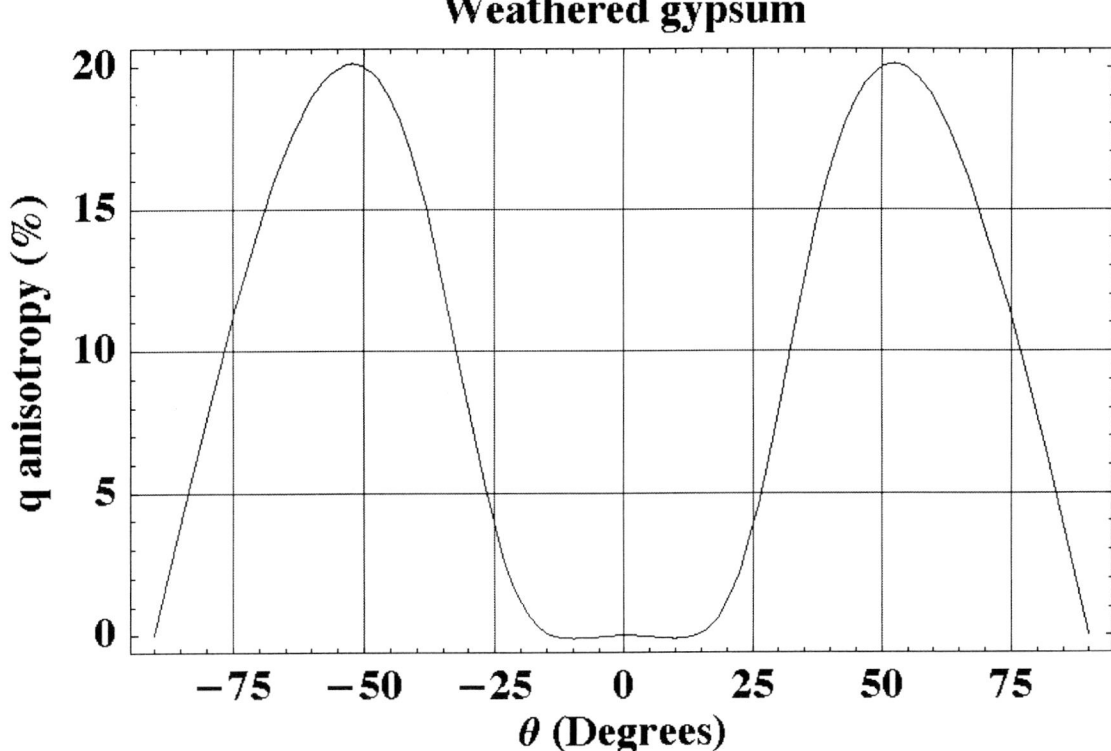

Fig. 5. *q* anisotropy for weathered gypsum. The maximum anisotropy is 20% at +/- 52 degrees (assuming vertical TI symmetry).

Impulse responses are generated by extrapolating a single spike through 300 m using a velocity model parameterized with no anisotropy $(\varepsilon = 0, \delta^* = 0)$, vertical TI, and TI with a -45 degree symmetry axis.

Discussion

The isotropic impulse response of Figure (6) is an upside down hyperbola showing numerical attenuation at the larger travel times. (This purpose of this effect is to suppress Fourier wraparound, and is achieved by adding a small imaginary component, calculated as a percentage, to the velocities). Comparison of this curve to the vertical TI impulse response of Figure (7) shows how the TI extrapolation is faster for higher dip angles and shows less numerical attenuation.

Figure (9) is a plot of *q* for the isotropic, vertical TI and TI with φ = -45 degrees, and is useful for understanding the impulse responses. We see that, due to anisotropy, the vertical TI extrapolation is computed with lower *q* than the isotropic example. Lower *q* means higher velocity and therefore shorter travel times through the same depth step. Shorter travel times can be seen in the impulse response as limbs, which have less dip. The reduction in dip implies increased wavelength in the *x* direction and therefore reduced numerical attenuation effects thus, the vertical TI impulse response is less dispersed.

The impulse response for φ = -45 degrees is shown in Figure (8). We see that, compared to the isotropic curve, the TI response is skewed down and to the left. This can be understood, again, with the help of Figure (9). The q curve for this non-vertical example is also skewed showing greater slowness for negative angles of incidence. If we think of a diffraction curve for this situation we see that waves incident along negatively dipping ray paths travel slower than those along positive ray paths thereby taking longer to traverse the same amount of depth. This changes as we follow the q curve from -90 degrees towards zero degrees where at about -20 degrees the trend reverses.

The diffraction curve for this example might be like the one pictured in Figure (10). If we then consider extrapolation as the cross correlation of the impulse with its diffraction response then a picture of the expected impulse response of the TI extrapolation emerges. It will look like a time and space reversed version of the diffraction response, and this is exactly what we see in Figure (8).

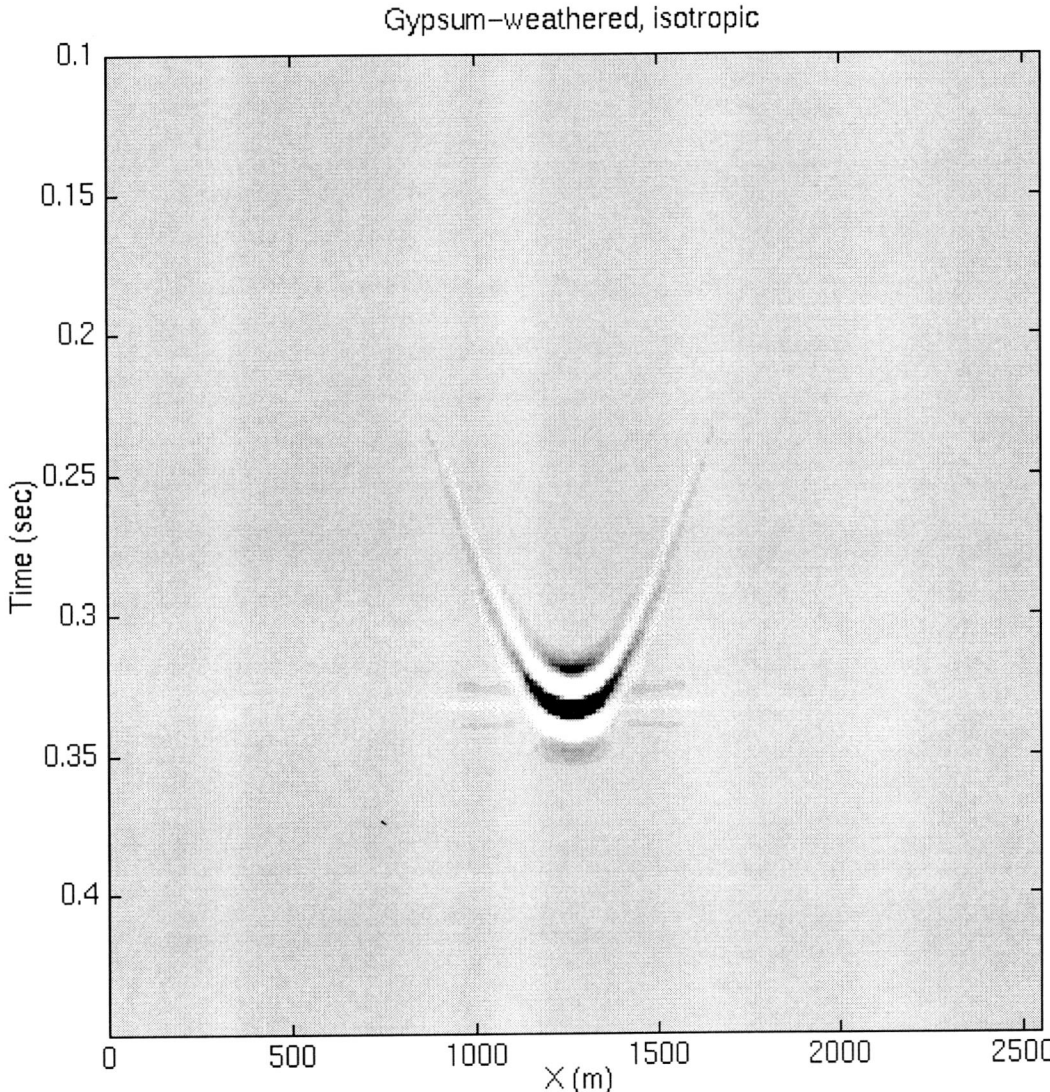

Fig 6. Isotropic impulse response for weathered gypsum.

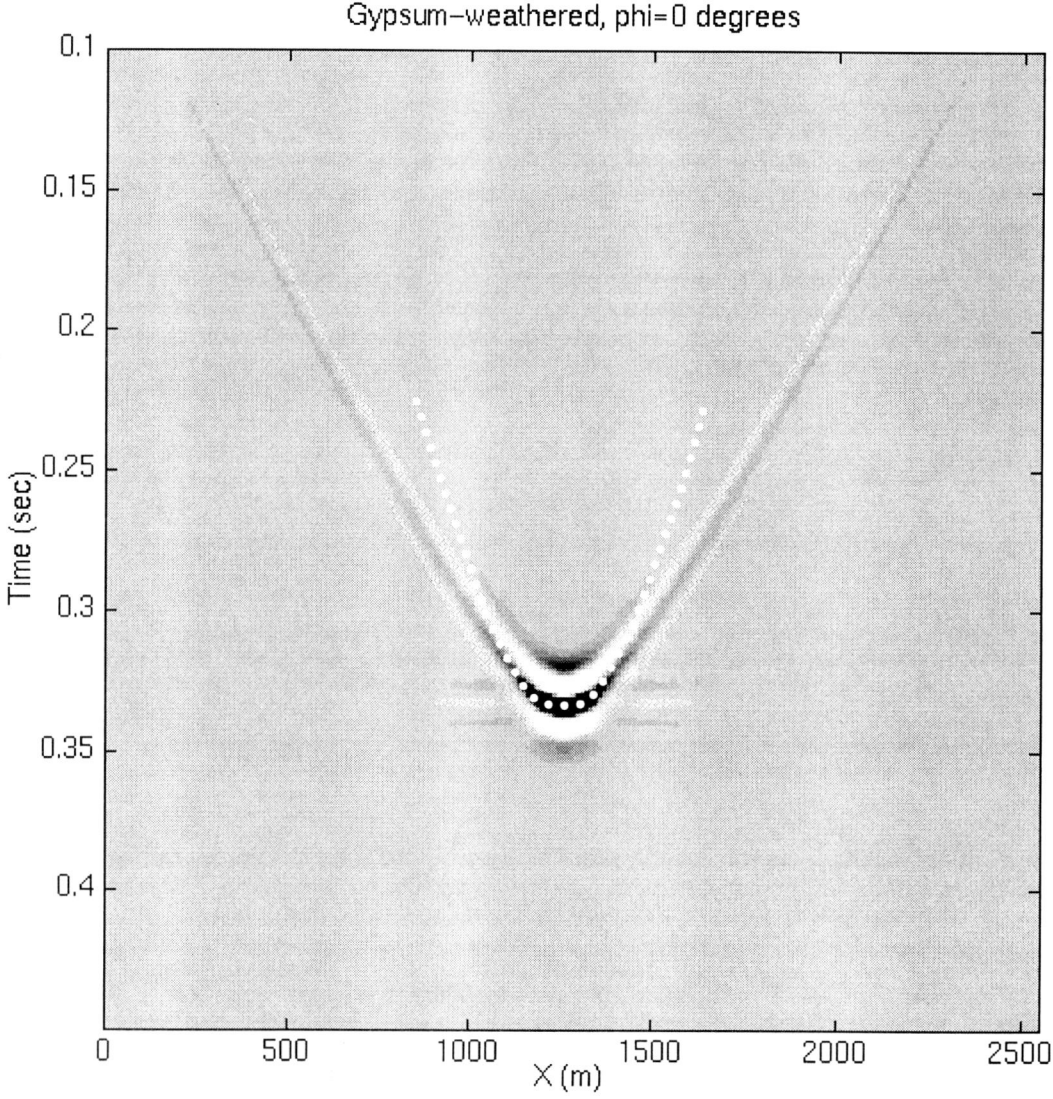

Fig. 7. Vertical TI impulse response for weathered gypsum. The isotropic response is overplotted in white.

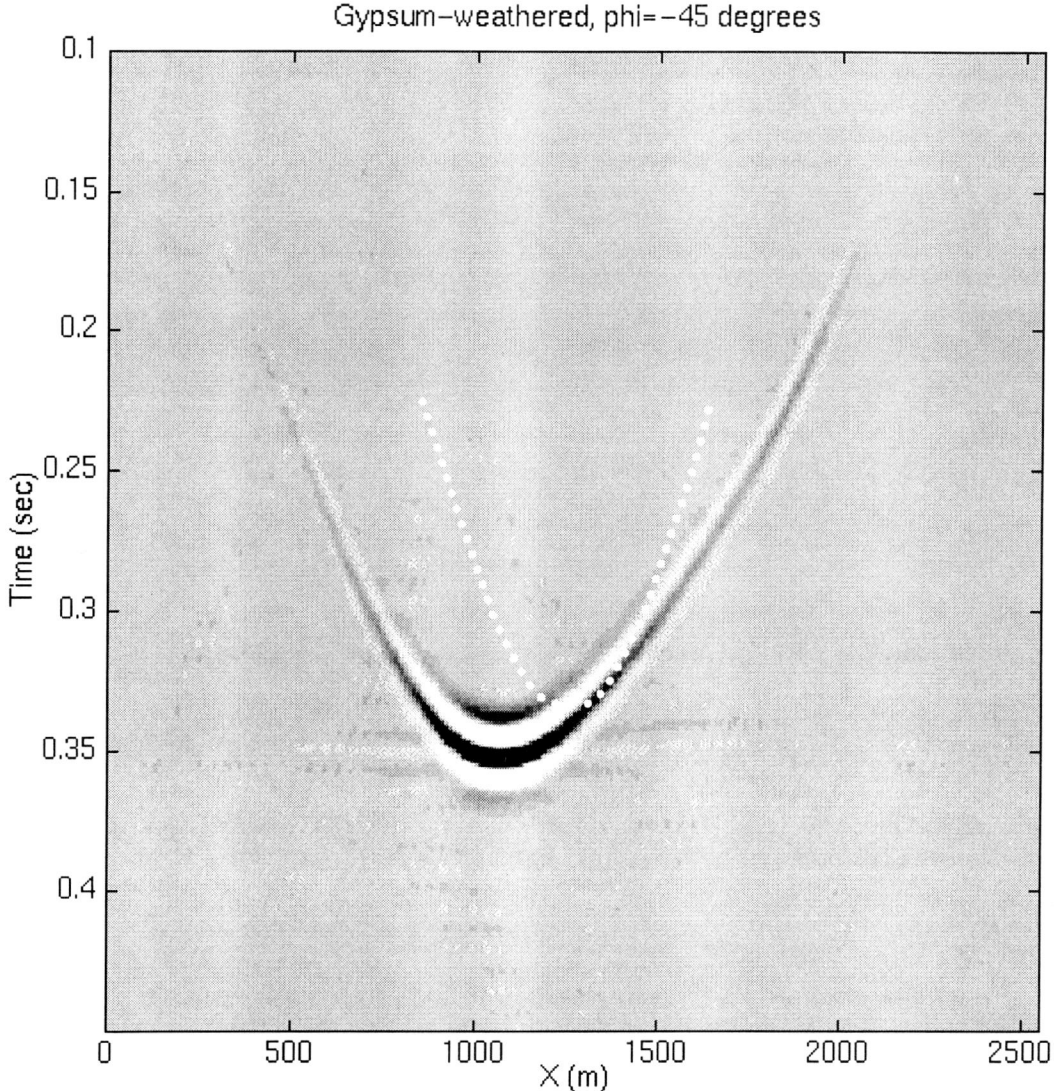

Fig. 8. TI impulse response for weathered gypsum. The isotropic response is overplotted in white.

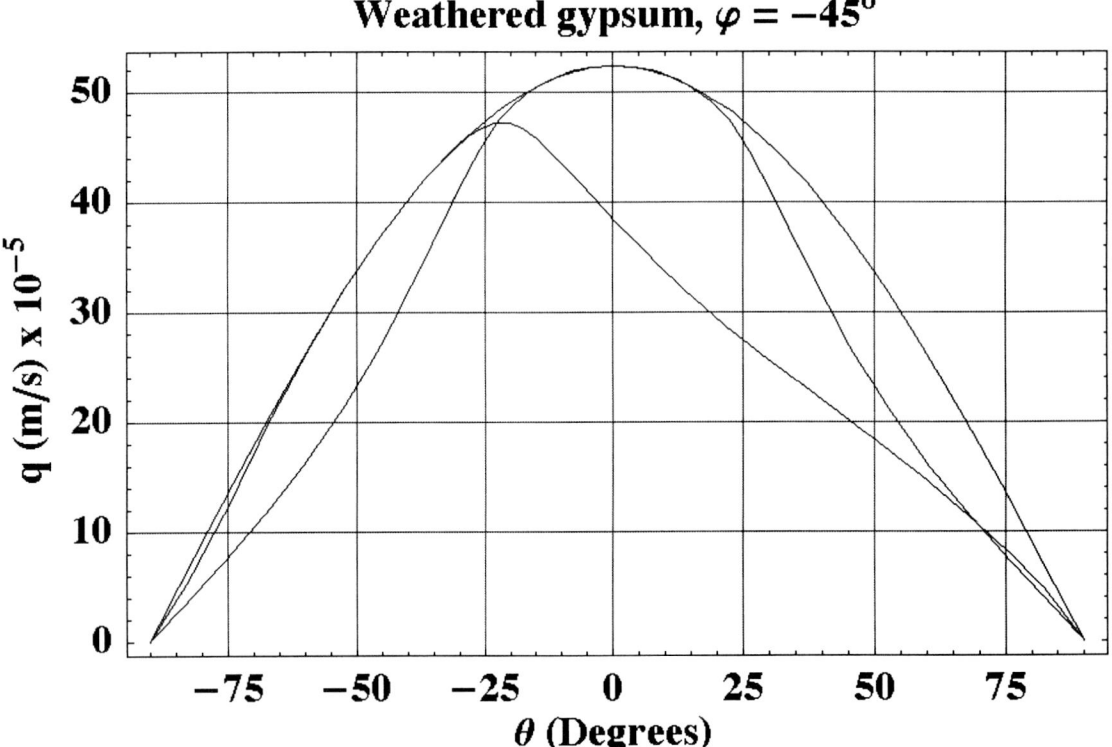

Fig. 9. Comparison of q, for different values of φ, as a function of θ. This plot is useful in understanding the skewed impulse response extrapolation through this medium (Figures 6, 7 and 8).

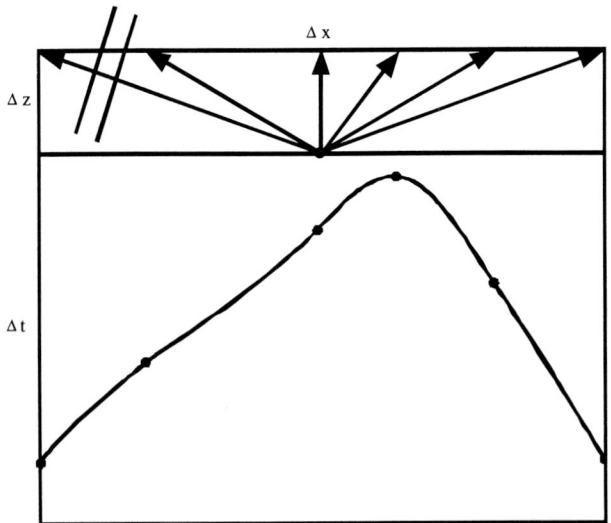

(a)

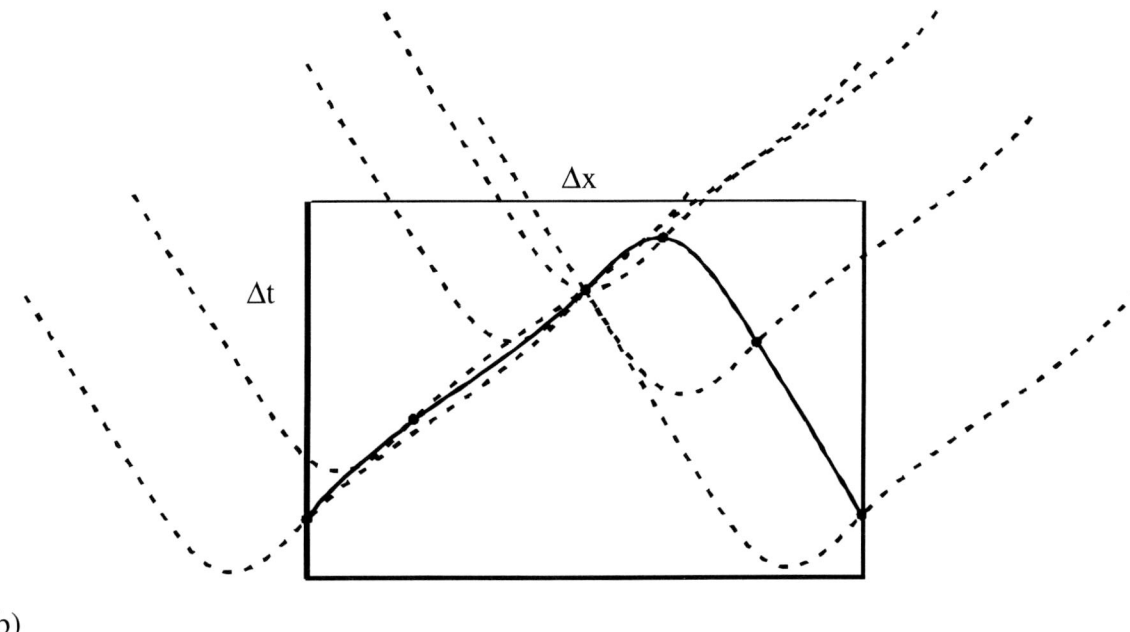

(b)

Fig. 10. (a) TI diffraction response of a diffactor in a TI medium. (The diffraction medium is weathered gypsum with a TI axis of symmetry of -45 degrees.) The arrows in (a) correspond to angles of incidence from a diffractor (the black dot) to the recording surface z = 0. Note that incident waves appear at 90 degrees to the rays (example: two parallel lines upper left corner). The corresponding travel times are represented by black dots on the diffraction curve. The extrapolation response (a, dashed line) is the time and space reversal of the diffraction curve. Focussing (b) is achieved by convolution of the extrapolation operator with the diffraction curve.

Mesaverde shale

The second example is of a much more isotropic material: Mesaverde shale. The anisotropic parameters for this material are:

$$\alpha_o = 3901, \beta_o = 2682, \varepsilon = 0.137, \delta^* = -0.078$$

The maximum anisotropy of this material is about 13% at +/-90 incidence. Figure (11) shows a plot of P-wave anisotropy. Figures (12) through (14) show 3D plots of the P-wave velocity, p and q of this material as a function of θ and φ. Again we see that the p and q surfaces are relatively simple compared to the velocity surface, and much simpler than those for weathered gypsum. Figure (15) is a plot of q anisotropy, which achieves a maximum of 3.2% at +/- 60 degrees.

Again, impulse responses are generated by extrapolating a single spike through 300 m of the material using first an isotropic velocity model $(\varepsilon = 0, \delta^* = 0)$, then with vertical TI, and then a 60 degree symmetry axis.

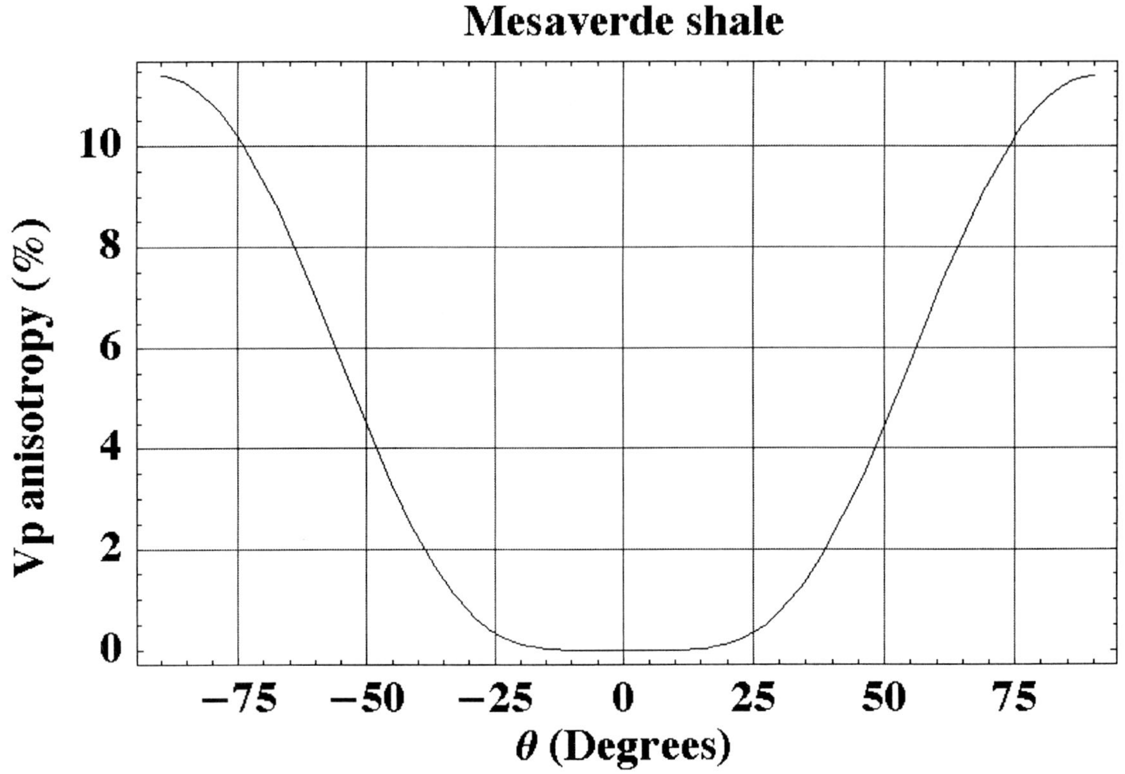

Mesaverde shale

Fig. 11. P-wave anisotropy for Mesaverde shale. The maximum anisotropy is 12.8% at +/- 90 degrees (assuming vertical TI symmetry).

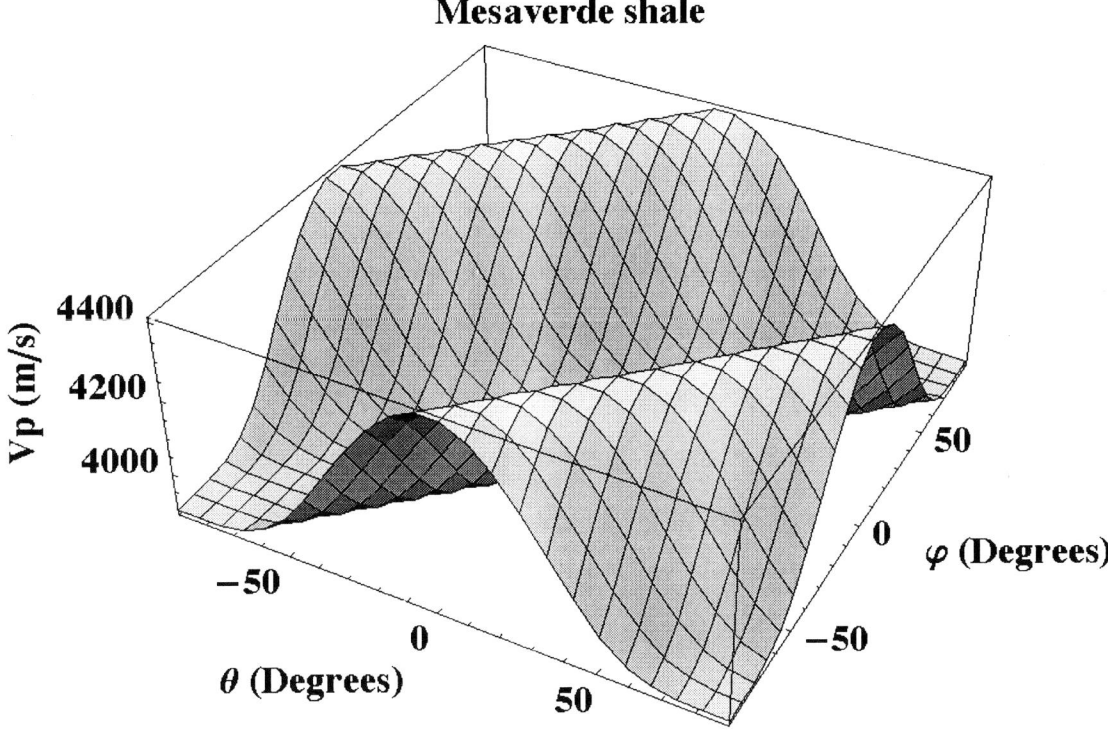

Fig. 12. P-wave velocity as a function of angle of incidence (θ) and TI symmetry axis (φ) for Mesaverde shale.

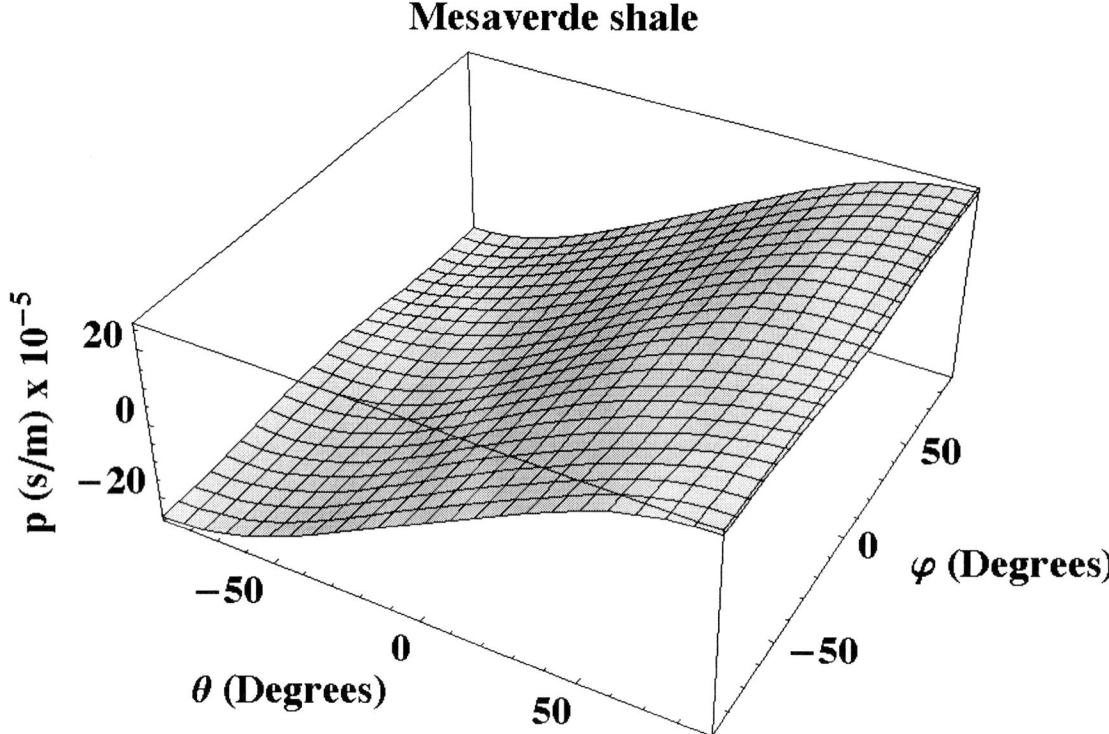

Fig. 13. Horizontal slowness (p) as a function of angle of incidence (θ) and TI symmetry axis (φ) for Mesaverde shale.

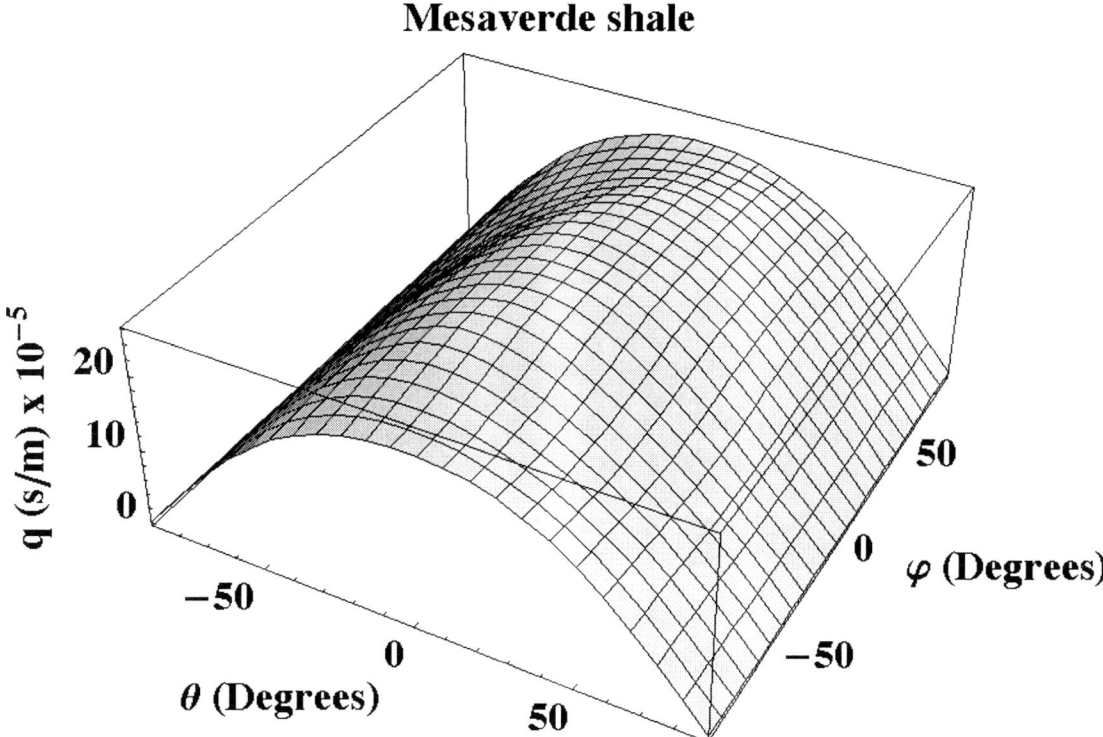

Fig. 14. Vertical slowness (q) as a function of angle of incidence (θ) and TI symmetry axis (φ) for Mesaverde shale.

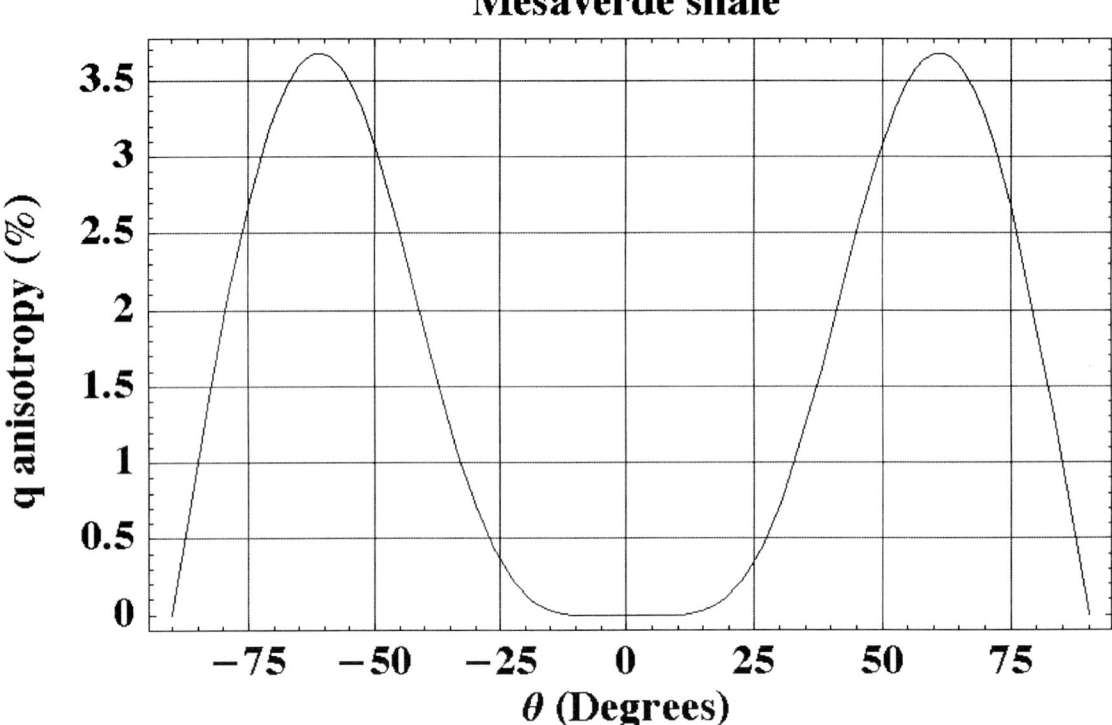

Fig. 15. q anisotropy for Mesaverde shale. The maximum anisotropy is 3.7% at +/- 61 degrees (assuming vertical TI symmetry).

Discussion

The isotropic, vertical TI and TI ($\varphi = 60$ degrees) impulse responses are given in Figures (16) to (18), and the corresponding q plot is shown in Figure (19). Comparison of Figures (16) and (17) shows that the vertical TI extrapolation is occurring faster for the higher angles of incidence, as in the gypsum example. From our previous discussion regarding the skew of the TI impulse response and its relation axis of symmetry and q (Figure 19) we expect it to be skewed down and to the right. The expected response is evident in Figure (18) though the effect is subtle compared to gypsum.

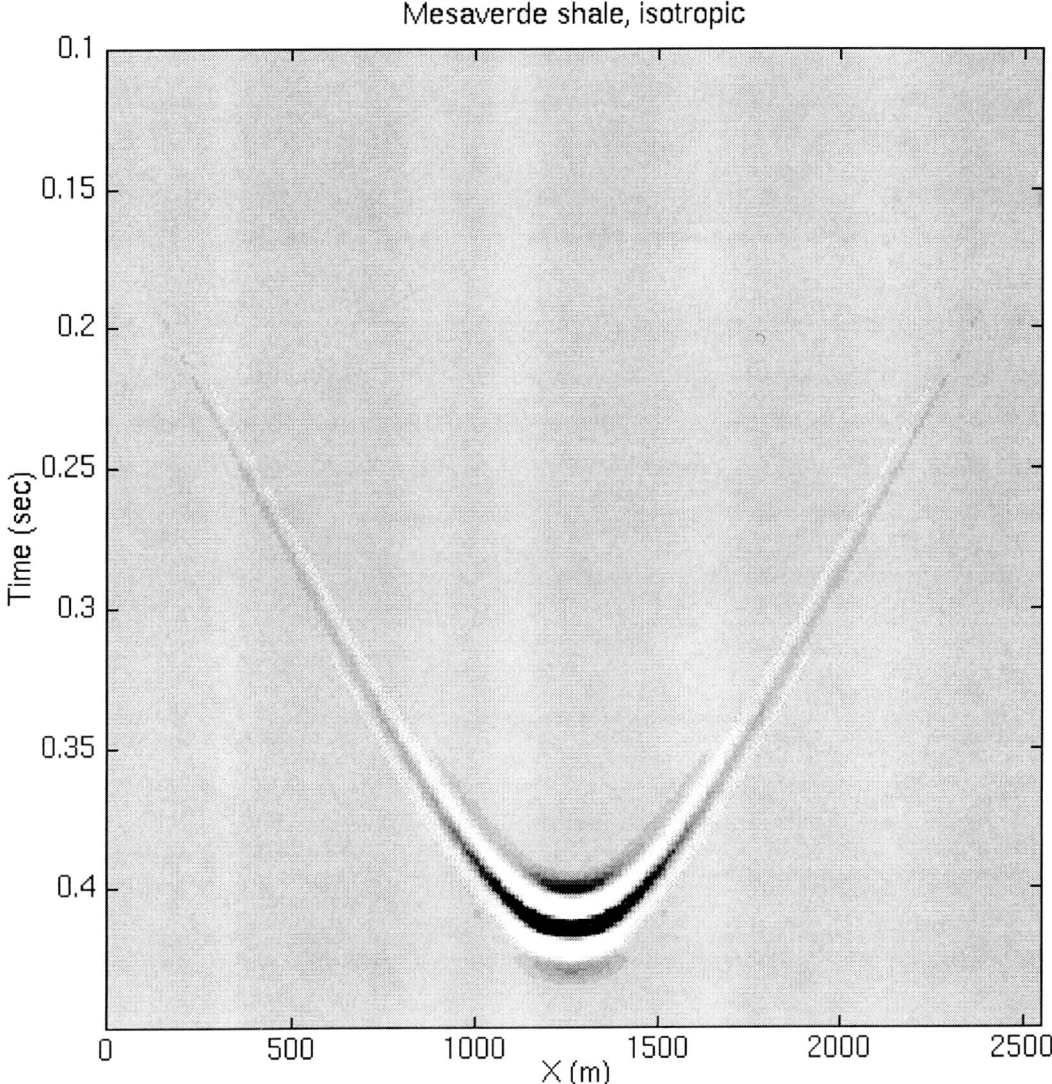

Fig 16. Isotropic impulse response for Mesaverde shale.

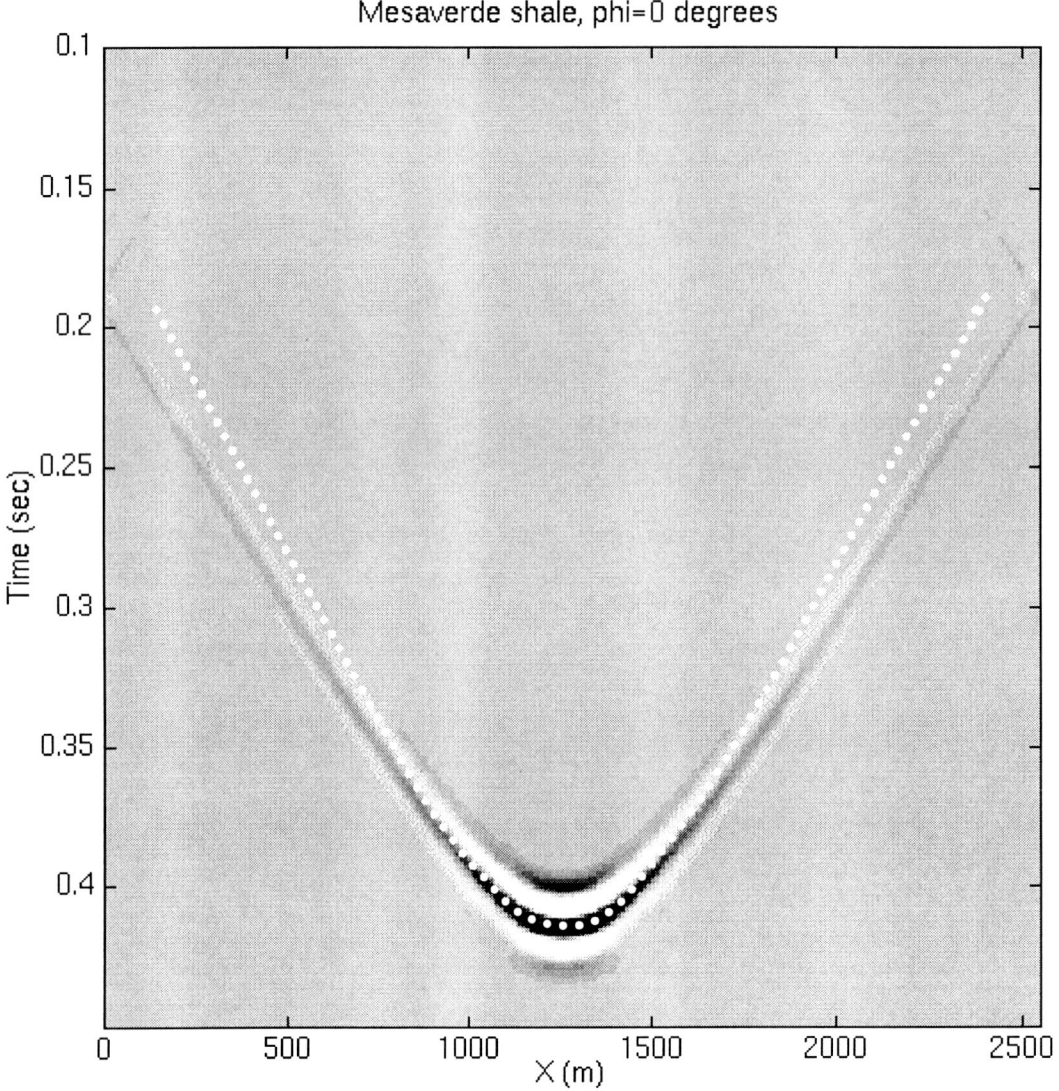

Fig. 17. Vertical TI impulse response for Mesaverde shale. The isotropic response is overplotted in white.

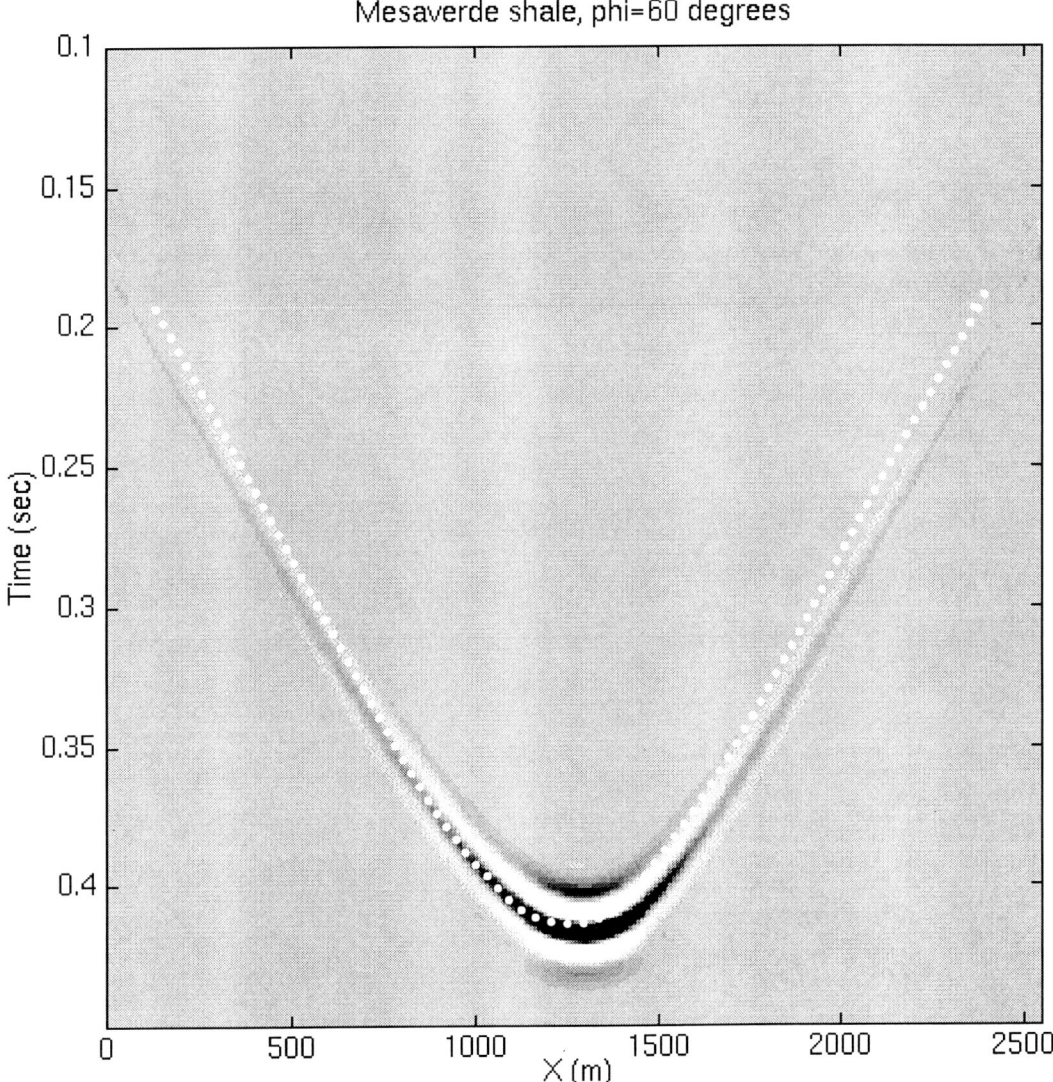

Fig. 18. TI impulse response for Mesaverde shale. The isotropic response is overplotted in white.

Weathered gypsum/Mesaverde shale

The final example is of a material that is a combination of the above to TI media. Figure (20) describes the geometry of this model in which 300m of Mesaverde shale meets abruptly with 300m of weathered gypsum. The anisotropyies of the two media are parameterized exactly as in the above two examples including the axes of symmetry.

For this TI model we extrapolate two spatially separate impulses simultaneously through the medium described in Figure (20). As indicated, the impulses are positioned such that, during translation through the medium, the one on the left encounters dipping weathered gypsum and the other encounters dipping Mesaverde shale.

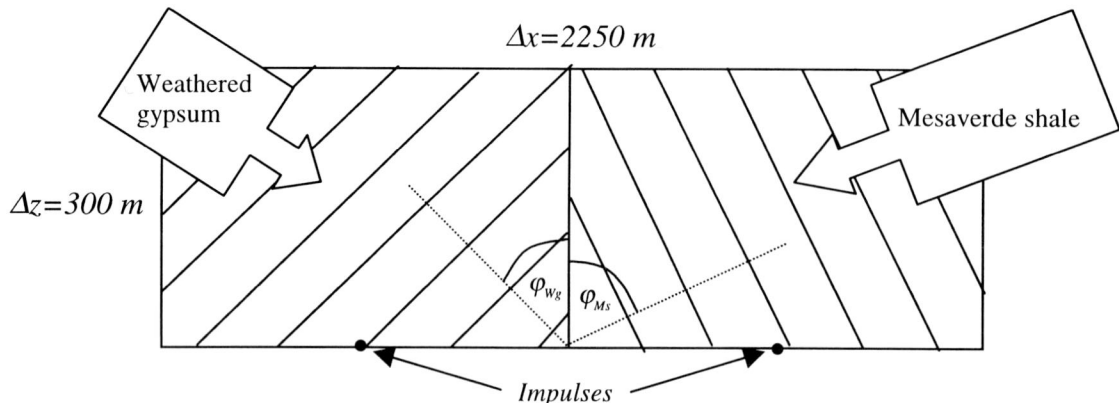

Fig. 20. Model of weathered gypsum and Mesaverde shale, $\varphi_{Wg} = -45$ degrees, $\varphi_{Ms} = 60$ degrees.

Discussion

As Figure (21) shows, we have achieved an extrapolated wavefield trough a media which is anisotropic and whose parameters vary with spatial position. The apparent boundary reflections are Fourier wraparound.

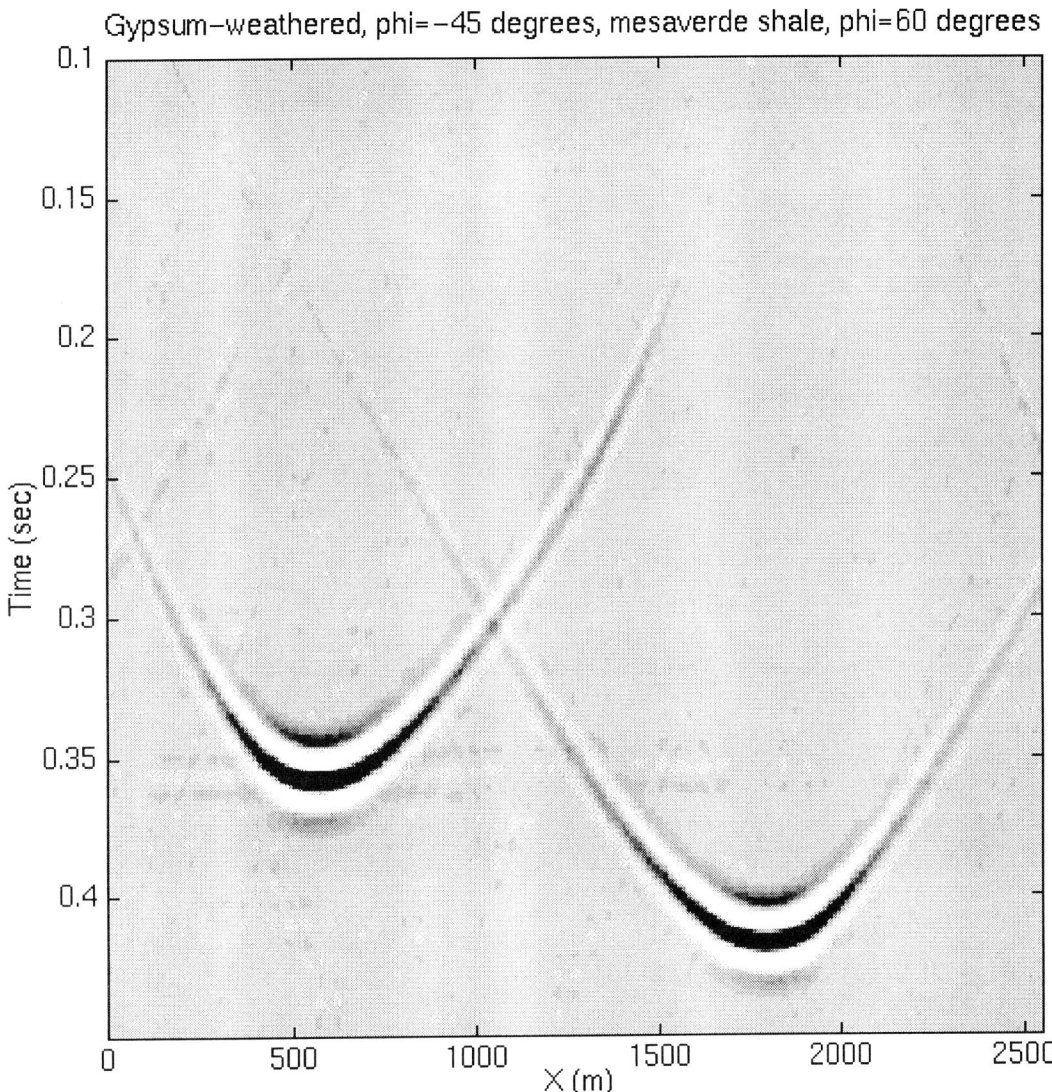

Fig. 21. TI impulse response for weathered gypsum/Mesaverde shale. A model of this material is given in Figure (20).

Conclusions

The theory of nonstationary filters, as applied to wavefield extrapolation, can describe extrapolation operators that vary laterally. The extension of this extrapolator to embrace variation with angle of incidence is presented in the context of transverse P-wave isotropy (TI). We show how standard nonstationary phase shift (NSPS) is modified, using vertical and horizontal slowness, to handle TI media with no restriction on the angle of TI symmetry. Thompson parameters are used to describe the anisotropy of the media. An advantage of our approach is that migration or modeling by TI NSPS requires knowledge of phase velocity alone.

We give P-wave impulse responses for NSPS extrapolation through different models of TI media. The first impulse response is for 300 meters of weathered Gypsum having a TI symmetry axis of –45 degrees. The second is for extrapolation through 300 meters of Mesaverde shale having an axis of 60 degrees. These two impulse responses are compared to their isotropic and zero degree counterparts. The Gypsum impulse response is found to show significant variation in shape, where the Mesaverde shale example is less so. These results are in agreement with analysis of velocities and horizontal slownesses which show Gypsum to have maximum anisotropy's respectively of 82% and 20% compared to 13% and 3% for Mesaverde shale.

We then present an example where two spatially separate impulses are extrapolated simultaneously through a 300-meter thick model consisting of Gypsum and Mesaverde shale with symmetry axes as described above. The two media are arranged in such a way that Gypsum grades abruptly into Mesaverde shale. This last example demonstrates the ability of NSPS to cope with lateral velocity variation simultaneously with angle of incidence.

References

Black, J. L., Su, C. B., Wason, C. B., 1984, Steep-D-p depth migration: Expanded abstracts, **54**th Annual Mtg. Soc. Expl. Geophy., 456-457.

Byun, B. B., 1984, Seismic parameters for transversely isotropic media: Geophysics, **49**, 1908-1914.

Crampin, S., 1984, Anisotropy in exploration seismics: First Break, 19-21.

Ferguson, R. J., and Margrave, G. F., 1997, Comparison of algorithms for nonstationary phase shift: CREWES Research Report, Vol. **9**, 31-1 - 31-19.

Isaac, J. H. and Lawton, D. C., 1997, Anisotropic physical modeling: Foothills Research Report, Vol. **3**, 15-1 - 15-26.

Kitchenside, P. W., 1991, Phase shift-based migration for transverse isotropy: **67**st Mtg., Soc. Expl. Geophys., Expanded Abstracts, 993-996.

Margrave, G. F., 1997, Theory of nonstationary linear filtering in the Fourier domain with application to time variant filtering: Geophysics, *in press*.

Margrave, G. F., and Ferguson R. J., 1997, Wavefield extrapolation by nonstationary phase shift: Margrave, G. F., and Ferguson, R. J., 1997, Wavefield extrapolation by nonstationary phase shift: Expanded abstracts, **67**th Annual Mtg. Soc. Expl. Geophy., 1599-1602.

Postma, G. W., 1955, Wave propagation in a stratified medium: Geophysics, **20**, 780-807.

Thomsen, L., 1986, Weak elastic anisotropy: Geophysics, **51**, 1954-1966.

Uzcategui, O., 1995, 2-D depth migration in transversely isotropic media using explicit operators: Geophysics, **60**, 1819-1829.

Wapenaar, C. P. A., 1992, Wave equation based seismic processing: In which domain?: EAGE **54**th Conference Extended Abstracts, B019.

Examples of prestack depth migration in TI media

Robert J. Ferguson and Gary F. Margrave

ABSTRACT

Wave field extrapolation by nonstationary phase shift can be formulated to allow velocity variation with both wavenumber and position, and so can accommodate anisotropic effects in highly inhomogeneous media. The examples presented use physical data, acquired at The University of Calgary in the Department of Geology and Geophysics modeling facility. The first represents a zero-offset recording in which the target geology is a reef edge underlying an anisotropic medium with a non-vertical axis of symmetry. Isotropic migration incorrectly positions the reef edge by 350 meters laterally. Anisotropic migration correctly positions the reef.

The second example is an anisotropic thrust sheet embedded in an isotropic medium. The sheet is made up of 4 separate blocks of a common anisotropic material with different axes of TI symmetry. The model is anisotropic and strongly heterogeneous. Two seismic lines were acquired; a constant source/geophone geometry (approximately zero offset), and a prestack geometry (split spread). Using isotropic migration neither data set could correctly image the model - especially the base of the model and reflections internal to the anisotropic thrust. Using anisotropic migration by nonstationary phase shift both data sets correctly image the model.

The extra cost in computer run time relative to isotropic migration is about 20%. Distribution of the TI thrust migration (prestack) over a number of computers reduced the run time by a factor proportional to the number of PCs.

INTRODUCTION

The notion that seismic anisotropy is intrinsic to wavefield propagation in the subsurface is swiftly gaining recognition as a factor in imaging. Though anisotropy is reported throughout the earth's' crust and upper mantle (Crampin, 1984) the special case of transverse isotropy (TI) is sufficiently common to warrant consideration (Byun, 1984). One of the most mathematically tractable forms of anisotropy, TI describes a medium in which a large number of homogeneous, isotropic layers are arranged in alternating planes. Five elastic parameters provide a complete description of the larger scale elastic properties of the medium (Postma, 1955).

Thomsen (1986) points out the fundamental inconsistency of trying to image an anisotropic subsurface using the assumption of isotropy. Authors such as Uzcategui (1995) and Kitchenside (1991) propose algorithms to image TI media. Their algorithms are based on $\omega - x$ extrapolation techniques which Margrave and Ferguson (1997) and Black et al. (1984) point out are space domain approximations to a nonstationary Fourier operator. Le Rousseau (1997) has extended Kitchenside's concepts to laterally varying media using a modified PSPI technique. Authors such as Margrave and Ferguson (1997), Wapenaar (1994) and Black et al. (1984) show that a wavefield extrapolation scheme can be formulated in Fourier space, that can cope with lateral velocity variation. Margrave and Ferguson (1997) refer to this scheme as NSPS (nonstationary phase shift).

This paper presents an extension of nonstationary extrapolators to handle TI media for both zero-offset and prestack acquisition geometries. This is done without restriction on velocity gradient or TI axis of symmetry, and without restricting the strength of the anisotropy. Thomsen parameters are used to describe the anisotropy. Examples are presented that illustrate the ability of anisotropic migration to improve seismic images relative to isotropic migration.

NONSTATIONARY PHASE SHIFT

Using the model of exploding reflectors, a zero-offset equation for phase shift migration is a direct outcome of the linear theory of nonstationary filters (Margrave, 1998, Ferguson and Margrave, 1998) is

$$r(x_{\Delta z}) = \int \psi(x) W(x_{\Delta z}, x_{\Delta z} - x) dx \tag{1}$$

where $r(x_{\Delta_a})$ is an estimate of the reflectivity at depth Δz, x_{Δ_z} and x are respectively the horizontal coordinate at depth Δz and the horizontal at depth 0. Wavefield $\psi(x)$ is the coincident source/geophone input, and $W(x_{\Delta z}, x_{\Delta z} - x)$ is a wavefield extrapolator. A reflectivity estimate can be made at Δz by averaging all frequencies (Ferguson and Margrave, 1998). (Quantities ψ and W depend on temporal frequency ω though this notation is suppressed for simplicity.) Equation (1) corresponds to a spatial domain *combination* filter which Margrave and Ferguson (1997) call the limiting form of PSPI. There is an alternate form where the extrapolator W has dependence on x in it's first coordinate and is, what Margrave (1998) calls, a *convolution* filter. (Ferguson and Margrave (1998) derive a hybrid operator which Margrave and Ferguson (1998) show, due to it's symmetry, has greater stability than either of it's component forms.)

Migration of a single source gather $\psi(x, x_s)$ is detailed in Ferguson and Margrave (1998) and is given by

$$r(x_{\Delta z}, x_s) = W(x_s, x_{\Delta z} - x_s) \int \psi(x, x_s) W(x_{\Delta z}, x_{\Delta z} - x) dx \tag{2}$$

The reflectivity r is unique for each source coordinate x_s, similarly for the input wavefield ψ. Equation (2) corresponds to the migration of a single source gather, and the migration of many sources is simple to implement on a large number of processors.

Both zero-offset and prestack migration (equations 1 and 2) share the same extrapolator W. In the following examples W was applied in the Fourier domain using

$$\int \psi(x) W(x_{\Delta z}, x_{\Delta z} - x) dx = \int \left[\int \varphi(k_x) A(k_{\Delta z} - k_x, k_x) dk_x \right] \exp(-ik_{\Delta z} x_{\Delta z}) dk_{\Delta z} \tag{3}$$

where, $\varphi(k_x)$ is the forward Fourier transform of $\psi(x)$.

$$\varphi(k_x) = \int \psi(x)\exp(ik_x x)dx \tag{4}$$

The Fourier domain extrapolator A is related to W by

$$A(\delta k_x, k_x) = \frac{1}{2\pi}\int \alpha(x_{\Delta z}, k_x)\exp(i\delta k_x x_{\Delta z})dx_{\Delta z} \tag{5}$$

where $\delta k_x = k_{\Delta z} - k_x$. The nonevanescent form of α is

$$\alpha(x_{\Delta z}, k_x) = \int W(x_{\Delta z}, \delta x)\exp(ik_x \delta x)d\delta x$$
$$= \exp\left(i\Delta z \sqrt{\frac{\omega^2}{v(x_{\Delta z})^2} - k_x^2}\right) \tag{6}$$

where $\delta x = x_{\Delta z} - x$.

In zero-offset migration the velocity term v in equation (6) is halved to conform to the model of exploding reflectors.

W in equation (6) is a function of laterally variable velocity $v(x)$ thus it is nonstationary. We propose that velocity variation be extended by allowing $v(x)$ to become $v(x, k_x)$. Then equation (6) becomes a wavefield extrapolator that varies velocity with position and phase angle

$$\alpha(k_x, x) = \exp\left[i\Delta z\sqrt{\frac{\omega^2}{v(x)^2} - k_x^2}\right] \Rightarrow \exp\left[i\Delta z\sqrt{\frac{\omega^2}{v(x,k_x)^2} - k_x^2}\right]. \tag{7}$$

In the following source gather migrations the source extrapolator $W(x_s, x_{\Delta z}\text{-}x_s)$ is approximated using an NMO correction applied as a temporal phase shift. The resulting reflectivity estimates in each migrated source gather are most accurate for positions directly beneath the source and decrease in accuracy for positions further away. In isotropic migrations the NMO corrections are calculated using rms velocities computed from surface at the location beneath the source. A more accurate source wavefield extrapolator is under development.

TI NONSTATIONARY PHASE SHIFT

Velocity $v(x,k_x)$ in equation (7) is the phase velocity for a Fourier plane wave determined by k_x and ω. A major strength of the phase-shift approach, compared to time domain Kirchhoff methods, is that only phase velocity is required. A complicating factor is the determination of the horizontal wavenumber dependence of velocity from theoretical descriptions that depend on angular direction.

The angle dependant velocities of a TI media can be completely described by five elastic constants whose mathematical forms are well established. However, these constants are difficult to interpret physically. Thomsen (1986) derives the five elastic constants such that two of them are the physically realizable values of P-wave velocity (α_0) and S-wave velocity (β_0), measured at zero offset, plus three additional constants ε, δ^* and γ. For P-wave propagation the number of parameters reduces to four. We adopt the Thomsen (1986) derivations as input parameters for wavefield extrapolation (we do not use the weak form of Thomsen's equations).

P-wave velocity as a function of angle of incidence θ (Thomsen, 1986) is

$$v_p(\theta) = \alpha_o^2 \left[1 + \varepsilon \sin^2 \theta + D^*(\theta) \right]$$ (8)

with

$$D^* = \frac{1}{2}\left[1 - \frac{\beta_o^2}{\alpha_o^2}\right]\left\{\left(1 + \frac{4\delta^*}{1-\frac{\beta_o^2}{\alpha_o^2}}\sin^2\theta\cos^2\theta + \frac{4\left[1-\frac{\beta_o^2}{\alpha_o^2}+\varepsilon\right]\varepsilon}{\left(1-\frac{\beta_o^2}{\alpha_o^2}\right)^2}\sin^4\theta\right)^{\frac{1}{2}} - 1\right\}.$$ (9)

Each of the anisotropic parameters in equations (8) and (9) are functions of input position x, thus, we have five numerical fields of anisotropic parameters to specify, plus a field for the axes of symmetry.

TI MIGRATION

Wavefield extrapolation by phase shift through one depth step, Δz, is an operation that loops over frequency; one monochromatic wavefield is extrapolated per loop. At all times during extrapolation k_x and ω are known; but in the anisotropic case, it is difficult to directly measure the angle of incidence of the incoming waves. A way to relate k_x and ω to angle of incidence θ is to compute a set of horizontal slownesses (p) for a range of evenly spaced angles (θ) using anisotropic velocities (Kitchenside, 1991). (For a non-zero symmetry axis, care must be taken to relate the angle of incidence of waves to the rotated axis of anisotropy.) From Snell's law,

$$p(x,\theta) = \frac{\sin(\theta)}{v(x,\theta)}.$$ (10)

Then θ as a function of p can be established using an interpolating polynomial,

$$\theta(p) = a_0 + a_1 p + a_2 p^2 + \ldots + a_n p^n$$ (11)

where a_i are the coefficients of an interpolating polynomial of order n, and p^i are increasing powers of horizontal slowness (computed using equation 10). Note that p can be given in terms of k_x and ω

$$p = \frac{k_x}{\omega}.$$ (12)

Given the coefficients a_i, equation (10) can be substituted into equation (11):

$$\theta\left(\frac{k_x}{\omega}\right) = a_0 + a_1 \frac{k_x}{\omega} + a_2 \left[\frac{k_x}{\omega}\right]^2 + \ldots + a_n \left[\frac{k_x}{\omega}\right]^n. \tag{13}$$

Thus, since k_x and ω are known at all points during the extrapolation process, an angle of incidence can be computed for each wave using equation (13). The corresponding P-wave velocities can then be computed using equations (8) and (9). Vertical slowness q is then given by

$$q(x,\theta) = \frac{\cos(\theta)}{v(x,\theta)}. \tag{14}$$

For the non-evanescent region the q values from equation (14) are used to calculate the values of the vertical wavenumbers

$$k_z = \omega q. \tag{15}$$

The form of the nonstationary extrapolator from equation (4) then becomes

$$\alpha(k_x, x, \omega) = \exp ik_z \Delta z \tag{16}$$

where,

$$k_z = \omega q(x, \theta(k_x, \omega)). \tag{17}$$

Anisotropic depth migration for zero-offset data then proceeds in the usual recursive manner with the anisotropic extrapolator (equation 7) and, for the Fourier domain expression of combination migration, using equation (3). Anisotropic depth migration for source gathers proceeds, again in the Fourier domain, by applying the anisotropic combination filter to the traces recursively, then shifting them in an offset dependent manner after recursion. In the following examples the simple NMO process for equivalent media described previously is used. Note that the source correction is not applied recursively, but is applied directly from the surface for each depth level.

EXAMPLES

Isotropic reef with an anisotropic overburden

This section presents depth migrations of two physical model experiments – both acquired at the physical modeling facility at The University of Calgary. The first (figure 1a) was designed to represent a reef edge underlying a TI media. The TI media has a single axis of symmetry, 45 degrees measured from normal to the bedding of the material, and the media containing the reef is isotropic. Though not a nonstationary problem (the anisotropy is laterally invariant) the reef model is a good example of the need for anisotropic imaging. Figure (1b) shows a zoomed image of the zero-offset section data. Migration of the zero-offset section by isotropic phase shift yields an image of the reef edge which is displaced by about 350 m to the left of it's true position (figure 2a). Migration by TI phase shift correctly positions the edge of the reef (figure 2b).

Anisotropic thrust sheet in an isotropic background

The second migration example is that of a flat reflector overlain by a TI thrust sheet embedded in an isotropic background, and represents a true anisotropic/nonstationary problem (the anisotropic parameters vary in the lateral coordinate). The thrust sheet is composed of four blocks (figure 3a) - each with a unique axis of symmetry (the zero-offset section is given in figure 3b). Migration of the zero-offset data by nonstationary phase shift (isotropic), yields a migrated image where reflectors within and below the TI material are incorrectly positioned (figure 4b). In particular the basement reflector exhibits substantial pull up. Migration by nonstationary phase shift (TI) does a better job of positioning the reflectors (Figure 4a), and has nearly flattened the basement reflector.

Notable in the migrated image is the apparent no data zone beneath the steepest two blocks of the thrust. The crossing energy at the base reflector between 2000 and 3300m is believed to indicate a shadow zone caused by the high-velocity thrust sheet overlying slower material. The shadow zone is a result of the zero-offset geometry of the recording. A normal incidence ray from the flat lying reflectors beneath the thrust tends to strike the hanging wall at an angle greater than the critical angle, so that zero-offset reflections from the area beneath the steep thrust blocks are not possible for nonevanescent energy.

Migration of the prestack data by anisotropic source-gather migration correctly positions the base reflector and fills in the shadow zone due to the multiplicity of ray paths afforded by the prestack geometry (figure 5a). Again, as can be seen in figure 5b, error due to using an isotropic migration (this time isotropic prestack) is evident in the incorrect positioning of the base reflector.

An example of the migration of a single source gather is given in figures 6a through 6d. The intermediate wavefields (figures 6b and c) show the process by which reflectors gradually focus on their way to the top of the section (t = 0) where they are output to image space.

Using a simple NMO correction for the downward continuation of the source causes an offset dependent error in the migrated source gather. The reflectivities directly beneath the source will be correctly imaged but less so with increased offset. On the geophone side, downward continuation proceeds as a nonstationary filter. Accuracy is limited only by run time, very fast for smooth models, very slow for complex models (Ferguson and Margrave, 1997). If the velocity variation is extremely strong there may also be stability problems (Margrave and Ferguson, 1998). Future source-gather migrations will incorporate an improved source extrapolator.

COMPUTER CONSIDERATIONS

Source-gather migrations were computed in a new computer lab at The University of Calgary. Available are 24 Pentium PCs with 128 megabytes of memory each. Fourteen PCs have 233 MHz processors while 10 have 300 MHz processors. One source gather in the anisotropic thrust data set above consists of, after padding, 256 traces and 512 samples per trace (4 ms sample rate). There are 86 source gathers acquired at 60 m intervals along the line. The required TI model is first constructed with 5 anisotropic parameters specified every 20 m in lateral coordinate, and every 10 m in depth. To ensure reasonable run times the model is smoothed and desampled in lateral coordinate and depth. (Nonstationary filters applied entirely in the Fourier domain run faster with smooth models).

Desampling is accomplished using RMS calculations in both spatial dimensions. Here each output coordinate, in the desampled model, is an RMS value computed using the eight neighboring samples laterally and four neighboring samples vertically. Only Vp and Vs are smoothed in this way. The other TI parameters are desampled without smoothing. The error caused by this use of the RMS method is not yet quantified. The resulting vertical sample rate of the model (figure 3a) is 40 m (split-step phase-shift migration is used to interpolate the output back to 10 m in depth). The horizontal sample rate of the model is 160 meters. The resulting smooth model consists of 56 by 32 samples.

The above rationale for desampling comes from access restrictions to the computer lab. Nighttime blocks of about 8 hours are generally available. A 300 MHz PC requires about 75 minutes and 267 megaflops to TI migrate one source gather, while a 233 MHz PC requires about 97 minutes. Thus the 86 source gather migrations require 5.3 days of run time. The extra cost of TI migration over isotropic migration is about 15 minutes per source gather on the 300 MHz PC and about 20 minutes on the 233 MHz PC (20%).

To meet the 8-hour limit, migration of the anisotropic thrust model was distributed over the 24 PCs in the new lab. Unfortunately, distribution of source gathers did not consider speed differences in the processors (they were assumed to be the same) thus optimal use of the lab was not made. Each machine was given 3 or 4 source gathers and a set of the required software (Matlab source code) by FTP. Distributing the 5.3-day run time reduced the actual wait time to about 6 hours. Interpretation of the migrated source gathers can be done individually or by binning into common offset sections.

For comparison a similarly designed and distributed stationary/isotropic source-gather migration required only 3.4 hours to migrate all 86 gathers. However, each nonstationary migration (run with the 56 by 32 model) is equivalent to computing 32 stationary migrations/source gather - needing 4.5 days of run time. Each stationary phase-shift migration can then be modified to accept anisotropic velocities at an extra cost of about 20%. The required run time then increases to 5.4 days - not including the overhead required to assemble 32 migrations/source into single images.

CONCLUSIONS

Depth migration by nonstationary phase shift allows velocity variation with both wavenumber and position, and so can image highly inhomogeneous media. Depth migration of TI physical models demonstrates the utility basing an imaging algorithm on nonstationary filter theory. The extra cost of allowing TI anisotropy compared to a similarly implemented isotropic migration is about 20%. Distribution of a prestack depth migration over a number of processors reduces run time by a factor proportional to the number of processors.

ACKNOWLEDGEMENTS

The authors would like to thank Dr. Don Lawton, Dr. Helen Isaac, Jennifer Leslie, Eric Gallant and the sponsors of CREWES for their assistance in this work.

REFERENCES

Black, J. L., Su, C. B., Wason, C. B., 1984, Steep-Dip depth migration: Expanded abstracts, 54th Annual Mtg. Soc. Expl. Geophy., 456-457.

Byun, B. B., 1984, Seismic parameters for transversely isotropic media: Geophysics, **49**, 1908-1914.

Crampin, S., 1984, Anisotropy in exploration seismics: First Break, 19-21.

Ferguson R. J., and Margrave, 1998, A nonstationary description of depth migration: CREWES Research Report, **10**, Chapter 41.

Ferguson R. J., and Margrave, 1997, Algorithm comparisons for nonstationary phase shift: CREWES Research Report, **9**, Chapter 31.

Kitchenside, P. W., 1991, Phase shift-based migration for transverse isotropy: 61st Mtg., Soc. Expl. Geophys., Expanded Abstracts, 993-996.

Le Rousseau, J. H., 1997, Depth migration in heterogeneous, transversely isotropic media with the phase-shift-plus-interpolation method: 67th Mtg., Soc. Expl. Geophys., Expanded Abstracts, 1703-1706.

Margrave, G. F., 1998, Theory of nonstationary linear filtering in the Fourier domain with application to time variant filtering: Geophysics, **63**, 244-259.

Margrave, G. F., and Ferguson, R. J., 1997, Wavefield extrapolation by nonstationary phase shift: 67th Mtg., Soc. Expl. Geophys., Expanded Abstracts, 1599-1602.

Margrave, G. F., and Ferguson, R. J., 1998Explicti Fourier wavefield extrapolators: CREWES Report, Vol., **10**, chapter 39.

Postma, G. W., 1955, Wave propagation in a stratified medium: Geophysics, **20**, 780-807.

Thomsen, L., 1986, Weak elastic anisotropy: Geophysics, **51**, 1954-1966.

Uzcategui, O., 1995, 2-D depth migration in transversely isotropic media using explicit operators: Geophysics, **60**, 1819-1829.

Wapenaar, C. P. A., 1992, Wave equation based seismic processing: In which domain?: EAGE 54th Conference Extended Abstracts, B019.

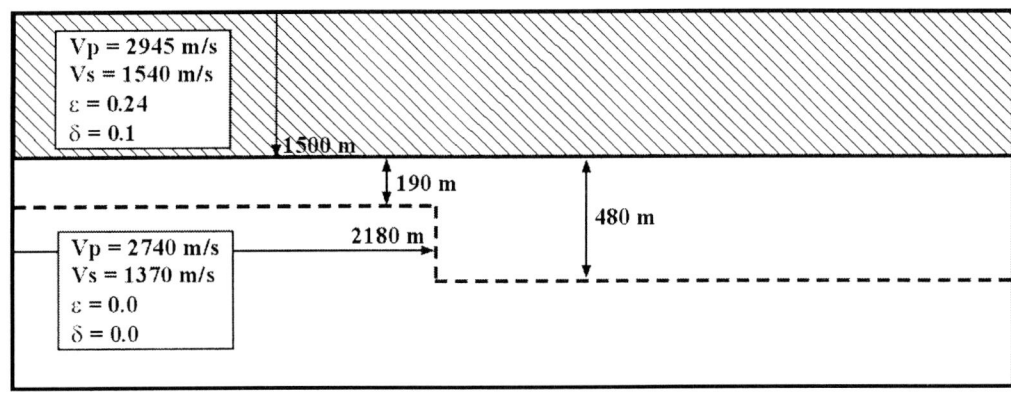

Fig.1. Model of an isotropic reef with an anisotropic overburden. (a) Dimensions and elastic parameters of the reef model. The axis of TI symmetry is +45 degrees. (b) Seismic data acquired with coincident source and receiver (the true offset scales to about 200 m).

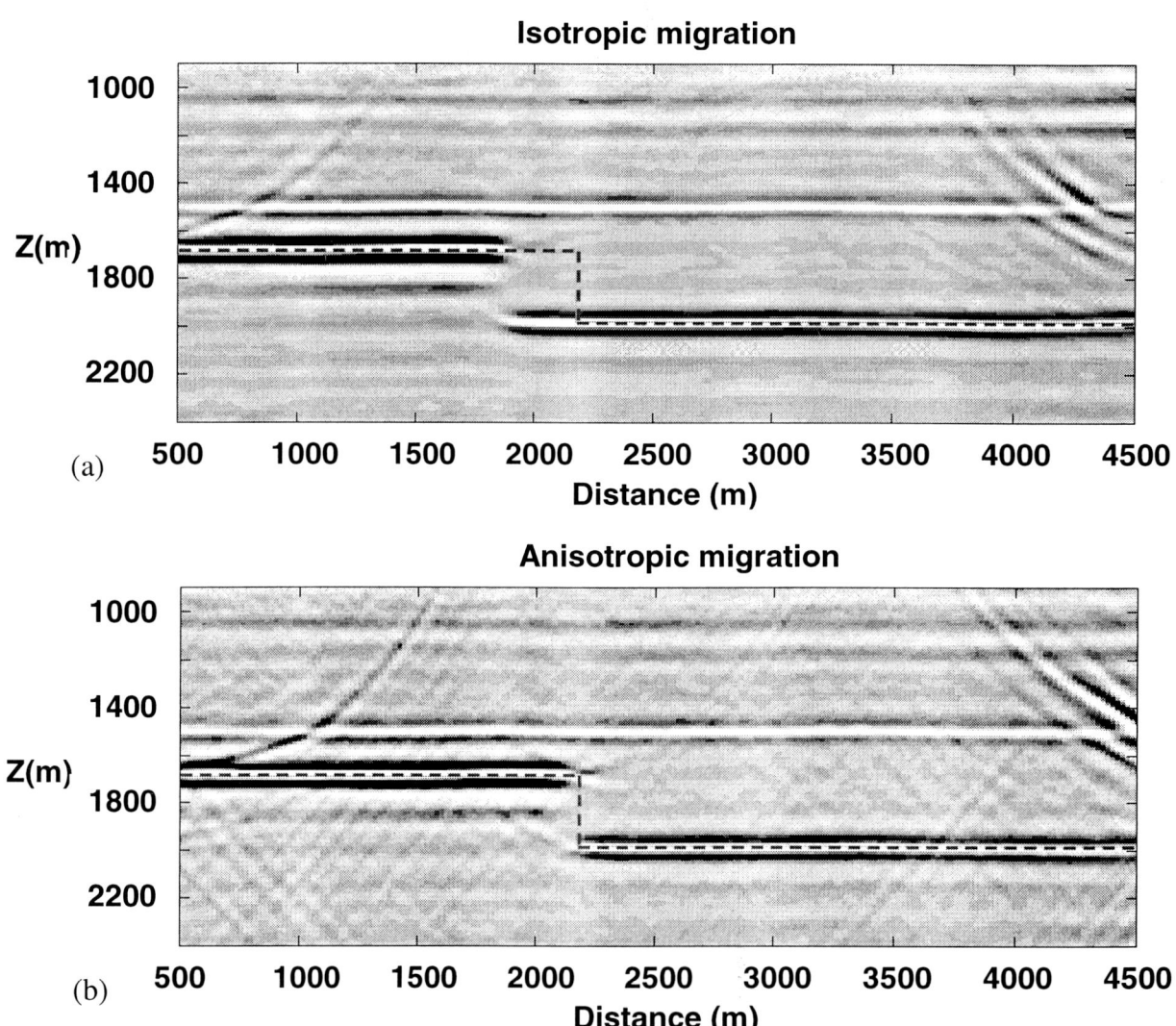

Fig.2. Migrations of figure 1a. (a) Isotropic phase-shift migration images the reef edge ~ 350 m left of its true position. (b) Anisotropic phase shift provides a correct image.

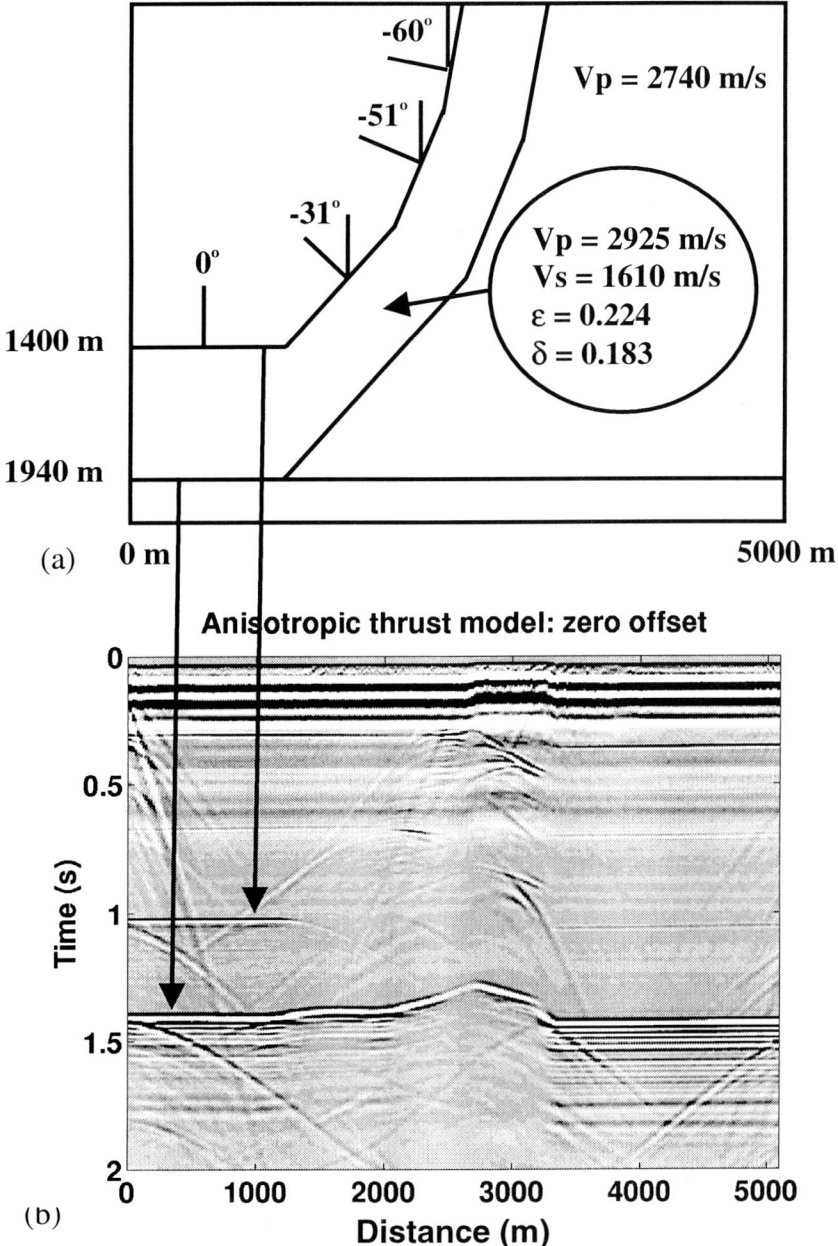

Fig. 3. Model of an anisotropic thrust sheet embedded in an isotropic background. (a) The thrust sheet is composed of 4 blocks of phenolic material (not to scale). The corresponding anisotropic parameters and axes of TI symmetry are labeled. (b) The zero-offset data. The arrows indicate the top and bottom reflections of the flat lying anisotropic block. Note the pull up of the base reflector between 1000 and 3500 m.

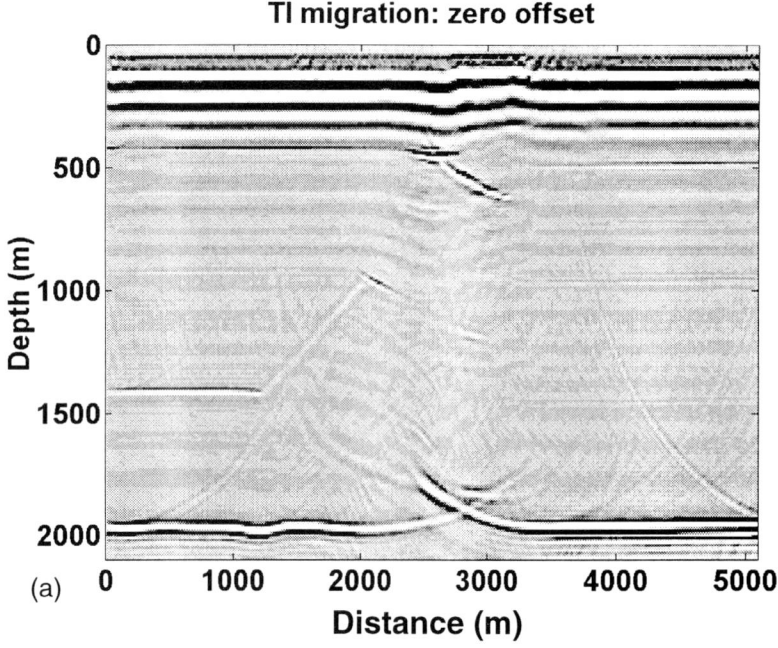

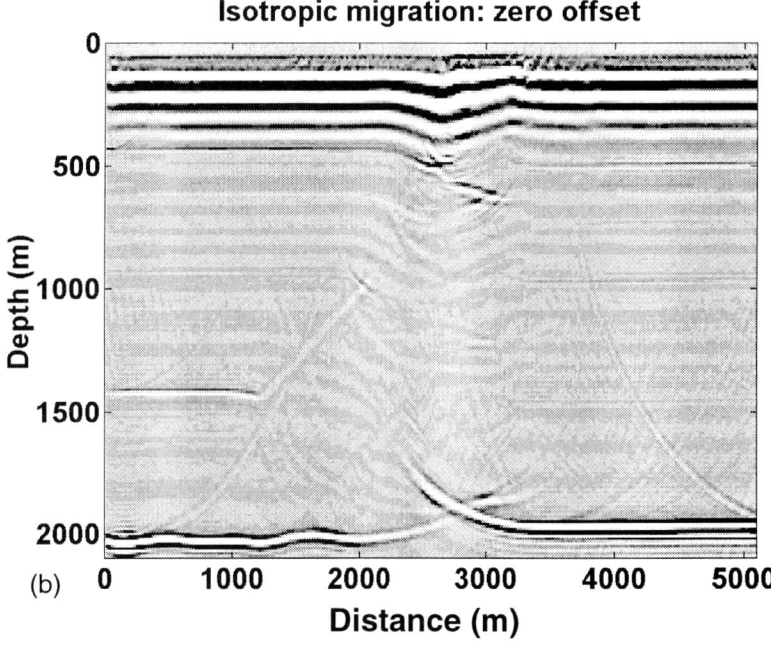

Fig. 4. Zero-offset migrations of the anisotropic thrust model (the depth scale was chosen to exaggerate differences). (a) TI migration flattens the base reflector and indicates a no data zone between 2000 and 3300 m. Isotropic incorrectly positions the base reflector (vp was chosen to be the average of the fast and slow directions = 3250 m/s). A difference in depth images (about 160 m) is apparent between base reflector under the anisotropic block and on the right.

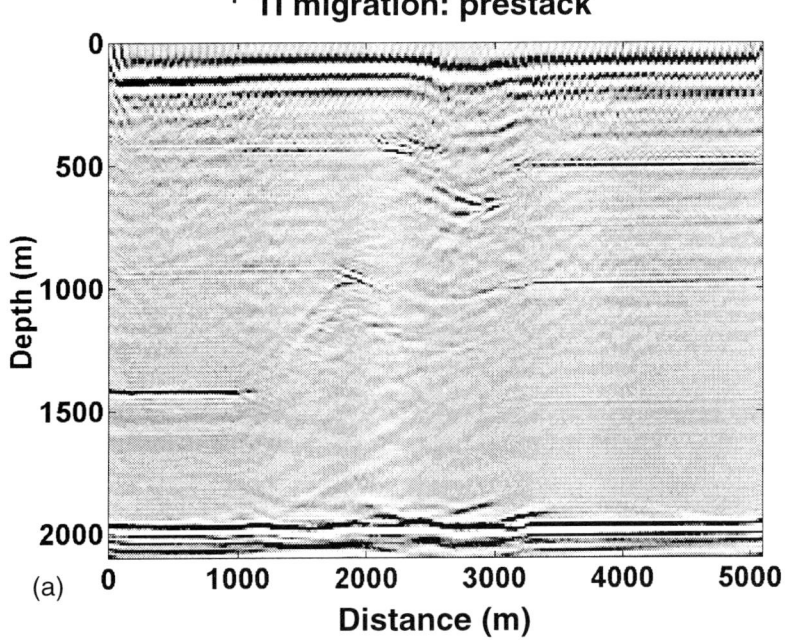

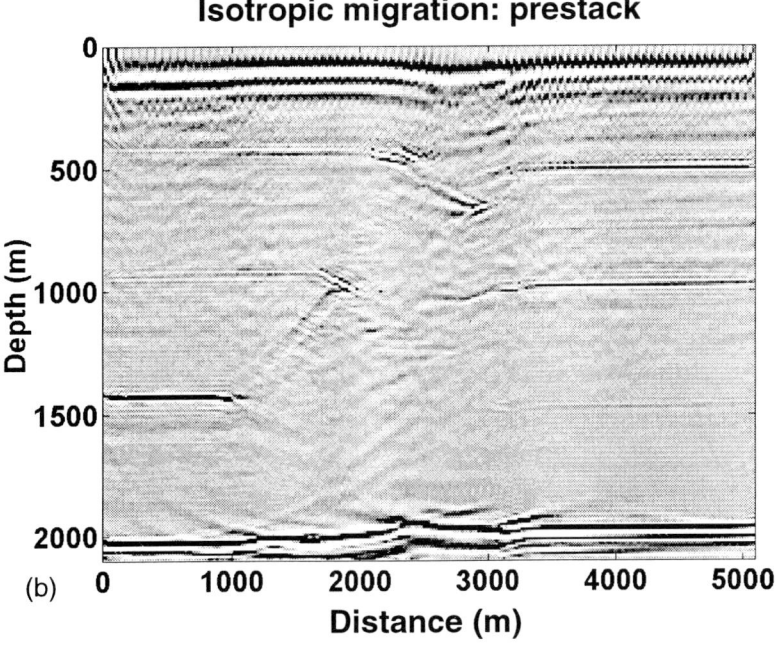

Fig. 5. Prestack migrations of the anisotropic thrust model. (a) TI migration flattens the base reflector. The scattered energy below the steepest part of the thrust sheet is due to model smoothing which was used to reduce run time. A less-smooth model would produce an even more coherent result. (b) Isotropic migration results in a stepped down image (about 160 m) of the base reflector similar to that of the zero-offset example. The base reflector beneath the steepest part of the thrust (2000 – 3000 m) is poorly imaged relative to the TI migration. The TI and isotropic migration models were equally smooth and the isotropic Vp was the average of the fast and slow directions (3250 m/s).

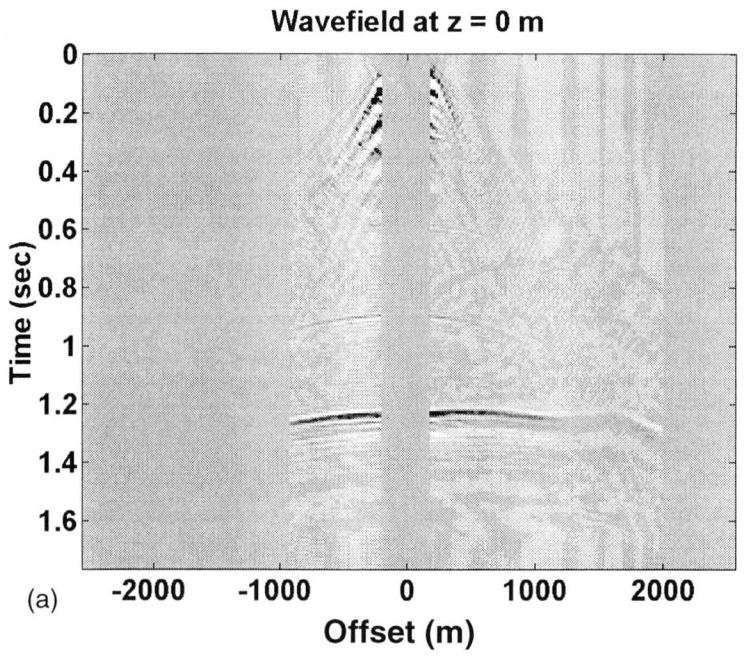

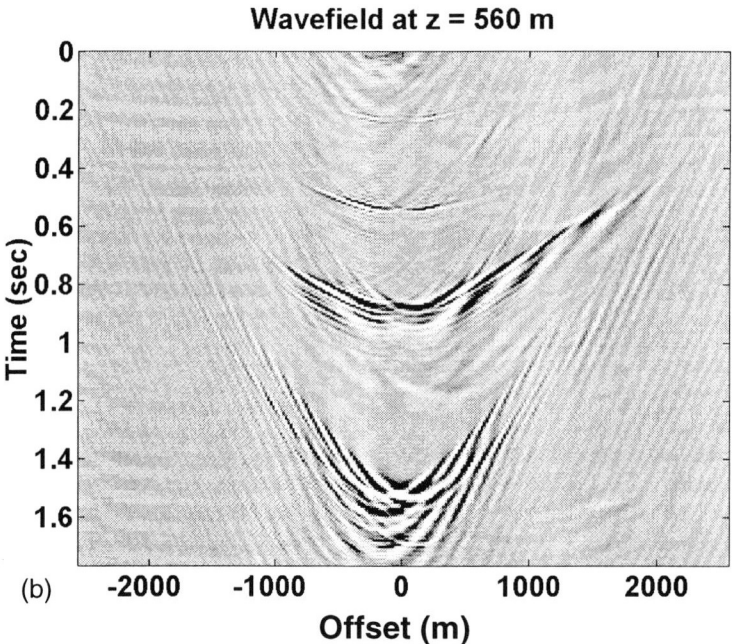

Fig. 6. Migration of a source gather. (a) A gather is input to TI source gather migration. (b) The wavefield downward continued to 560 m.

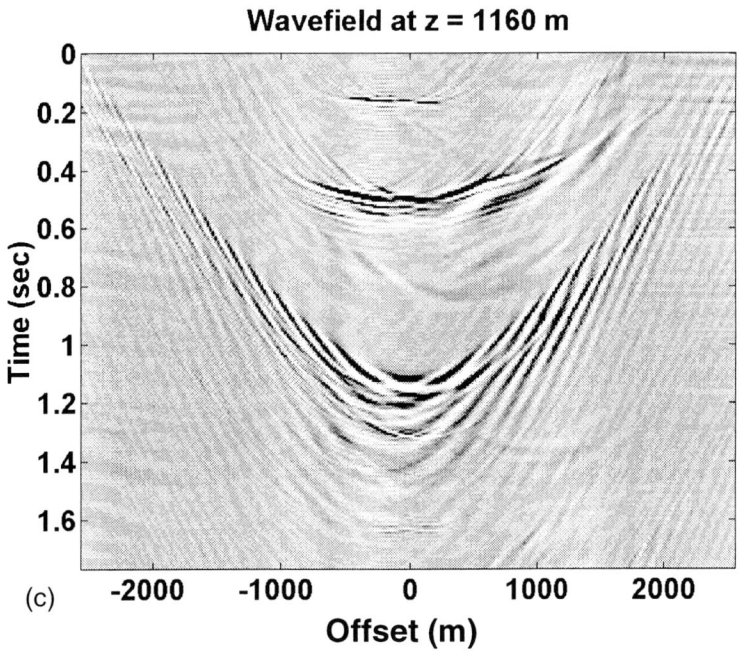

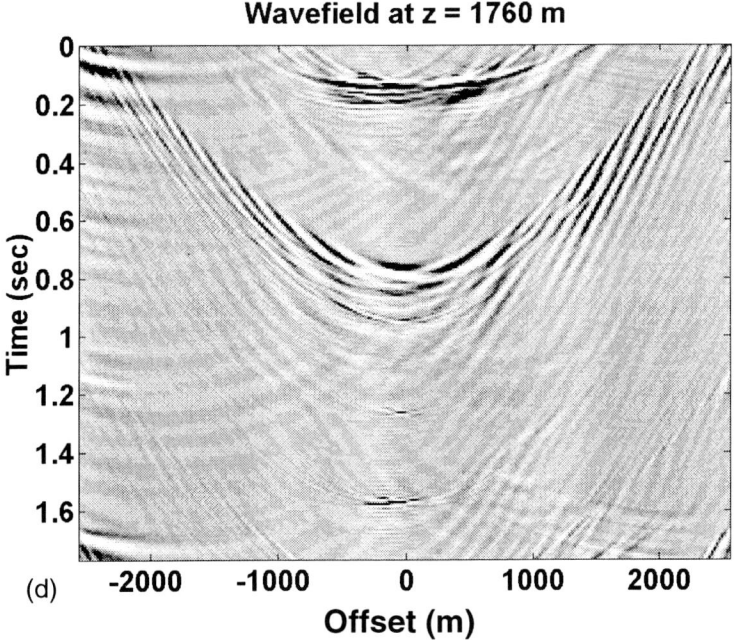

Fig. 6 cont. (c) The wavefield at 1160 m. The base reflector (t = 0.44) is continuing to focus. (d) The wavefield at 1760 m. the base reflector (t = 0.1 sec) is nearly completely focussed as it approaches t = 0 where it will be sent to output space.

Chapter 6 – Conclusions and Future Directions

Depth imaging of seismic data from the Canadian foothills can rarely be done effectively with conventional processing. An effective job generally requires prestack depth migration, preferably with anisotropic corrections. The main reasons that anisotropic migrations are only recently becoming used are twofold. The algorithms are more complex than isotropic algorithms, and also the necessary anisotropy parameters in these algorithms have only been measured for a few formations.

Finally, the reader may have noticed that we have not emphasized 3-D depth imaging. This is not because we feel 3-D imaging is unimportant – nothing could be further from the truth in foothills imaging. However, the cost of acquiring 3-D seismic data in the Canadian foothills is very expensive, and these data were not available for us to publish. We anticipate that as more 3-D data sets become available, we will see an increased number of successful applications published.